Remote Sensing by Satellite Gravimetry

Remote Sensing by Satellite Gravimetry

Editors

Thomas Gruber
Annette Eicker
Frank Flechtner

MDPI • Basel • Beijing • Wuhan • Barcelona • Belgrade • Manchester • Tokyo • Cluj • Tianjin

Editors
Thomas Gruber
Technische Universität München
Germany

Annette Eicker
HafenCity University Hamburg
Germany

Frank Flechtner
Helmholtz-Zentrum Potsdam Deutsches GeoForschungsZentrum—GFZ
Germany

Editorial Office
MDPI
St. Alban-Anlage 66
4052 Basel, Switzerland

This is a reprint of articles from the Special Issue published online in the open access journal *Remote Sensing* (ISSN 2072-4292) (available at: https://www.mdpi.com/journal/remotesensing/special_issues/satellite_gravimetry).

For citation purposes, cite each article independently as indicated on the article page online and as indicated below:

LastName, A.A.; LastName, B.B.; LastName, C.C. Article Title. *Journal Name* **Year**, *Volume Number*, Page Range.

ISBN 978-3-0365-0008-9 (Hbk)
ISBN 978-3-0365-0009-6 (PDF)

Contents

About the Editors . . . . **vii**

Preface to "Remote Sensing by Satellite Gravimetry" . . . . **ix**

Christoph Dahle, Michael Murböck, Frank Flechtner, Henryk Dobslaw, Grzegorz Michalak, Karl Hans Neumayer, Oleh Abrykosov, Anton Reinhold, Rolf König, Roman Sulzbach and Christoph Förste
The GFZ GRACE RL06 Monthly Gravity Field Time Series: Processing Details and Quality Assessment
Reprinted from: *Remote Sens.* **2019**, *11*, 2116, doi:10.3390/rs11182116 . . . . **1**

Ulrich Meyer, Krzysztof Sosnica, Daniel Arnold, Christoph Dahle, Daniela Thaller, Rolf Dach and Adrian Jäggi
SLR, GRACE and Swarm Gravity Field Determination and Combination
Reprinted from: *Remote Sens.* **2019**, *11*, 956, doi:10.3390/rs11080956 . . . . **23**

Xiang Guo and Qile Zhao
A New Approach to Earth's Gravity Field Modeling Using GPS-Derived Kinematic Orbits and Baselines
Reprinted from: *Remote Sens.* **2019**, *11*, 1728, doi:10.3390/rs11141728 . . . . **45**

Tyler Sutterley and Isabella Velicogna
Improved Estimates of Geocenter Variability from Time-Variable Gravity and Ocean Model Outputs
Reprinted from: *Remote Sens.* **2019**, *11*, 2108, doi:10.3390/rs11182108 . . . . **59**

Bihter Erol, Mustafa Serkan Işık and Serdar Erol
An Assessment of the GOCE High-Level Processing Facility (HPF) Released Global Geopotential Models with Regional Test Results in Turkey
Reprinted from: *Remote Sens.* **2020**, *12*, 586, doi:10.3390/rs12030586 . . . . **79**

Roland Pail, Hsien-Chi Yeh, Wei Feng, Markus Hauk, Anna Purkhauser, Changqing Wang, Min Zhong, Yunzhong Shen, Qiujie Chen, Zhicai Luo, Hao Zhou, Bingshi Liu, Yongqi Zhao, Xiancai Zou, Xinyu Xu, Bo Zhong, Roger Haagmans and Houze Xu
Next-Generation Gravity Missions: Sino-European Numerical Simulation Comparison Exercise
Reprinted from: *Remote Sens.* **2019**, *11*, 2654, doi:10.3390/rs11222654 . . . . **107**

Yufeng Nie, Yunzhong Shen and Qiujie Chen
Combination Analysis of Future Polar-Type Gravity Mission and GRACE Follow-On
Reprinted from: *Remote Sens.* **2019**, *11*, 200, doi:10.3390/rs11020200 . . . . **127**

Markus Hauk and Roland Pail
Gravity Field Recovery Using High-Precision, High–Low Inter-Satellite Links
Reprinted from: *Remote Sens.* **2019**, *11*, 537, doi:10.3390/rs11050537 . . . . **145**

Ingo Sasgen, Hannes Konrad, Veit Helm and Klaus Grosfeld
High-Resolution Mass Trends of the Antarctic Ice Sheet through a Spectral Combination of Satellite Gravimetry and Radar Altimetry Observations
Reprinted from: *Remote Sens.* **2019**, *11*, 144, doi:10.3390/rs11020144 - **163**

**Andreas Richter, Andreas Groh, Martin Horwath, Erik Ivins, Eric Marderwald,
José Luis Hormaechea, Raúl Perdomo and Reinhard Dietrich**
The Rapid and Steady Mass Loss of the Patagonian Icefields throughout the GRACE
Era: 2002–2017
Reprinted from: *Remote Sens.* **2019**, *11*, 909, doi:10.3390/rs11080909 . . . . **187**

Wondwosen M. Seyoum, Dongjae Kwon and Adam M. Milewski
Downscaling GRACE TWSA Data into High-Resolution Groundwater Level Anomaly Using
Machine Learning-Based Models in a Glacial Aquifer System
Reprinted from: *Remote Sens.* **2019**, *11*, 824, doi:10.3390/rs11070824 . . . . **207**

Gonca Okay Ahi and Shuanggen Jin
Hydrologic Mass Changes and Their Implications in Mediterranean-Climate Turkey from
GRACE Measurements
Reprinted from: *Remote Sens.* **2019**, *11*, 120, doi:10.3390/rs11020120 . . . . **231**

Xinyu Xu, Hao Ding, Yongqi Zhao, Jin Li and Minzhang Hu
GOCE-Derived Coseismic Gravity Gradient Changes Caused by the 2011
Tohoku-Oki Earthquake
Reprinted from: *Remote Sens.* **2019**, *11*, 1295, doi:10.3390/rs11111295 . . . . **255**

About the Editors

Thomas Gruber (Dr.) is a Senior Scientist at the Institute of Astronomical and Physical Geodesy at the Technical University of Munich. He coordinated the level-2 processing team for the GOCE mission and led several studies for future gravity field missions.

Annette Eicker (Prof. Dr.) is a Professor of "Geodesy and Adjustment Theory" at HafenCity University in Hamburg, Germany. She works on GRACE data analysis and geophysical applications and is currently the president of the Inter-Commission Committee for "Geodesy and Climate Research" of the International Association of Geodesy (IAG).

Frank Flechtner (Prof. Dr.) is Head of Section "Global Geomonitoring and Gravity Field" at the Hemholtz-Zentrum Potsdam Deutsches GeoForschungsZentrum and Professor of "Physical Geodesy" at the Technical University of Berlin. He was the GRACE Co-PI and is currently responsible for the operation of German-contributed mission elements for GRACE Follow-On.

Preface to "Remote Sensing by Satellite Gravimetry"

During the last two decades, satellite gravimetry has become a new remote sensing technique providing a detailed global picture of the physical structure of the Earth. With the CHAMP, GRACE, and GOCE missions, spatial mass irregularities and the transport of mass in the Earth system could be systematically observed and monitored from space. With the GRACE Follow-On mission, launched in 2018, the time series of mass transport observations is continued for another number of years, which enables the disentanglement of anthropogenic and natural sources of climate change impact on the Earth system. A wide range of Earth science disciplines and operational observing systems benefit from these observations and have been enabled to improve their models and to get new insights into processes of the Earth system (e.g., water cycle, continental hydrology, ocean modelling, ice sheet and glacier melting, lithosphere modelling). The value of satellite gravimetry has been acknowledged by the international Earth science community in various resolutions, and in the meantime it is regarded as a new remote sensing tool, providing information that is complementary to other remote sensing techniques. Therefore, the Earth science community and space agencies are currently planning future satellite gravimetry missions in order to secure sustained observations of mass distribution and mass transport on a long-term basis with higher accuracy and better spatial and temporal resolution. This book is a collection of papers reporting results and research on various aspects of satellite gravimetry. It includes contributions about modelling the static and time-variable Earth gravity field using satellite gravimetry data; about design and capabilities of next-generation satellite gravimetry missions; and about results obtained in Earth science applications, namely for the cryosphere, the hydrosphere, and earthquake research.

Thomas Gruber, Annette Eicker, Frank Flechtner
Editors

Article

The GFZ GRACE RL06 Monthly Gravity Field Time Series: Processing Details and Quality Assessment

Christoph Dahle [1,*], Michael Murböck [1,2], Frank Flechtner [1,2], Henryk Dobslaw [1], Grzegorz Michalak [1], Karl Hans Neumayer [1], Oleh Abrykosov [1,†], Anton Reinhold [1], Rolf König [1], Roman Sulzbach [1,3] and Christoph Förste [1]

[1] Department 1: Geodesy, GFZ German Research Centre for Geosciences, 14473 Potsdam, Germany; michael.murboeck@gfz-potsdam.de (M.M.); frank.flechtner@gfz-potsdam.de (F.F.); henryk.dobslaw@gfz-potsdam.de (H.D.); grzegorz.michalak@gfz-potsdam.de (G.M.); karl.neumayer@gfz-potsdam.de (K.H.N.); oleg.abrikosov@gfz-potsdam.de (O.A.); contact@antoncouper.com (A.R.); rolf.koenig@gfz-potsdam.de (R.K.); roman.lucas.sulzbach@gfz-potsdam.de (R.S.); christoph.foerste@gfz-potsdam.de (C.F.)
[2] Institute of Geodesy and Geoinformation Science, Technische Universität Berlin, 10623 Berlin, Germany
[3] Institute of Meteorology, Freie Universität Berlin, 12165 Berlin, Germany
* Correspondence: christoph.dahle@gfz-potsdam.de
† Current address: SpaceTech GmbH Immenstaad, 88090 Immenstaad, Germany.

Received: 10 August 2019; Accepted: 6 September 2019; Published: 11 September 2019

Abstract: Time-variable gravity field models derived from observations of the Gravity Recovery and Climate Experiment (GRACE) mission, whose science operations phase ended in June 2017 after more than 15 years, enabled a multitude of studies of Earth's surface mass transport processes and climate change. The German Research Centre for Geosciences (GFZ), routinely processing such monthly gravity fields as part of the GRACE Science Data System, has reprocessed the complete GRACE mission and released an improved GFZ GRACE RL06 monthly gravity field time series. This study provides an insight into the processing strategy of GFZ RL06 which has been considerably changed with respect to previous GFZ GRACE releases, and modifications relative to the precursor GFZ RL05a are described. The quality of the RL06 gravity field models is analyzed and discussed both in the spectral and spatial domain in comparison to the RL05a time series. All results indicate significant improvements of about 40% in terms of reduced noise. It is also shown that the GFZ RL06 time series is a step forward in terms of consistency, and that errors of the gravity field coefficients are more realistic. These findings are confirmed as well by independent validation of the monthly GRACE models, as done in this work by means of ocean bottom pressure in situ observations and orbit tests with the GOCE satellite. Thus, the GFZ GRACE RL06 time series allows for a better quantification of mass changes in the Earth system.

Keywords: satellite gravimetry; GRACE; Level-2 processing; time-variable gravity field; mass change monitoring

---

1. Introduction

During more than 15 years (April 2002 through June 2017) of successful science operations phase, the Gravity Recovery and Climate Experiment (GRACE) mission enabled breakthroughs in monitoring the terrestrial water cycle (e.g., [1,2]), ice sheet and glacier mass balances (e.g., [3,4]), sea-level change (e.g., [5,6]) and ocean bottom pressure variations (e.g., [7,8]). A comprehensive overview of numerous other GRACE-related studies and their contributions to understanding changes in the global climate system is reviewed by Tapley et al. [9]. These results are based on time-variable, in general monthly, global gravity field models.

Such models, so-called GRACE Level-2 products, are routinely generated by the joint US-German GRACE Science Data System (SDS) consisting of the Center for Space Research at the University of Texas at Austin (CSR), NASA's Jet Propulsion Laboratory (JPL), and the German Research Centre for Geosciences (GFZ). To provide consistent long-term gravity field time series of highest possible quality to the user community the SDS has recently reprocessed its gravity field solutions over the complete GRACE mission duration. This latest reprocessing, referred to as release 06 (RL06) [10–12], comprises updated background models and processing standards which are also applied to consistently process the first release of gravity field solutions based on data from the GRACE Follow-on (GRACE-FO) mission [13]. The GRACE-FO satellites were successfully launched in May 2018 and are designed to continue the unique GRACE data record for at least five additional years.

Apart from the SDS, recent GRACE gravity field time series are provided by other processing centers, e.g., ITSG-Grace2018 [14], CNES/GRGS RL04 [15], AIUB RL02 [16], or Tongji-Grace2018 [17].

The purpose of this work is to demonstrate the progress that has been achieved with the current GFZ GRACE RL06 time series compared to its precursor GFZ RL05a [18]. First, an overview of the GRACE gravity field processing procedure at GFZ is given and modifications implemented within the RL06 reprocessing are described (Section 2). Thereafter, results are discussed with focus on the internal quality in the spectral and spatial domain comparing RL06 with RL05a (Section 3). Additionally, both time series are validated by external data (Section 4). Finally, conclusions are drawn, and the main findings of this work are summarized (Section 5). All results confirm that GFZ's RL06 time series is a significant step forward in terms of accuracy, consistency and reliability and thus enables a better quantification of climate change related phenomena in the Earth system.

2. GRACE Level-2 Gravity Field Processing at GFZ

GRACE global monthly gravity field recovery at GFZ is based on the so-called "dynamical approach" using GFZ's Earth Parameter and Orbit System (EPOS) software package (https://www.gfz-potsdam.de/en/section/global-geomonitoring-and-gravity-field/topics/earth-system-parameters-and-orbit-dynamics/earth-parameter-and-orbit-system-software-epos/). Underlying satellite orbit perturbations rely on a precise numerical orbit integration taking into account all reference system and force model related quantities [19]. The integrated orbit is then fitted to the GRACE tracking observations, i.e., GPS code and carrier phase observations and K-band inter-satellite ranging data between the two GRACE satellites. This step is done in a least-squares adjustment process solving iteratively for both satellites' state vector at the beginning of each arc, observation-specific parameters, in particular GPS receiver clock offsets, GPS carrier phase ambiguities and calibration parameters for the accelerometers, and other arc-specific parameters such as empirical accelerations. The term "arc" refers to the time length of the integrated orbit starting with one initial state vector which is typically one day. After convergence of the initial orbit adjustment with the a priori force models, the observation equations are extended by partial derivatives for the unknown global parameters describing the gravitational potential, represented by spherical harmonic (SH) gravity field coefficients. Arc-by-arc normal equation (NEQ) systems are generated in this way from the observation equations and accumulated over nominally one month to one overall system which is then solved by matrix inversion. For the complete GRACE mission 163 of these NEQ systems and corresponding Level-2 gravity field products have been derived.

In the following subsections, specific aspects relevant for GRACE gravity field processing are discussed and modifications in GFZ's RL06 processing relative to the processing of the previous release GFZ RL05a are described.

2.1. GPS Constellation

Since GPS tracking data are processed, the precise orbits and sender clock offsets of the GPS constellation, nominally consisting of 32 satellites, must be known. Using GFZ's EPOS software, it has been demonstrated by König [20] that an integrated processing of both GPS satellites and the GRACE satellites, i.e., simultaneous orbit determination and parameter adjustment at the observation level, is beneficial for the determination of the terrestrial reference frame (TRF). However, the dynamic part of the TRF in König [20] is limited to the SH gravity field coefficients of degrees one and two and all parameters have been estimated daily. When expanding the parameter space of the gravity field to a maximum degree and order (d/o) of 90 or higher, typically for monthly GRACE gravity field solutions, which requires that daily NEQs need to be stacked to obtain a monthly solution, such an integrated approach becomes quite demanding in terms of computational efficiency and proper separation of daily and monthly Earth system and other parameters. Thus, for GRACE gravity field recovery, it is common practice to determine precise GPS orbits and clock offsets beforehand using GPS tracking data from a globally distributed ground station network, and then keep them fixed in the subsequent GRACE orbit and gravity field adjustment process. At GFZ, the GPS constellation used for gravity field determination is traditionally generated in-house allowing for best possible consistency.

For processing GRACE RL06, the GPS constellation has been reprocessed as well. Compared to the previous RL05 GPS constellation, the following changes have been implemented: (1) application of the ITRF2014 reference frame realized by the IGS14 ground station network (https://mediatum.ub.tum.de/doc/1341338/1341338.pdf) instead of ITRF2008/IGS08; (2) increase in the number of GPS ground stations from approx. 70 to approx. 120 to 140; (3) improved solar radiation pressure parameterization; and (4) adaption of the background models according to GRACE RL06 standards (see Section 2.3). To assess the level of accuracy of GFZ's RL05 and RL06 GPS constellations, daily root mean square (RMS) values of position differences regarding final orbits provided by the International GNSS Service (IGS) for all available GPS satellites are calculated. During the GRACE mission period, these 3D RMS values are typically in the range of 3 to 5 cm in the first years until end of 2006 and around 3 cm for all years thereafter for the RL06 constellation. The corresponding global 1D RMS over the whole 15 years is 1.96 cm for RL06 and 2.28 cm for RL05 revealing that the current RL06 constellation is closer to and more consistent with the official IGS products compared to its predecessor.

2.2. GRACE Observations

Both GFZ RL05a and RL06 are based on official GRACE Level-1B (L1B) instrument data processed and provided by JPL [21]. In particular, the following L1B observations are used:

- K-band range-rate (KRR) observations (KBR1B product) as primary observations to retrieve monthly gravity field estimates.
- GPS code and carrier phase observations (GPS1B product) used for precise orbit determination of the GRACE satellites. Please note that inside GFZ's EPOS software, zero-difference ionosphere-free (L3) linear combinations of the measurements are generated and processed.
- Linear accelerations (ACC1B product) to model non-conservative forces acting on the GRACE satellites. Please note that these onboard accelerometer (ACC) observations are not treated as classical observations in a least-squares sense, but only as part of the right-hand side force model.
- Star camera observations (SCA1B product) describing the GRACE satellites' attitude, required for the rotation from the satellite reference frame (SRF) to the inertial frame.

For GFZ RL05a, L1B RL02 data were used. The same RL02 data are used for RL06 in case of ACC1B and GPS1B. Regarding KBR1B and SCA1B, an improved L1B RL03 dataset [22] has been made available by JPL and is used for RL06. At the end of the mission, i.e., during the period November 2016 through June 2017, the ACC instrument aboard GRACE-B was turned off due to battery issues, and thus transplanted ACC observations from GRACE-A have to be used (in the following, this period is denoted by "GRACE single ACC"). For RL05a, a simple ACC data transplant was used, which had only attitude and time corrections applied. As part of JPL's L1B RL03 dataset, an improved transplant version [23], additionally corrected for thruster spikes, has been made available and is used for RL06.

Before orbit and gravity field determination with EPOS, the L1B data are preprocessed as follows for RL05a: (1) KRR observations are modified by adding the light time correction and the antenna offset correction (both also taken from the KBR1B products); (2) GPS observations are cleaned, phase cycle slips are detected, and the data are downsampled to 30 s; (3) ACC observations are downsampled from 1 Hz to 0.2 Hz by simple decimating and data gaps < 100 s are interpolated; and (4) data gaps < 500 s in the SCA1B data are interpolated. For RL06, the only modification concerns (4): SCA1B data are downsampled from 1 Hz to 0.2 Hz, smoothed and data gaps are filled, all done simultaneously using spherical quadrangle interpolation (SQUAD). After preprocessing, the L1B observations are analyzed for data gaps caused by satellite-specific events such as, e.g., maneuvers and reboots of the Instrument Processing Unit (IPU). In case of larger gaps, the nominally one day long arcs are split into two or more shorter arcs over partial days. The minimum length of an arc, however, is set to three hours. To avoid any unwanted effects in the observations around such events, a margin of ten minutes is applied before and after data gaps when defining the final arcs.

2.3. Background Models

By definition, GRACE gravity field models represent geophysical signals caused by variations in the terrestrial water storage, mass loss in polar ice sheets and inland glacier systems, ocean mass variations, global isostatic adjustment and large earthquakes. Consequently, gravity variations caused by solid Earth and pole tides, atmosphere and ocean tides or short-term non-tidal atmospheric and oceanic mass variations are not supposed to be included in the gravity field solutions and are therefore taken into account during the data processing via background models. On the other hand, any error contained in the background models degrades the quality of the gravity field solutions, especially at low frequencies [24]. Thus, some of these background models have been updated for processing the GFZ RL06 time series (Table 1).

Table 1. Overview of background models used for GFZ RL05a and RL06 processing.

Background Model	GFZ RL05a	GFZ RL06
Static a priori gravity field	EIGEN-6C [25] (up to d/o 200)	EIGEN-6C4 [26] (up to d/o 200)
Time-variable a priori gravity field	Trend, annual and semi-annual coefficients of EIGEN-6C (up to d/o 50)	GFZ RL05a (DDK1 smoothed, up to d/o 50), only used during data editing
Ocean tides	EOT11a [27]	FES2014 [28]
Atmospheric tides	Biancale & Bode [29]	same as RL05a
Non-tidal atmospheric and oceanic mass variations	AOD1B RL05 [30]	AOD1B RL06 [31]
Ocean pole tide	Desai [32]	same as RL05a
Solid Earth and pole tides	IERS2010 [33]	same as RL05a
3rd body ephemerides	JPL DE421	JPL DE430

Among the background models listed in Table 1, the choice of the Atmosphere and Ocean De-aliasing (AOD) and the ocean tide model has the most impact on the quality of the monthly GRACE solutions and hence their geophysical interpretation [24]. The AOD1B model is an official GRACE L1B product routinely generated at GFZ. Compared to its precursor AOD1B RL05, the most recent release AOD1B RL06 has higher temporal (3-hourly vs. 6-hourly) and spatial (maximum SH d/o 180 vs. 100) resolution. Further details are described by Dobslaw et al. [31], where also improvements in GRACE gravity field processing by means of variance reduction of K-band range-acceleration residuals are already reported when using AOD1B RL06 instead of RL05. Regarding ocean tides, similar improvements are observed when using FES2014 instead of EOT11a. Generally, the choice of the ocean tide model does not significantly affect the quality of individual monthly GRACE solutions, but regional deficiencies, especially at high latitudes, can become visible in GRACE time series analysis. In this context, Ray et al. [34] reported that the FES2014 model, used for GFZ RL06, shows improvements in most of the polar regions compared to its precursor and performs comparable to other state-of-the-art ocean tide models.

2.4. Processing Strategy

The processing scheme of GRACE gravity field processing at GFZ consists of the following steps: (1) GPS data editing; (2) KRR data editing; (3) generation of a priori orbits; (4) generation of arc-wise NEQs; and (5) accumulation to and solving of monthly NEQs. For all previous GFZ GRACE releases from RL01 to RL05a, the strategy how these steps are performed is more or less identical and has been described by Schmidt [35]. For the current GFZ GRACE RL06 processing, the strategy has been considerably changed and is described in this section. An overview of the GFZ RL05a and RL06 processing strategies is given in Table 2.

Table 2. Overview of GFZ RL05a and RL06 processing strategies.

		GFZ RL05a	GFZ RL06
GPS data editing			
Remarks		GRACE-A & -B jointly processed	GRACE-A & -B independently processed
Time-variable a priori gravity field		yes	yes
	$\sigma_{GPSphase}$	0.7 cm	0.3 cm
Observation weights	$\sigma_{GPScode}$	70.0 cm	40.0 cm
	σ_{KRR}	50 µm/s	no KRR observations
KRR data editing			
Remarks		further GPS data editing still possible automated editing based on 8-sigma elimination	no further GPS data editing no automated editing, instead visual inspection of residuals
Time-variable a priori gravity field		yes	yes
	$\sigma_{GPSphase}$	0.7 cm	0.3 cm
Observation weights	$\sigma_{GPScode}$	70.0 cm	40.0 cm
	σ_{KRR}	0.1 µm/s	0.3 µm/s
Generation of a priori orbits / generation of arc-wise NEQs			
Time-variable a priori gravity field		yes	no
	$\sigma_{GPSphase}$	0.7 cm	arc-wise based on residuals scaled by empirical factor of 7
Observation weights	$\sigma_{GPScode}$	70.0 cm	
	σ_{KRR}	0.1 µm/s	arc-wise based on residuals

Editing of GPS data is done by automated elimination during an iterative precise orbit determination (POD). The elimination is based both on an n-sigma criterion, with varying values for n in the different iterations, and additionally an absolute threshold for the size of GPS residuals. The main difference in the GPS data editing step is that for RL06 the POD for both GRACE satellites is done completely independently from each other in contrast to RL05a, where down-weighted KRR observations were included and a common POD for GRACE-A and -B was applied. The goal for RL06 is to obtain best possible absolute orbit accuracy for each of the two spacecraft and thus to avoid that the elimination of GPS observations for one satellite is influenced by the other. Another difference concerns the empirical corrections for GPS phase center variations (PCV): For RL05a, a certain unvaried PCV correction based on one month (April 2008) of GPS residuals has been applied for the whole time series, whereas for RL06, monthly PCV corrections are computed from the corresponding monthly residuals. A novelty of RL06 is that also GPS code residual variation maps per month are computed and applied. In general, these empirical corrections are very stable over time, but they are significantly affected by systematic patterns whenever radio occultation measurements are activated onboard one of the GRACE satellites (Figure 1). Because activation and deactivation of these measurements has occurred several times throughout the GRACE mission (mostly related to satellite swap maneuvers), a monthly computation of these corrections has been chosen as the preferable option. Finally, the a priori weights for GPS phase and code observations have been changed from RL05a to RL06 to better reflect the actual level of the RMS of the corresponding residuals (see Table 2).

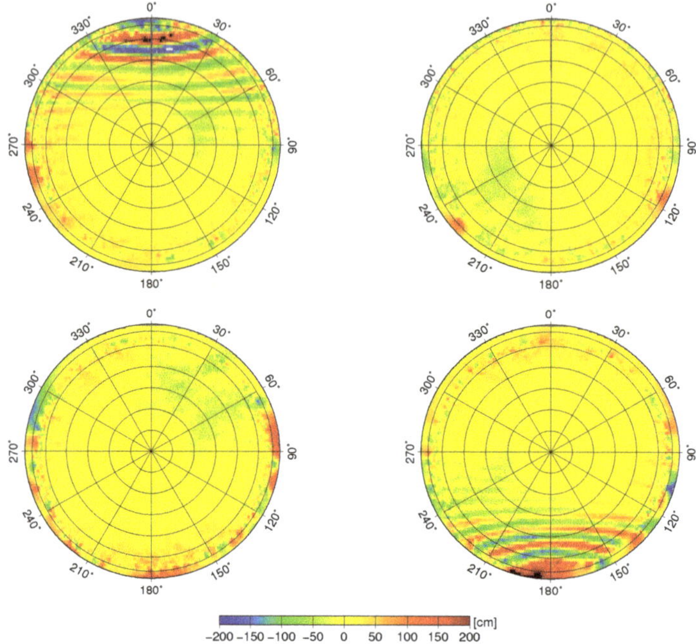

Figure 1. Code residual variation maps of GRACE-A (**top**) and GRACE-B (**bottom**) for June 2014 (**left**) and August 2014 (**right**). Radio occultation measurements on GRACE-A were activated in June 2014 and deactivated in August 2014 and vice versa on GRACE-B, after a satellite swap in July 2014.

For KRR data editing, a common POD for GRACE-A and -B is performed. In case of RL05a, where down-weighted KRR observations were already included in the previous step, the KRR weight was increased by setting the a priori standard deviation to 0.1 μm/s, and elimination of KRR observations was automatically done based on an 8-sigma criterion. Further editing of GPS observations during this step was not explicitly turned off. In contrast, now for RL06 KRR observations are only introduced in this step with a slightly increased a priori standard deviation of 0.3 μm/s which better matches—as done similarly for GPS—the corresponding orbit fit RMS. No automated elimination at all is applied; instead, possible elimination of KRR observations-if needed-is based on visual inspection of the KRR residuals. Usually, observations are edited if the residuals exceed a threshold of 3 μm/s, but this threshold is not fixed and other editing criteria such as anomalous behavior of the KRR residuals over a certain period of time within one arc might be applied. By modifying the KRR data editing as described, inadvertent elimination of observations over areas with large geophysical signals is avoided and about 1% more KRR observations (approx. 0.3 days accumulated over one month) remain and are used for gravity field determination in the case of RL06.

The purpose of generating the a priori orbit is to assure that convergence of a POD using the edited GPS and KRR observations and iterated orbit parameter estimates is reached after one iteration. If this condition is fulfilled, arc-wise NEQs are set up starting with the observations and initial parameters from the a priori orbit run. In case of RL05a, the observation weights from the previous data editing steps remained unchanged, whereas for RL06 individual arc-wise weighting for GPS and KRR observations has been introduced. Arc-wise a priori standard deviations are based on the RMS of the corresponding residuals after the KRR editing run: In case of KRR, they are equal to the RMS values; for GPS, the RMS values are multiplied by an empirical factor of 7. This intentional down-weighting of GPS observations relative to the K-band observations is necessary to obtain better gravity field solutions and is also done similarly by other processing centers (see, e.g., [10,16]).

Another modification from RL05a to RL06 regards the time-variable a priori gravity field which in case of RL05a was used as background model also during these two last-mentioned processing steps. This required restoring of the monthly mean of these a priori fields to the estimated monthly GRACE solution to provide users a Level-2 gravity field product which contains the full time-variable signal as expected per definition. For RL06, the time-variable gravity field background model is only used during the data editing steps, where background modeling is desired to be as realistic as possible, but not during the steps where gravity field determination takes place. Thus, any possible bias that might be introduced by applying a remove-restore procedure as described for RL05a is avoided for RL06.

2.5. Parametrization

In conjunction with the processing strategy, the orbit and instrument parametrization has also been significantly changed from GFZ RL05a to RL06 (see Table 3).

In RL05a processing, empirical accelerations were only set up and estimated during the GPS data editing step. In the subsequent steps, these parameters were completely removed, and empirical K-band parameters were introduced [35]. For RL06, empirical accelerations are set up more frequently (once per orbital revolution) and remain as parameters throughout all processing steps to assure a consistent parametrization from GPS editing until gravity field estimation. To avoid that these parameters absorb too much gravity field signal, an a priori standard deviation of $1\text{E-}8\,\text{m/s}^2$ is applied as constraint. In contrast to RL05a, no K-band parameters at all are set up in RL06. Internal tests have shown that the additional estimation of K-band parameters does not significantly impact the estimated gravity field solution, but leads to a degradation in orbit accuracy.

Further modifications from RL05a to RL06 affect the ACC instrument parameters. Regarding the ACC biases, their number has been decreased. Now for RL06, usually three biases per arc are estimated in along-track and radial direction (at the beginning, middle, and end of the arc), and nine in cross-track direction (at the beginning and end, and equally spaced in between). The number of biases can become less in case the arc length is shorter than 24 h, as the minimum spacing between two biases is set to three hours. Between the epochs where biases are estimated, the bias is modeled as a natural cubic spline function (RL05a: linear interpolation). Regarding ACC scale factors, one per arc and direction is estimated in case of RL06. This is a lesson learned from RL05a processing, where for most of the time series the scales were fixed to one, which turned out to be not optimal especially in the later years of the GRACE mission. Therefore, this was modified within RL05a to estimating 3-hourly scales, which helped to improve the quality of the RL05a solutions, but on the other hand tends to over-parameterize the gravity field estimation process. Another notable difference between RL05a and RL06 is the parametrization during the "GRACE single ACC" period: Whereas for RL05a the parametrization was the same as for the rest of the GRACE mission, a fully populated scale factor matrix is estimated per arc in case of RL06. This has been proposed first by Klinger and Mayer-Gürr [36] and is applied to the complete ITSG-Grace2016/2018 time series, as well as to other recent releases such as the CSR RL06 [10] and JPL RL06 [11] time series. For GFZ RL06, the additional six parameters, i.e., the off-diagonal elements of the scale factor matrix, are constrained with an a priori standard deviation of 1E-3 since otherwise, the inversion of the NEQ system becomes unstable and in many cases would fail.

In summary, the total number of orbit and instrument parameters described above has decreased for RL06 compared to most of the RL05a solutions, and is nearly identical compared to the RL05a solutions with modified ACC parametrization. Moreover, as already mentioned, the parametrization is now consistent during all processing steps. It has to be mentioned as well that there are other parameters not listed in Table 3, namely GPS receiver clock offsets (2880 per 1-day arc and satellite) and GPS phase ambiguities (approx. 400 per 1-day arc and satellite). However, these two parameter groups are pre-eliminated before gravity field estimation and there are no differences in their treatment between RL05a and RL06 processing.

Finally, regarding the main parameters of interest, i.e., the gravity field parameters, the maximum SH d/o estimated was slightly increased from 90 (RL05a) to 96 (RL06). New with GFZ RL06, an independently estimated GRACE time series only up to d/o 60 is provided additionally. These two decisions have been made jointly within the GRACE SDS to ensure better consistency between the three SDS time series than it was the case with RL05. Furthermore, both KRR and GPS observations contribute to the full spectrum of the monthly gravity field estimates in GFZ RL06, whereas in GFZ RL05a the contribution of GPS was limited to SH d/o 80.

Table 3. Number and properties of GFZ RL05a and RL06 orbit and instrument parameters (numbers are per arc representative for the nominal arc length of 24 h).

	GFZ RL05a		GFZ RL06	
	GPS editing step	*subsequent steps*	*GPS editing step*	*subsequent steps*
Orbital elements		6/6 (GRACE-A/GRACE-B)		
Empirical accel.	20/20	none	64/64	
Details		cos/sin coefficients of 1/rev periodical model		
	every 4.8 h in TN	-	1/rev in TN	
	no constraint	-	constraint: $\sigma = $ 1E-8 m/s^2	
K-band param.	none	48	none	
Details	-	range-rate bias & drift every 90 min; cos/sin coeff. of range bias every 180 min	-	
ACC param.	75/75 [1]		18/18 [3]	
	54/54 [2]		18/18 [4]	24/24 [4]
Details	1-hourly biases in RTN; scale factors fixed to 1 [1]		3 biases per arc in RT; 9 in N 1 scale factor per arc in RTN	
	3-hourly biases in RTN; 3-hourly scale factors in RTN[2]		6 off-diagonal elements of scale factor matrix; constraint: $\sigma = $ 1E-3 [4]	

R, T, N: radial, along-track and cross-track direction in SRF. [1] period from 2003/01 through 2013/05; [2] period from 2002/04 through 2002/12 and 2013/06 through 2017/06; [3] period from 2002/04 through 2016/08; [4] period from 2016/11 through 2017/06 ("GRACE single ACC").

2.6. Orbit Quality

The quality of the GRACE orbits determined prior to gravity field estimation was not in the focus during GFZ RL05a processing. For GFZ RL06, the modifications in processing strategy and parametrization are motivated not only to obtain improved gravity fields, but also orbits of high quality.

A first indication that this goal is reached is given by GPS phase and code residuals of the GPS editing runs which are extremely stable during the whole mission with arc-wise RMS values of about 3 mm and 40 cm, respectively (for the "GRACE single ACC" period, only a slight increase is observed). In contrast to RL05a, the size of GPS residuals does not increase at all when adding KRR observations in the subsequent processing steps which can be attributed to the consistent parametrization.

Another independent orbit validation during RL06 processing is routinely done by means of satellite laser ranging (SLR) observations from ground stations to the GRACE satellites, provided by the International Laser Ranging Service (ILRS). Coarse outliers in the SLR observations are eliminated by a 20 cm threshold. Mean values and standard deviations of GRACE-A and -B SLR residuals per year are shown in Figure 2 for all available stations and a subset of high-quality stations. The definition of such a subset has been outlined by Arnold et al. [37] and the same 12 stations are used here for better comparability. These high-quality stations contribute between 50% and 75% of all available observations. It can be seen that the standard deviations are in the range of 20 mm to 25 mm for all stations and about 15 mm for the high-quality stations. For the year 2010, the standard deviations for GRACE-A are 24.5 mm and 13.7 mm, respectively, which agrees very well with values of 24.4 mm and 12.3 mm, respectively, reported in [37]. As for the GPS residuals, the values are the same whether only GPS observations are used or KRR observations are added, and are also very similar for both GRACE satellites. Increased standard deviations, in particular for recent years, are observed for the a priori orbits which can be explained by the fact that these orbits are determined without a time-variable gravity background model and GPS observations are down-weighted. The mean values of GRACE-A and -B as are also very consistent, independent of the processing step or whether all or the high-quality stations are evaluated. They are mostly in the range of -10 mm to -15 mm which is relatively large. However, this is not necessarily due to the orbit quality, but may also be caused by incorrect values for the GPS phase center offset (PCO) of the GRACE GPS navigation antennas. For GFZ RL06, PCOs relative to the antenna reference point provided by Montenbruck et al. [38] are used. A geometrical offset (distance between the satellites' center of mass and the antenna reference point) of -444 mm is added–only for the z-component in the SRF–resulting in a total L3 PCO in z-direction of -391.7 mm. This value is approx. 22 mm larger than the corresponding value derived from the official GRACE L1B vector offset product for the GPS main antenna (VGN1B, see [21]) which might at least partly explain the relatively large negative offsets reported here.

Overall, the quality of the GFZ GRACE RL06 orbits is satisfyingly well confirming that the processing changes relative to GFZ RL05a have been a step into the right direction.

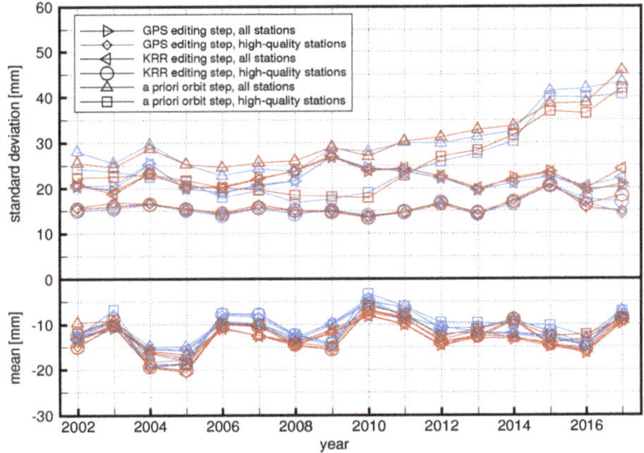

Figure 2. Mean and standard deviation per year of GRACE-A (blue) and -B (red) SLR residuals during the different GRACE processing steps for all available stations and a subset of high-quality stations.

3. Results

GFZ GRACE RL06 gravity field results are analyzed and discussed in comparison to GFZ's RL05a GRACE time series in the following subsections. Most of the results shown are relative to a climatology model (individually derived for each time series) which has been estimated as follows: The dominating signal content of the time series is approximated by fitting a proper parameter model coefficient-wise to the monthly solutions. Here, eight parameters describing the constant and linear part as well as periodic sine and cosine amplitudes for annual, semi-annual and 161-days (GRACE aliasing period for the ocean tide S2) periods are used. Furthermore, the formal errors of the monthly SH coefficients are used as a priori information to weight each individual monthly coefficient when estimating the climatology. Months with short repeat cycles (i.e., solutions which were regularized in RL05a) as well as the seven "GRACE single ACC" solutions are excluded.

3.1. Formal and Empirical Errors

In this subsection formal and empirical errors of the GFZ RL06 time series are analyzed in the spectral domain and compared to GFZ RL05a. The formal errors are the standard deviations of the gravity field parameters estimated in the least-squares adjustment process, i.e., the square root of the diagonal of the gravity field parameter part of the variance-covariance matrices. In principle, these errors should give a good indication of the real errors of the estimated parameters. However, as variance-covariance information of all the input data is insufficiently (observations) or not all (background models) applied, the formal errors are typically too optimistic.

Another quantification of the errors of such a gravity field time series are empirically derived values. Here, residuals regarding the climatology described above are defined as empirical errors of the time series.

Figure 3 shows the RMS over the whole time series of the empirical and formal errors for RL05a and RL06. The main differences between empirical and formal errors can be seen in the very low SH degrees and around so-called resonance orders. Due to residual signals in the very low degrees, which are not covered by the eight parameters, the empirical errors show much higher values than the formal ones here. Around the resonance orders, which are integer multiples of approximately 15, it is well known that GRACE errors are larger due to systematic effects from temporal aliasing caused by background model errors. This effect is not very well represented in the formal errors.

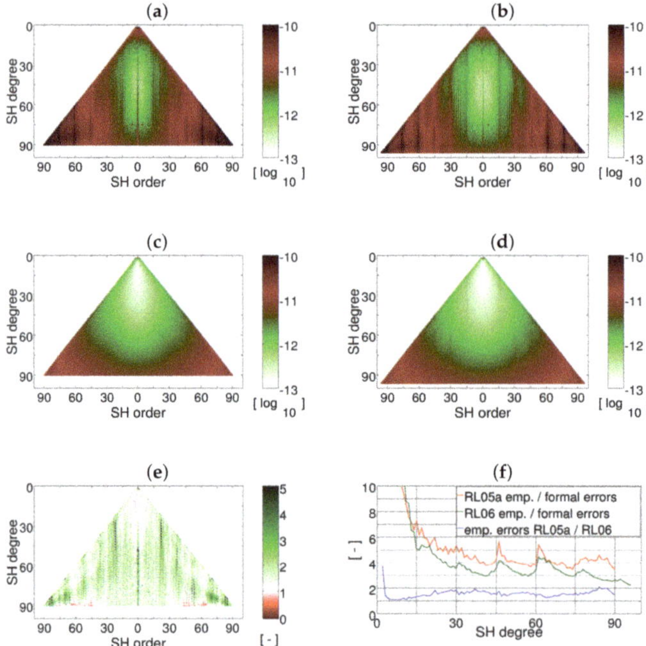

Figure 3. (**a–d**) SH spectra of empirical error RMS for RL05a (**a**) and RL06 (**b**); and of formal error RMS for RL05a (**c**) and RL06 (**d**); (**e**) Ratio of SH spectra of empirical error RMS "RL05a/RL06"; (**f**) SH degree amplitudes of the ratio of empirical error RMS "RL05a/RL06" (blue), and the ratios "empirical/formal error RMS" for RL05a (red), and RL06 (green).

Compared to RL05a (Figure 3a,c), Figure 3b,d reveal smaller empirical errors and more realistic formal errors for RL06. This becomes even more clear when looking at the ratio of the empirical error

RMS values between RL05a and RL06 (Figure 3e) which is > 1 for nearly all coefficients and also when plotted as amplitudes per SH degree (Figure 3f). Also, the degree amplitudes of the ratio between empirical and formal errors indicate that the GFZ RL06 formal errors are more realistic as smaller variations than for GFZ RL05a are visible and the curve is closer to the value of one.

3.2. Degree Amplitudes

Difference degree amplitudes relative to climatology are shown in Figure 4. The spread of monthly degree amplitudes is less for RL06 than for RL05a which illustrates that RL06 is a more homogeneous time series. Significant differences between the RL05a and RL06 median degree amplitudes are already visible at approx. SH degree 15, and almost all monthly RL06 degree amplitude curves are well below the median RL05a curve (or vice versa) for medium and high degrees indicating a notably improved signal-to-noise ratio for RL06. When only looking at the "GRACE single ACC" period, large improvements from RL05a to RL06 are achieved according to the corresponding median degree amplitudes. These improvements are not only present for medium and high degrees, but also for the very low degrees. However, the quality of the "GRACE single ACC" solutions is still significantly worse than for the rest of the GRACE mission also in case of RL06.

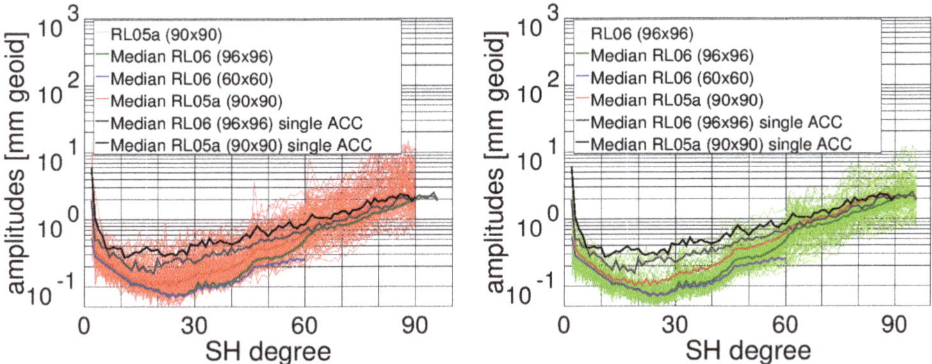

Figure 4. Degree amplitudes relative to a climatology model for GFZ RL05a (**left**) and GFZ RL06 (**right**); thin lines represent monthly solutions (without "GRACE single ACC" solutions and those regularized in RL05a), and bold lines represent the median curves (the same curves are shown in both plots) for RL05a (red), RL06 (96 × 96, green), RL06 (60 × 60, blue), and the "GRACE single ACC" months only for RL05a (black) and RL06 (96 × 96, grey).

3.3. RMS of Residuals in the Spatial Domain

To assess the quality of the GFZ RL06 time series in the spatial domain, monthly residual SH coefficients relative to climatology are converted to gridded mass anomalies in terms of equivalent water height (EWH). To reduce the impact of spatially correlated noise, the solutions are de-correlated and smoothed by applying the non-isotropic DDK filter [39]. Then, RMS values of the time series per grid point are calculated and shown in Figure 5. For the period until August 2016, i.e., without the "GRACE single ACC" months, a clear reduction of RMS variability for RL06 has been achieved (Figure 5a,b). Geophysical signals over continental areas, where they are much larger than over the oceans, are less superimposed by the typical GRACE striping pattern and thus better detectable. Variability over the oceans is generally expected to be rather small and is therefore often interpreted as upper error bound for monthly global GRACE gravity field models. Latitude-dependent weighted RMS (wRMS) values over the oceans decrease from 6.8 cm (RL05a) to 4.0 cm (RL06, 41% relative improvement) when DDK5 filtered, and from 3.4 cm (RL05a) to 2.1 cm (RL06, 38% relative improvement) when DDK3 filtered. For the "GRACE single ACC" period,

Figure 5c shows DDK3 filtered RMS variability to allow a direct comparison with the period before, but it becomes obvious that much stronger decorrelation and smoothing would be required here to extract geophysical signals. Nevertheless, the corresponding wRMS values over ocean decrease again significantly from 9.5 cm (RL05a) to 6.1 cm (RL06, 36% relative improvement).

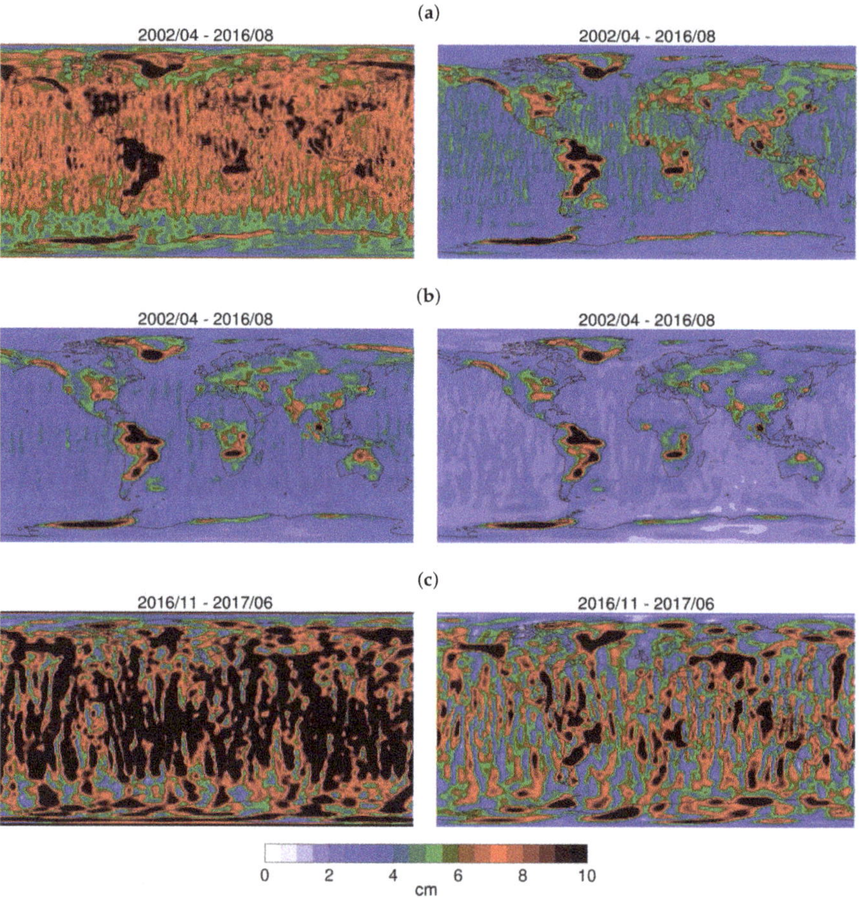

Figure 5. RMS of the time series of residuals (cm EWH) relative to a climatology model (without months regularized in RL05a) for GFZ RL05a (left) and GFZ RL06 (right); the following different cases are shown: period from 2002/04 through 2016/08, DDK5 filtered (**a**) and DDK3 filtered (**b**); and "GRACE single ACC" period, DDK3 filtered (**c**).

Monthly wRMS values over the oceans for the complete GRACE time series (DDK5 filtered) are shown in Figure 6. Again, it becomes visible that GFZ RL06 is a clear improvement over RL05a in terms of noise reduction and homogeneity. Some months where RL06, particularly for the 96 × 96 time series, exhibits larger wRMS values than RL05a can be attributed to short period repeat orbit cycles (the most harmful repeat orbits during the GRACE mission are: 61/4 around September 2004, 46/3 around May 2012, 77/5 around December 2013, 31/2 around February 2015). RL05a solutions for these months were regularized which is not the case anymore for RL06 as the additionally provided RL06 60 × 60 time series, which is less sensitive to these short period repeat orbits, might be analyzed instead. Apart from the repeat cycles just mentioned before, the wRMS values of the RL06 96 × 96 and

60×60 time series are mostly almost identical. Periods where these values are notably larger are again related to less harmful repeat cycles such as the long-lasting 107/7 repeat orbit around December 2009. Finally, also Figure 6 shows that the "GRACE single ACC" months are of much less quality than the rest of the time series. At least, the RL06 solution for May 2017 is now of comparable quality (it must be noted that for this solution GRACE-B ACC data is actually available and used).

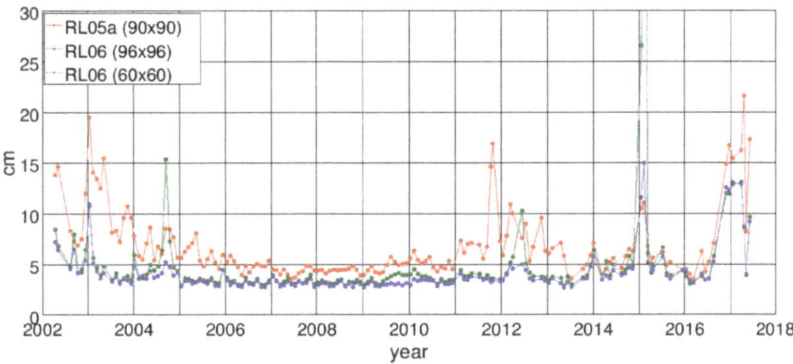

Figure 6. wRMS over the oceans (cm EWH) of DDK5 filtered residuals relative to a climatology model for the complete GFZ RL05a (red), GFZ RL06 (96 $\times$ 96, green), and GFZ RL06 (60 $\times$ 60, blue) time series.

3.4. Low Degree Harmonics

In this subsection, time series of selected low degree SH coefficients are analyzed, starting with C_{20}. This coefficient is known to be poorly estimated from GRACE (see, e.g., [40]), and it is common practice to replace it, e.g., with estimates derived from SLR observations to geodetic satellites. Despite the fact that the GFZ GRACE RL06 C_{20} values have significantly improved compared to GFZ RL05a (Figure 7a), a replacement of C_{20} is still recommended for RL06 before using the time series for geophysical interpretation. Available SLR-based replacement time series which are consistent with RL06 standards are, e.g., GRACE Technical Note TN-11 generated by CSR [41], or a similar time series provided by GFZ [42] which is also shown in Figure 7a.

Two other coefficients requiring special attention are C_{21} and S_{21}. When analyzing surface mass variations from the GRACE SDS RL05 time series, Wahr et al. [43] recommended corrections to these coefficients to account for effects of the applied mean pole model. Since all three SDS RL06 time series including GFZ RL06 are processed based on a linear mean pole model which is conform to the updated IERS2010 mean pole convention (http://iers-conventions.obspm.fr/chapter7.php), this recommendation is not applicable anymore to these reprocessed time series. Looking at the GFZ RL06 C_{21} time series in comparison to GFZ RL05a (Figure 7b), however, one can see an anomalous behavior during the "GRACE single ACC" period. Although already the RL05a time series shows larger amplitudes in that period, this is even more pronounced in RL06. A similar behavior is visible also for S_{21} (Figure 7c). The reason for these anomalies in C_{21} and S_{21} is not yet fully understood and subject to further investigation. As it is clearly correlated with the use of ACC data transplant, a possible explanation would be that it is due to inaccurate modeling of surface forces, potentially in conjunction with an inappropriate parametrization. First experiments at GFZ combining GRACE and SLR on NEQ level have revealed promising results and might lead to a replacement time series, similar to C_{20}, to overcome these deficiencies in the near future. It should be mentioned here that a GRACE+SLR combination would not be a novelty as, e.g., the GRACE solutions provided by the CNES/GRGS group are in fact already based on a combination with SLR [15].

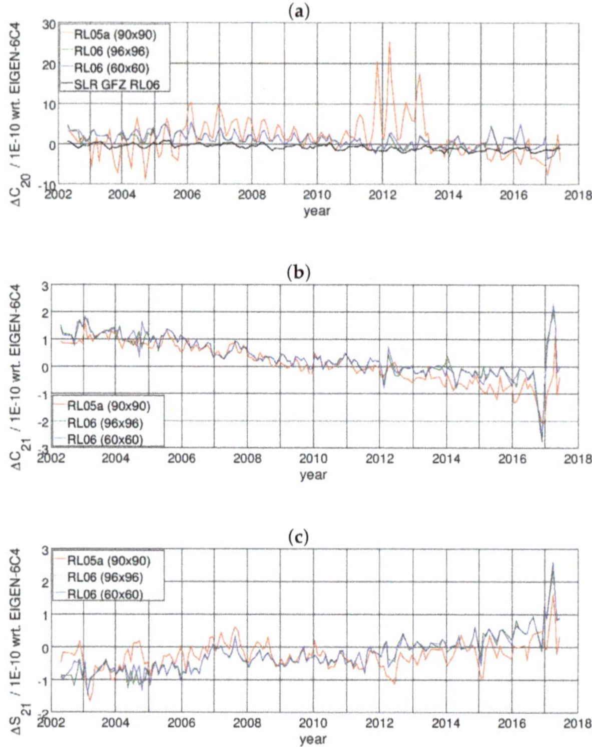

Figure 7. Time series of SH coefficients C_{20} (**a**); C_{21} (**b**); and S_{21} (**c**); each plot shows values of GFZ RL05a (red), GFZ RL06 (96 × 96, green), and GFZ RL06 (60 × 60, blue); for C_{20}, the SLR-based time series König et al. [42] is shown additionally (black).

4. External Validation

Due to the uniqueness of GRACE Level-2 products as observable for studies of Earth surface mass transport and climate change, it is nontrivial to validate them against independent data or models, and thus to reliably assess the quality of different GRACE time series in terms of signal content rather than only assessing their internal noise level. In the following subsections, two methods to evaluate the quality of the GFZ RL06 and RL05a time series by external data are presented.

4.1. OBP Validation

First, the GFZ GRACE RL06 and RL05a solutions are independently validated by comparing them with ocean bottom pressure (OBP) in situ observations.

The OBP database used here was initially compiled by Macrander et al. [44] and consists of 167 stations which are irregularly scattered over the oceans covering the time period from 2002 through 2010 with observation lengths for individual stations of up to eight years. The station data are preprocessed as outlined by Poropat et al. [7] to obtain time series of OBP observations with removed trends, tidal variability, outliers, and discontinuities at certain dates related to instrument issues including maintenance and battery replacement.

For the OBP validation, the GFZ RL06 and RL05a Level-2 solutions are post-processed as follows: The C_{20} coefficients are replaced with GRACE Technical Notes TN-11 and TN-07 [40], for RL06 and RL05a, respectively, the effects of glacial isostatic adjustment are corrected by subtracting the model by A et al. [45], co-seismic signatures from three megathrust earthquakes are removed with estimates from the GOCO06s model [46], and approximated degree-1 coefficients according to Bergmann-Wolf et al. [47] are added. The DDK filter is applied to de-correlate the solutions: DDK4 is used for the long-term trend component, whereas DDK2 is used for the annual and semi-annual components and for the remaining residual monthly signals. Please note that for five monthly solutions with particularly poor signal-to-noise ratio, the DDK1 filter is used. Finally, the monthly GAD background model [48] including atmospheric surface pressure and non-tidal OBP is added back.

These post-processed GRACE data are evaluated at the locations of the OBP in situ recorders. Regionally different linear trends, specifically caused by changing sea level, are removed. Since GRACE data do not represent a point-measurement, but an average over a large area, areas of coherent OBP variability are identified and the GRACE OBP data are averaged over these areas [49]. The selection of these areas follows [7].

To compare GRACE and in situ OBP variations at the OBP in situ sites, relative explained variances (defined as $\sigma_r^2 = (\sigma_{\text{in situ OBP}}^2 - \sigma_{\text{in situ OBP - GRACE OBP}}^2)/\sigma_{\text{in situ OBP}}^2$) and correlation coefficients are calculated from both time series. Generally, positive relative explained variances are observed for about 35% of the OBP in situ stations (Figure 8a), indicating that GRACE-derived and observed OBP variations correspond rather poorly in many regions. However, improvements in relative explained variance for GFZ RL06 compared to GFZ RL05a become visible in most regions (Figure 8c). The same conclusion can be drawn for the correlations, where a slight increase for GFZ RL06 can be seen as well again in most regions (Figure 8d). Generally, correlation coefficients between GRACE and in situ OBP are within the range of 0.1 to 0.7 for most of the stations (Figure 8b); the corresponding 25th, 50th, and 75th percentiles are 0.21, 0.34, and 0.54, respectively. Overall, a slightly better performance of GFZ RL06 over GFZ RL05a in explaining OBP variability over wide regions is achieved.

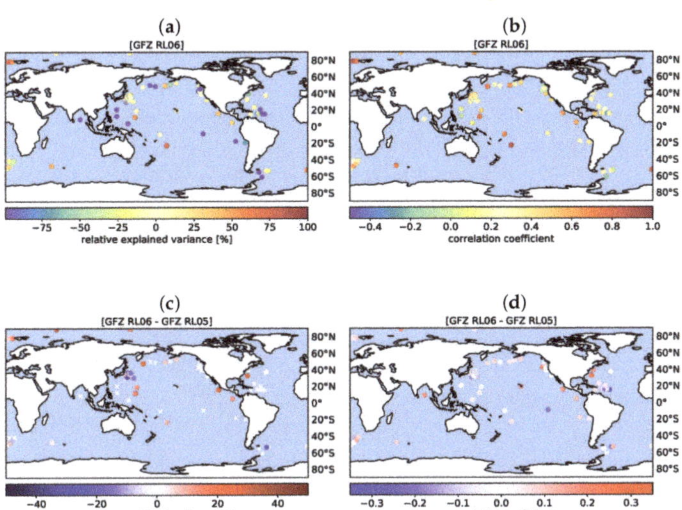

Figure 8. (**a**) Relative explained variances σ_r^2 at in situ OBP stations for GFZ RL06; (**b**) Correlation coefficients between in situ OBP and GRACE OBP for GFZ RL06; (**c**) Difference of relative explained variances for GFZ RL06 and GFZ RL05a; (**d**) Difference of correlation coefficients for GFZ RL06 and GFZ RL05a. For (**c**) and (**d**), red colors indicate improvements of GFZ RL06 over GFZ RL05a; stations with relative explained variances or correlation coefficients < 0 for both RL06 and RL05a are marked with white crosses.

4.2. GOCE Orbit Tests

As another independent validation, orbits of ESA's Gravity field and steady-state Ocean Circulation Explorer (GOCE) mission are used to compare the quality of monthly GRACE gravity solutions. The GOCE satellite [50], in orbit from March 2009 until November 2013, had a very low orbital altitude of about 255 km and thus shows a rather high sensitivity to the Earth's gravity field.

For these orbit tests, dynamic orbits are fitted to GOCE kinematic 3D orbit positions which are taken as observations (i.e., not directly the GPS tracking data). These kinematic orbits (Precise Science Orbits) have been generated at AIUB [51] and are provided within the GOCE High Level Processing Facility. For this study, GOCE orbit tests have been carried out for four months (November and December 2009, October and November 2010), each consisting of 30 individual GOCE arcs with a length of 1.25 days. The ocean tide model used here is FES2014 [28] to SH d/o 100. The reference system and gravitational force modeling is done applying the IERS 2010 [33] conventions. GOCE common mode accelerations are used during orbit computation instead of non-gravitational force models. The scale factors of the common mode accelerations cannot be accurately estimated per arc due to the drag-free control system which has compensated most of the signal and are therefore fixed to one [52]. This value is accurate within 3%, as demonstrated by Visser and van den IJssel [53]. The GOCE gradiometer works best in the measurement bandwidth of 10 to 200 s [54], and consequently the common mode accelerations include an instrumental bias. Therefore, three common mode acceleration biases per arc are estimated, one in each direction, in addition to the initial state vector.

For each month, two versions of orbit fits are computed which are identical except for the background gravity field model: The first version uses the corresponding GFZ RL06 monthly solutions, the second one uses the GFZ RL05a solutions instead, both up to SH d/o 90. Due to the high gravitational sensitivity of GOCE, the monthly GRACE models are filled up with SH coefficients from the long-term static GOCE model GO_CONS_GCF_2_DIR_R6 [55] up to d/o 240 to achieve reasonable orbit fits at the level of few centimeters.

The results of the GOCE orbit tests are listed in Table 4. When using GFZ RL06 instead of GFZ RL05a as background model, the RMS values of the orbit fits are clearly reduced for all four months, with relative improvements of GFZ RL06 over GFZ RL05a ranging from 12% to 25%. The significant differences in the orbit fits prove that such kind of orbit validation tests are an appropriate tool for the validation of monthly GRACE gravity field solutions. For future validation purposes, it is planned to extend the orbit validation to the complete GOCE mission period and to investigate whether orbits of other Low Earth Orbiting satellites such as, e.g., CHAMP and Swarm can be used as well.

Table 4. RMS of orbit fits [cm] for the time-variable GFZ RL05a and RL06 gravity field models and (only for reference) for the static model GO_CONS_GCF_2_DIR_R6. RMS values are based on 3D residuals and represent mean values of the 30 individual arcs within a particular month.

Gravity Field Model	Month			
	2009/11	2009/12	2010/10	2010/11
GFZ RL05a	8.39	9.14	7.53	7.40
GFZ RL06	7.39	6.84	6.24	6.21
GO_CONS_GCF_2_DIR_R6	3.56	3.37	3.82	3.76

5. Conclusions

GFZ has reprocessed an improved monthly gravity field time series for the complete GRACE mission consisting of 163 gravity field models (Level-2 products) in the period from April 2002 through June 2017. This GFZ GRACE RL06 time series incorporates a reprocessed in-house GPS constellation (orbits and clocks), reprocessed Level-1B K-band ranging, star camera and accelerometer (ACC) transplant observations (L1B RL03 dataset provided by JPL), updated background models for tidal (FES2014) and non-tidal (AOD1B RL06) mass variations, and a considerably modified processing strategy including a different parametrization with (for most months) even less parameters compared to the precursor GFZ RL05a. Key features of the new RL06 processing strategy are a strict separation of GPS and K-band data editing, manual instead of automated sigma-based K-band data editing, and omittance of a time-variable gravity background model during the gravity field estimation step. Main differences in RL06 parametrization compared to RL05a are consistently estimated parameters throughout all processing steps, less ACC parameters (biases and scale factors), and –exclusively for the last months where ACC transplant data has to be used for GRACE-B ("GRACE single ACC" period)–the estimation of a fully populated ACC scale factor matrix. Independent validation by satellite laser ranging (SLR) observations reveals a satisfying quality of the GRACE orbits prior to gravity field adjustment with standard deviations of SLR residuals $< 20\,\text{mm}$ for selected high-quality stations.

With the new GFZ RL06 time series significant improvements have been achieved: The noise is considerably reduced and, consequently, geophysical signals are better detectable and can be analyzed at smaller spatial scales. Relative improvements over GFZ RL05a in terms of residual RMS variability are about 40% for both DDK3 and DDK5 filtered solutions. Furthermore, the complete time series is more homogeneous. Although the GFZ RL06 formal errors are still too optimistic for most of the gravity field coefficients, they exhibit a more realistic behavior, and also empirical errors in terms of residuals relative to a climatology model are smaller than for GFZ RL05a. The quality of the gravity fields within the "GRACE single ACC" period is clearly worse compared to the rest of the time series, but relative to RL05a, the RL06 solutions are also significantly improved here. Special attention needs to be paid to the C_{21} and S_{21} coefficients showing unrealistic amplitudes during that period. A combined GRACE+SLR replacement time series for these coefficients might help to mitigate this issue, as first investigations at GFZ have indicated. Regarding the C_{20} coefficient, known to be poorly estimated from GRACE, it is still advised to replace the values by external time series, e.g., derived from SLR. Such a time series that is consistently processed with GRACE RL06 standards, is also provided by GFZ [42].

External validation by means of comparison with in situ ocean bottom pressure observations as well as orbit tests with the GOCE satellite confirm that improvements have been achieved with GFZ RL06 over RL05a, enabling thus a better understanding of phenomena in the Earth system related to climate change. To put the relative improvement from RL05a to RL06 in context with the relative improvements between all GFZ GRACE releases since RL01, Figure 9 shows gravity field anomalies for all previous GFZ GRACE releases exemplarily for the month August 2003. The corresponding relative improvements in terms of wRMS over ocean are as follows: RL01 to RL02: 14%, RL02 to RL03: 24%, RL03 to RL04: 4%, RL04 to RL05a: 0%, RL05a to RL06: 41%. This is another clear indication of the remarkable improvements achieved with the GFZ RL06 reprocessing and also depicts that even after more than 15 years of the first instrument data release a substantial gain in the quality of monthly GRACE gravity field products is possible thanks to reprocessing efforts regarding Level-1 and Level-2 products, but also improved background models and enhanced processing strategies. Hence, reprocessing of a GRACE RL07 time series is already planned, for which a final release of Level-1 products will be available. Apart from using these new Level-1 data and possible background model updates, the specific focus at GFZ for RL07–or other likely upcoming releases–will be on the reported C_{21}/S_{21} issue, as well as on a further reduction of noise as achieved by other groups (see, e.g., [14]). Whereas modification or fine tuning of the parametrization is always a promising option in view of improvements, in particular the application of an improved stochastic modeling of errors in observations and background models is envisaged.

A comparison between GFZ RL06 and recently published GRACE time series by other processing centers is not the purpose of this work; however, such comparisons were already done in several other studies: Göttl et al. [56] report an increased consistency of the SDS (CSR, JPL, GFZ) RL06 and ITSG-Grace2018 solutions compared to the SDS RL05 and ITSG-Grace2016 solutions. Kvas et al. [14] investigated the signal content of the SDS RL06 and ITSG-Grace2018 solutions by evaluating river basin averages and conclude that all four solutions exhibit the same signal content. Adhikari et al. [57] calculated sea-level fingerprints using the SDS RL06 time series and find that differences between these three solutions are within 1-sigma uncertainties.

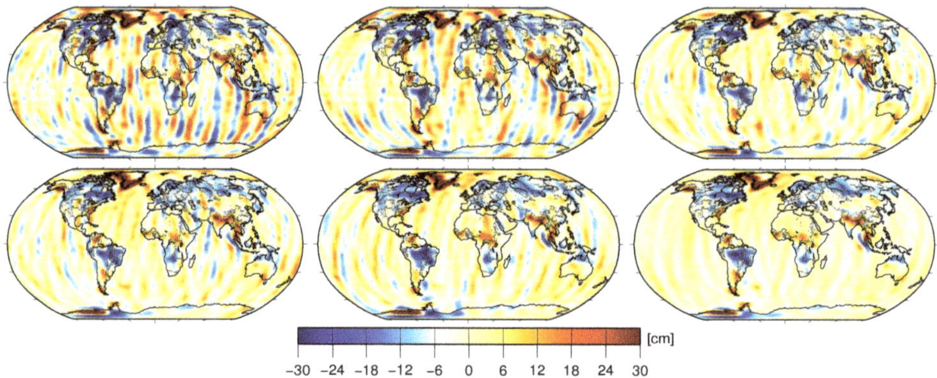

Figure 9. Gravity field anomalies in terms of cm EWH (DDK3 filtered) for the month 2003/08 for all GFZ GRACE releases so far: RL01 (**top left**), RL02 (**top middle**), RL03 (**top right**), RL04 (**bottom left**), RL05a (**bottom middle**), and RL06 (**bottom right**).

The GFZ GRACE RL06 monthly gravity field time series consists of fully unconstrained spherical harmonic (SH) Level-2 products, i.e., no regularization at all is applied, and is provided in two versions as agreed upon within the GRACE SDS: (1) up to SH degree and order 96; and (2) up to SH degree and order 60. GFZ GRACE RL06 is available at GFZ's Information System and Data Center (ISDC) archive (https://isdc.gfz-potsdam.de/grace-isdc/) along with related documentation ([12]; Release Notes for GFZ RL06 Level-2 products (ftp://isdcftp.gfz-potsdam.de/grace/DOCUMENTS/RELEASE_NOTES/GRACE_GFZ_L2_Release_Notes_for_RL06.pdf); GRACE Level-2 User Handbook (ftp://isdcftp.gfz-potsdam.de/grace/DOCUMENTS/Level-2/GRACE_L2_Gravity_Field_Product_User_Handbook_v4.0.pdf)).

GFZ GRACE RL06 processing standards and background models are also used for the initial GFZ GRACE-FO Level-2 product release [58]. Moreover, GFZ GRACE/GRACE-FO RL06 Level-2 products are the basis for GFZ's web portal GravIS (Gravity Information Service, http://gravis.gfz-potsdam.de), jointly developed with the Alfred-Wegener-Institut and TU Dresden, where dedicated Level-3 products for hydrological, oceanic and polar ice-sheet applications are visualized and offered for download. Finally, the GFZ GRACE RL06 time series contributes to the newly established International Combination Service for Time-variable Gravity Fields (COST-G), a product center of the International Gravity Field Service (IGFS).

Author Contributions: Conceptualization, C.D.; Processing of GRACE orbits and gravity fields, C.D.; Processing of GPS orbits, A.R.; Software, C.D., M.M., G.M., K.H.N. and O.A.; Validation of results, C.D., M.M., H.D., R.S. and C.F.; Writing—Original Draft Preparation, C.D.; Writing—Review and Editing, M.M., F.F., H.D., R.S. and C.F.; Visualization, C.D., M.M. and R.S.; Supervision, F.F., H.D. and R.K.; Project Administration, F.F.; Funding Acquisition, F.F., H.D. and R.K.

Funding: This research was partly funded by the German Ministry for Education and Research (BMBF) with FKZ 03F0654A, and by the German Research Foundation (DFG) within Research Group 2736 NEROGRAV (New Refined Observations of Climate Change from Spaceborne Gravity Missions).

Acknowledgments: We would like to thank the German Space Operations Center (GSOC) of the German Aerospace Center (DLR) for providing continuously and nearly 100% of the raw telemetry data of the twin GRACE satellites. Valuable comments by four anonymous reviewers helped to improve the manuscript and are highly appreciated.

Conflicts of Interest: The authors declare no conflict of interest. The funders had no role in the design of the study; in the collection, analyses, or interpretation of data; in the writing of the manuscript, or in the decision to publish the results.

References

1. Rodell, M.; Famiglietti, J.S.; Wiese, D.N.; Reager, J.T.; Beaudoing, H.K.; Landerer, F.W.; Lo, M.H. Emerging trends in global freshwater availability. *Nature* **2018**, *557*, 651–659. [CrossRef] [PubMed]
2. Kusche, J.; Eicker, A.; Forootan, E.; Springer, A.; Longuevergne, L. Mapping probabilities of extreme continental water storage changes from space gravimetry. *Geophys. Res. Lett.* **2016**, *43*, 8026–8034. [CrossRef]
3. Sasgen, I.; Konrad, H.; Ivins, E.R.; Van den Broeke, M.R.; Bamber, J.L.; Martinec, Z.; Klemann, V. Antarctic ice-mass balance 2003 to 2012: Regional reanalysis of GRACE satellite gravimetry measurements with improved estimate of glacial-isostatic adjustment based on GPS uplift rates. *Cryosphere* **2013**, *7*, 1499–1512. [CrossRef]
4. Wouters, B.; Gardner, A.S.; Moholdt, G. Global Glacier Mass Loss During the GRACE Satellite Mission (2002–2016). *Front. Earth Sci.* **2019**, *7*, 96. [CrossRef]
5. Reager, J.T.; Gardner, A.S.; Famiglietti, J.S.; Wiese, D.N.; Eicker, A.; Lo, M.H. A decade of sea level rise slowed by climate-driven hydrology. *Science* **2016**, *351*, 699–703. [CrossRef] [PubMed]
6. Rietbroek, R.; Brunnabend, S.E.; Kusche, J.; Schröter, J; Dahle, C. Revisiting the contemporary sea-level budget on global and regional scales. *Proc. Natl. Acad. Sci. USA* **2016**, *113*, 1504–1509. [CrossRef] [PubMed]
7. Poropat, L.; Dobslaw, H.; Zhang, L.; Macrander, A.; Boebel, O.; Thomas, M. Time variations in ocean bottom pressure from a few hours to many years: In situ data, numerical models, and GRACE satellite gravimetry. *J. Geophys. Res. Ocean* **2018**, *123*, 5612–5623. [CrossRef]
8. Landerer, F.W.; Wiese, D.N.; Bentel, K.; Boening, C.; Watkins, M.M. North Atlantic meridional overturning circulation variations from GRACE ocean bottom pressure anomalies. *Geophys. Res. Lett.* **2015**, *42*, 8114–8121. [CrossRef]
9. Tapley, B.D.; Watkins, M.M.; Flechtner, F.; Reigber, C.; Bettadpur, S.; Rodell, M.; Sasgen, I.; Famiglietti, J.S.; Landerer, F.W.; Chambers, D.P.; et al. Contributions of GRACE to understanding climate change. *Nat. Clim. Chang.* **2019**, *9*, 358–369. [CrossRef]
10. Bettadpur, S. UTCSR Level-2 Processing Standards Document (For Level-2 Product Release 0006) (Rev. 5.0, April 18, 2018). GRACE Publication 327–742. 2018. Available online: ftp://isdcftp.gfz-potsdam.de/grace/DOCUMENTS/Level-2/ (accessed on 10 September 2019).
11. Yuan, D.N. JPL Level-2 Processing Standards Document For Level-2 Product Release 06 (Rev. 6.0, June 1, 2018). GRACE Publication 327–744. 2018. Available online: ftp://isdcftp.gfz-potsdam.de/grace/DOCUMENTS/Level-2/ (accessed on 10 September 2019).
12. Dahle, C.; Flechtner, F.; Murböck, M.; Michalak, G.; Neumayer, K.; Abrykosov, O.; Reinhold, A.; König, R. *GRACE 327-743 (Gravity Recovery and Climate Experiment): GFZ Level-2 Processing Standards Document for Level-2 Product Release 06 (Rev. 1.0, October 26, 2018)*; Scientific Technical Report STR - Data, 18/04; GFZ German Research Centre for Geosciences: Potsdam, Germany, 2018. [CrossRef]
13. Kornfeld, R.P.; Arnold, B.W.; Gross, M.A.; Dahya, N.T.; Klipstein, W.M.; Gath, P.F.; Bettadpur, S. GRACE-FO: The Gravity Recovery and Climate Experiment Follow-On Mission. *J. Spacecr. Rocket.* **2019**, *56*, 931–951. [CrossRef]
14. Kvas, A.; Behzadpour, S.; Ellmer, M.; Klinger, B.; Strasser, S.; Zehentner, N.; Mayer-Gürr, T. ITSG-Grace2018: Overview and Evaluation of a New GRACE-Only Gravity Field Time Series. *J. Geophys. Res. Solid Earth* **2019**, *124*. [CrossRef]
15. Lemoine, J.M.; Bourgogne, S.; Biancale, R.; Bruinsma, S. RL04 monthly gravity field solutions from CNES/GRGS. In Proceedings of the the GRACE/GRACE-FO Science Team Meeting, Potsdam, Germany, 9–11 October 2018.
16. Meyer, U.; Jäggi, A.; Jean, Y.; Beutler, G. AIUB-RL02: An improved time-series of monthly gravity fields from GRACE data. *Geophys. J. Int.* **2016**, *205*, 1196–1207. [CrossRef]

17. Chen, Q.; Shen, Y.; Chen, W.; Francis, O.; Zhang, X.; Chen, Q.; Li W.; Chen, T. An optimized short-arc approach: Methodology and application to develop refined time series of Tongji-Grace2018 GRACE monthly solutions. *J. Geophys. Res. Solid Earth* **2019**, *124*, 6010–6038. [CrossRef]

18. Dahle, C.; Flechtner, F.; König, R.; Michalak, G.; Neumayer, K.; Gruber, C.; König, D. GFZ RL05: An Improved Time-Series of Monthly GRACE Gravity Field Solutions. In *Observation of the System Earth from Space - CHAMP, GRACE, GOCE and Future Missions. Advanced Technologies in Earth Sciences*; Flechtner, F., Sneeuw, N., Schuh, W., Eds.; Springer: Berlin/Heidelberg, Germany, 2014; pp. 29–39, ISBN 978-3-642-32134-4. [CrossRef]

19. Reigber, C. Gravity field recovery from satellite tracking data. In *Theory of Satellite Geodesy and Gravity Field Determination*; Lecture Notes in Earth Sciences; Sanso, F., Rummel, R., Eds.; Springer: Berlin/Heidelberg, Germany, 1989; Volume 25, pp. 197–234, ISBN 3-540-51528-3.

20. König, D. A Terrestrial Reference Frame realised on the observation level using a GPS-LEO satellite constellation. *J. Geod.* **2018**, *92*, 1299–1312. [CrossRef]

21. Case, K.; Kruizinga, G.; Wu, S. GRACE Level 1B Data Product User Handbook (Rev. 1.3). JPL Publication D-22027. 2010. Available online: ftp://isdcftp.gfz-potsdam.de/grace/DOCUMENTS/Level-1/ (accessed on 10 September 2019).

22. GRACE. *GRACE Level 1B JPL Release 3.0*; Data Publication; PO.DAAC: CA, USA, 2018. [CrossRef]

23. Bandikova, T.; McCullough, C.; Kruizinga, G.L.; Save, H.; Christophe, B. GRACE accelerometer data transplant. *Adv. Space Res.* **2019**, *64*, 623–644. [CrossRef]

24. Flechtner, F.; Neumayer, K.H.; Dahle, C.; Dobslaw, H.; Fagiolini, E.; Raimondo, J.C., Güntner, A. What Can be Expected from the GRACE-FO Laser Ranging Interferometer for Earth Science Applications? *Surv. Geophys.* **2016**, *37*, 453–470. [CrossRef]

25. Förste, C.; Bruinsma, S.; Shako, R.; Marty, J.C.; Flechtner, F.; Abrykosov, O.; Dahle, C.; Lemoine, J.M.; Neumayer, K.H.; Biancale, R.; et al. *EIGEN-6—A New Combined Global Gravity Field Model Including GOCE Data from the Collaboration of GFZ-Potsdam and GRGS-Toulouse*; Geophysical Research Abstracts Vol. 13, EGU2011-3242-2; EGU General Assembly: Vienna, Austria, 2011.

26. Förste, C.; Bruinsma, S.; Abrykosov, O.; Lemoine, J.M.; Marty, J.C.; Flechtner, F.; Balmino, G.; Barthelmes, F.; Biancale, R. *EIGEN-6C4 The Latest Combined Global Gravity Field Model Including GOCE Data Up to Degree and Order 2190 of GFZ Potsdam and GRGS Toulouse*; Data Publication; GFZ Data Services: Potsdam, Germany, 2014. [CrossRef]

27. Savcenko, R.; Bosch, W. *EOT11a—Empirical Ocean Tide Model from Multi-Mission Satellite Altimetry*; Report No. 89; Deutsches Geodätisches Forschungsinstitut: München, Germany, 2012.

28. Carrere, L.; Lyard, F.; Cancet, M.; Guillot, A.; Picot, N. FES2014, a new tidal model—Validation results and perspectives for improvements. In Proceedings of the ESA Living Planet Symposium 2016, Prague, Czech Republic, 9–13 May 2016.

29. Biancale, R.; Bode, A. *Mean Annual and Seasonal Atmospheric Tide Models Based on 3-hourly and 6-hourly ECMWF Surface Pressure Data*; Scientific Technical Report STR, 06/01; GFZ German Research Centre for Geosciences: Potsdam, Germany, 2006. [CrossRef]

30. Dobslaw, H.; Flechtner, F.; Bergmann-Wolf, I.; Dahle, C.; Dill, R.; Esselborn, S.; Sasgen, I.; Thomas, M. Simulating high-frequency atmosphere-ocean mass variability for de-aliasing of satellite gravity observations: AOD1B RL05. *J. Geophys. Res. Ocean* **2013**, *118*, 3704–3711. [CrossRef]

31. Dobslaw, H.; Bergmann-Wolf, I.; Dill, R.; Poropat, L.; Thomas, M.; Dahle, C.; Esselborn, S.; König, R.; Flechtner, F. A new high-resolution model of non-tidal atmosphere and ocean mass variability for de-aliasing of satellite gravity observations: AOD1B RL06. *Geophys. J. Int.* **2017**, *211*, 263–269. [CrossRef]

32. Desai, S.D. Observing the pole tide with satellite altimetry. *J. Geophys. Res.* **2002**, *107*, 3186. [CrossRef]

33. Petit, G.; Luzum, B. *IERS Conventions (2010)*; IERS Technical Note No. 36; Verlag des Bundesamts für Kartographie und Geodäsie: Frankfurt am Main, Germany, 2010; p. 179, ISBN 3-89888-989-6.

34. Ray, R.D.; Loomis, B.D.; Luthcke, S.B.; Rachlin, K.E. Tests of ocean-tide models by analysis of satellite-to-satellite range measurements: An update. *Geophys. J. Int.* **2019**, *217*, 1174–1178. [CrossRef]

35. Schmidt, R. Zur Bestimmung des cm-Geoids und Dessen Zeitlicher Variationen mit GRACE. Scientific Technical Report STR, 07/04. Ph.D. Thesis, GFZ German Research Centre for Geosciences, Potsdam, Germany, 2007. [CrossRef]

36. Klinger, B.; Mayer-Gürr, T. The role of accelerometer data calibration within GRACE gravity field recovery: Results from ITSG-Grace2016. *Adv. Space Res.* **2016**, *58*, 1597–1609. [CrossRef]

37. Arnold, D.; Montenbruck, O.; Hackel, S.; Sosnica, K. Satellite laser ranging to low Earth orbiters: Orbit and network validation. *J. Geod.* **2018**. [CrossRef]

38. Montenbruck, O.; Garcia-Fernandez, M.; Yoon, Y.; Schön, S.; Jäggi, A. Antenna phase center calibration for precise positioning of LEO satellites. *GPS Solut.* **2009**, *13:23*. [CrossRef]

39. Kusche, J.; Schmidt, R.; Petrovic, S.; Rietbroek, R. Decorrelated GRACE time-variable gravity solutions by GFZ, and their validation using a hydrological model. *J. Geod.* **2009**, *83*, 903–913. [CrossRef]

40. Cheng, M.K.; Ries, J.C. The unexpected signal in GRACE estimates of C20. *J. Geod.* **2017**, *91*, 897–914. [CrossRef]

41. Cheng, M.K.; Ries, J.C. Monthly estimates of C20 from 5 SLR Satellites Based on GRACE RL06 Models. GRACE Technical Note TN-11. Available online: ftp://isdcftp.gfz-potsdam.de/grace/DOCUMENTS/TECHNICAL_NOTES/TN-11_C20_SLR_RL06.txt (accessed on 10 September 2019).

42. König, R.; Schreiner, P.; Dahle, C. *Monthly Estimates of C(2,0) Generated by GFZ from SLR Satellites Based on GFZ GRACE/GRACE-FO RL06 Background Models. V. 1.0*; Data Publication; GFZ Data Services: Potsdam, Germany, 2019. [CrossRef]

43. Wahr, J.; Nerem, R. S.; Bettadpur, S.V. The pole tide and its effect on GRACE time-variable gravity measurements: Implications for estimates of surface mass variations. *J. Geophys. Res.: Solid Earth* **2015**, *120*, 4597–4615. [CrossRef]

44. Macrander, A.; Böning, C.; Boebel, O.; Schröter, J. Validation of GRACE Gravity Fields by In-Situ Data of Ocean Bottom Pressure. In *System Earth via Geodetic-Geophysical Space Techniques. Advanced Technologies in Earth Sciences*; Flechtner, F., Gruber, T., Güntner, A., Mandea, M., Rothacher, M., Schöne, T., Wickert, J., Eds.; Springer: Berlin/Heidelberg, Germany, 2010; pp. 169–185, ISBN 978-3-642-10227-1. [CrossRef]

45. A, G.; Wahr, J.; Zhong, S. Computations of the viscoelastic response of a 3-D compressible Earth to surface loading: An application to glacial isostatic adjustment in Antarctica and Canada. *Geophys. J. Int.* **2013**, *192*, 557–572. [CrossRef]

46. Kvas, A.; Mayer-Gürr, T.; Krauss, S.; Brockmann, J.M.; Schubert, T.; Schuh, W.D.; Pail, R.; Gruber, T.; Jäggi, A.; Meyer, U. *The Satellite-Only Gravity Field Model GOCO06s*; Data Publication; GFZ Data Services: Potsdam, Germany, 2019. [CrossRef]

47. Bergmann-Wolf, I.; Zhang, L.; Dobslaw, H. Global eustatic sea-level variations for the approximation of geocenter motion from GRACE. *J. Geod. Sci.* **2014**, *4*, 37–48. [CrossRef]

48. Dobslaw, H.; Dill, R.; Dahle, C. *GRACE Geopotential GAD Coefficients GFZ RL06. V. 6.0*; Data Publication; GFZ Data Services: Potsdam, Germany, 2018. [CrossRef]

49. Böning, C.; Timmermann, R.; Macrander, A.; Schröter, J. A pattern-filtering method for the determination of ocean bottom pressure anomalies from GRACE solutions. *Geophys. Res. Lett.* **2008**, *35*, L18611. [CrossRef]

50. Drinkwater, M.; Haagmans, R.; Muzi, D.; Popescu, A.; Floberghagen, R.; Kern, M.; Fehringer, M. The GOCE gravity mission: ESA's first core explorer. In Proceedings of the 3rd International GOCE User Workshop, Frascati, Italy, 6–9 November 2006; ESA SP-627, pp. 1–8, ISBN 92-9092-938-3.

51. Bock, H.; Jäggi, A.; Beutler, G.; Meyer, U. GOCE: Precise orbit determination for the entire mission. *J. Geod.* **2014**, *88*, 1047–1060. [CrossRef]

52. Gruber, T.; Visser, P.N.A.M.; Ackermann, C.; Hosse, M. Validation of GOCE gravity field models by means of orbit residuals and geoid comparisons. *J. Geod.* **2011**, *85*, 845–860. [CrossRef]

53. Visser, P.N.A.M.; van den Ijssel, J. Calibration and validation of individual GOCE accelerometers by precise orbit determination. *J. Geod.* **2016**; *90*, 1–13. [CrossRef]

54. Pail, R.; Bruinsma, S.; Migliaccio, F.; Förste, C.; Goiginger, H.; Schuh, W.D.; Höck, E.; Reguzzoni, M.; Brockmann, J.M.; Abrikosov, O.; et al. First GOCE gravity field models derived by three different approaches. *J. Geod.* **2011**, *85*, 819. [CrossRef]

55. Förste, C.; Abrykosov, O.; Bruinsma, S.; Dahle, C.; König, R.; Lemoine, J.M. *ESA's Release 6 GOCE Gravity Field Model by Means of the Direct Approach Based on Improved Filtering of the Reprocessed Gradients of the Entire Mission*; Data Publication; GFZ Data Services: Potsdam, Germany, 2019. [CrossRef]

56. Göttl, F.; Schmidt, M.; Seitz, F. Mass-related excitation of polar motion: An assessment of the new RL06 GRACE gravity field models. *Earth Planets Space* **2018**, *70*, 195. [CrossRef]

57. Adhikari, S.; Ivins, E.R.; Frederikse, T.; Landerer, F. W.; Caron, L. Sea-level fingerprints emergent from GRACE mission data. *Earth Syst. Sci. Data* **2019**, *11*, 629–646. [CrossRef]
58. Dahle, C.; Flechtner, F.; Murböck, M.; Michalak, G.; Neumayer, K.; Abrykosov, O.; Reinhold, A.; König, R. *GRACE-FO D-103919 (Gravity Recovery and Climate Experiment Follow-On): GFZ Level-2 Processing Standards Document for Level-2 Product Release 06 (Rev. 1.0, June 3, 2019)*; Scientific Technical Report STR - Data, 19/09; GFZ German Research Centre for Geosciences: Potsdam, Germany, 2019. [CrossRef]

Article

SLR, GRACE and Swarm Gravity Field Determination and Combination

Ulrich Meyer [1,*], Krzysztof Sosnica [2], Daniel Arnold [1], Christoph Dahle [1,3], Daniela Thaller [4], Rolf Dach [1] and Adrian Jäggi [1]

[1] Astronomical Institute, University of Bern, 3012 Bern, Switzerland; daniel.arnold@aiub.unibe.ch (D.A.); dahle@gfz-potsdam.de (C.D.); rolf.dach@aiub.unibe.ch (R.D.); adrian.jaeggi@aiub.unibe.ch (A.J.)
[2] Institute of Geodesy and Geoinformatics, Wroclaw University of Environmental and Life Sciences, 50-375 Wroclaw, Poland; krzysztof.sosnica@igig.up.wroc.pl
[3] GFZ German Research Centre for Geosciences, 14473 Potsdam, Germany
[4] Bundesamt für Kartographie und Geodäsie, 60598 Frankfurt, Germany; daniela.thaller@bkg.bund.de
* Correspondence: ulrich.meyer@aiub.unibe.ch

Received: 28 February 2019; Accepted: 16 April 2019; Published: 22 April 2019

Abstract: Satellite gravimetry allows for determining large scale mass transport in the system Earth and to quantify ice mass change in polar regions. We provide, evaluate and compare a long time-series of monthly gravity field solutions derived either by satellite laser ranging (SLR) to geodetic satellites, by GPS and K-band observations of the GRACE mission, or by GPS observations of the three Swarm satellites. While GRACE provides gravity signal at the highest spatial resolution, SLR sheds light on mass transport in polar regions at larger scales also in the pre- and post-GRACE era. To bridge the gap between GRACE and GRACE Follow-On, we also derive monthly gravity fields using Swarm data and perform a combination with SLR. To correctly take all correlations into account, this combination is performed on the normal equation level. Validating the Swarm/SLR combination against GRACE during the overlapping period January 2015 to June 2016, the best fit is achieved when down-weighting Swarm compared to the weights determined by variance component estimation. While between 2014 and 2017 SLR alone slightly overestimates mass loss in Greenland compared to GRACE, the combined gravity fields match significantly better in the overlapping time period and the RMS of the differences is reduced by almost 100 Gt. After 2017, both SLR and Swarm indicate moderate mass gain in Greenland.

Keywords: satellite gravimetry; ice mass change; GRACE; SLR; swarm; normal equation combination

---

1. Introduction

The Gravity Recovery and Climate Experiment satellite mission (GRACE) [1] dedicated to the observation of temporal variations of the gravity field allows for the quantification of ice mass loss of glacier accumulations in polar and sub-polar regions (e.g., [2–4]). However, this high resolution information is limited to the life-time of the GRACE satellites (2002–2017) and of the GRACE-FO (Follow On) [5] mission that was launched in May 2018, but suffered a failure of the main instrument processing unit between July and October 2018. To date, no other single satellite mission or proxy is able to bridge the gap between GRACE and GRACE-FO with comparable quality (compare, e.g., [6,7] or [8]). We therefore present a combination on the normal equation level of alternative satellite gravimetric data considering several missions.

Satellites not dedicated to gravity field determination already collected data before GRACE and during the gap between GRACE and GRACE-FO. These are mainly the geodetic satellite laser ranging (SLR) missions, e.g., the Laser Geodynamics Satellites (LAGEOS) or, at a significantly lower orbit altitude, the Laser Relativity Satellite (LARES) as the youngest member of the SLR family, which also provide information about temporal variations of the Earth gravity field at the lowest degrees of the spherical harmonic spectrum. The geodetic SLR satellites are optimal for gravimetry by their spherical geometry and their favorable area-to-mass ratio that minimizes the effect of surface forces [9]. For studies of temporal gravity field variations derived by SLR in the pre-GRACE era, see Cheng et al. [10], Bianco et al. [11] and Cheng and Tapley [12]. For a focus on variations in Earth oblateness, where GRACE results are unreliable [13], compare Cox and Chao [14] Cheng and Tapley [15] and Bloßfeld et al. [16].

In addition, other Earth observing satellites at orbits below 1500 km altitude, so-called low Earth orbiters (LEOs), which are equipped with GPS receivers for precise orbit determination, may serve for deriving the mass distribution of the Earth and its temporal variations at low to medium resolution [17]. Due to the time-period covered and their low orbit altitude, the three satellites of the Swarm mission [18] are well suited to bridge the gap between GRACE and GRACE-FO.

We present a long-term series covering 1995–2018 of low resolution monthly gravity field coefficients derived at the Astronomical Institute of the University of Bern (AIUB) from a combined orbit determination of LAGEOS and the geodetic SLR satellites in low Earth orbits [19]. The derived gravity fields are compared to time-series of GRACE solutions spanning 2003–2016, which were also determined at AIUB [20]. Finally, also the GPS-derived Swarm kinematic orbits that are routinely computed at AIUB [21] are exploited to derive monthly gravity fields for the time-period 2014–2018.

Due to the different orbital altitudes and ground track patterns of the satellites, and due to the different observation techniques used, the gravity fields derived differ in their spherical harmonic and corresponding spatial resolution [22]. The truncation of the spherical harmonic expansion of the gravity field leads to spatial leakage that has to be taken into account when comparing the results (e.g., [4,23–26]). Furthermore, the high-resolution GRACE gravity fields commonly are smoothed to suppress noise in the high-degree coefficients (e.g., [27–29]). We shortly introduce the representation of the gravity field in a spherical harmonic expansion and exemplify the problem of signal loss due to leakage or filter attenuation in Section 2. In the following, all monthly gravity fields are truncated to the same spherical harmonic degree for the sake of comparison of the different techniques.

To illustrate the capability of the different satellite gravimetric techniques to track temporal variations of the mass distribution within the system Earth, we transform the gravity variations to variations in equivalent water height (EWH) [30] or mass variations in Greenland and at the coast of West Antarctica and compare their spatial distribution and development with time. Moreover, we fit deterministic signal models of secular and seasonal variations to consecutive five-year time-periods and compare the derived mass trends.

Comparable studies were already performed by Matsuo et al. [31], Talpe et al. [32] and Bonin et al. [33]. The latter concluded that truncated at spherical harmonic degree 5 gravity field models are not able to correctly separate mass loss in Greenland and Antarctica and that, while the inter-annual variations in none of the SLR time-series are realistic, long-time mass trends are well captured in the regularized AIUB SLR time-series truncated at degree 10 [19] as opposed to other series. Bonin et al. [33] suggest to either reduce the temporal resolution of the SLR-derived gravity fields to reduce their scatter, which in our eyes is counter-intuitive because it also reduces the separability of mass loss signal in Greenland and Antarctica, or to combine SLR with other data.

We do not continue our regularized SLR-only degree 10 time-series beyond 2014 because in combination with Swarm the full sensitivity of SLR can be exploited without regularization. The combination approach is favorable because it is independent from external information not based on the original observations. We provide an unconstrained SLR-only degree 6 time-series and a combined SLR + Swarm time-series where SLR entered to degree and order 10. The decorrelation of the individual spherical harmonic coefficients is possible by the Swarm data. Due to the consistent processing of Swarm and SLR, we are able to perform the combination on the normal equation level, taking correlations between reference frame (co-estimated in the case of SLR), force model and orbit parameters into account [34]. Combinations of SLR with CHAllenging Minisatellite Payload

(CHAMP), GRACE or Gravity field and steady-state Ocean Circulation Explorer (GOCE) for gravity field determination were also studied by Moore et al. [35], Cheng et al. [36], Maier et al. [37] and Haberkorn et al. [38].

2. Materials and Methods

The dominating force acting on a satellite is due to Earth gravity. Satellite orbits at low altitudes are sensitive to mass distribution and redistribution and therefore to mass transport in the Earth system. The long-term mean gravity field of the Earth has been determined by the dedicated gravimetric satellite mission GFZ-1 [39], by CHAMP [40] that also observed the magnetosphere, and by GRACE [41,42] and GOCE [43,44], again dedicated to gravimetry. Temporal variations in gravity have mainly been derived from GRACE (see Wouters et al. [45] for an overview), but also from CHAMP [46] and other Earth observing LEOs such as the three satellites of the Swarm mission [47] or from the fleet of geodetic SLR satellites (e.g., [10–12]).

Prerequisite for gravity field determination is the precise observation of the satellite orbits. Today this may either happen by laser ranging or by GPS- or Doris-tracking of the satellites (the latter being irrelevant here due to the choice of missions). Critical for orbit modeling and signal separation is the knowledge of all forces acting on the satellites, the so-called background force model, and the set of parameters estimated to represent the orbit, improve the a priori force model and absorb model or instrument errors, the so-called orbit parameterization. In detail, the force model consists of a priori models of the gravitational forces by the Earth and third bodies, ocean, atmosphere and solid Earth tides, satellite specific models of the surface forces such as air drag, solar radiation pressure and Earth albedo, and an empirical or pseudo-stochastic part [48]. In case of dedicated gravity missions, the surface forces are normally measured by accelerometers onboard the satellites [49] (due to technical reasons only the sum of all forces plus an unknown but constant bias can be observed).

At AIUB, the gravity field determination from satellite observations is treated as a generalized orbit determination problem [48]. The satellites' orbits are co-estimated with the instrument specific parameters like accelerometer scale factors and with the parameters updating the a priori background forces, i.e., the weight coefficients C_{lm} and S_{lm} of the expansion of the Earth's gravitational potential V in spherical harmonics [50]

$$V(r, \phi, \lambda) = \frac{GM}{R} \sum_{l=0}^{\infty} \left(\frac{R}{r} \right)^{l+1} \sum_{m=0}^{l} P_{lm}(\sin \phi) \left(C_{lm} \cos m\lambda + S_{lm} \sin m\lambda \right), \tag{1}$$

where r, ϕ, λ are the spherical coordinates in an Earth-fixed reference frame, GM is the product of the gravitational constant and the Earth's mass, R is the semi-major axis of the Earth and P_{lm} are the fully normalized Legendre functions of degree l and order m. In case of SLR, air drag scale factors, station and geocenter coordinates, and Earth orientation parameters are also co-estimated [19].

Deficiencies of the force model are mitigated by pseudo-stochastic accelerations or pulses [51]. In order to reduce the absorption of gravity field signal by these parameters, their frequency has to be tailored to the orbit altitude and observation sampling of the specific satellite and their magnitude has to be limited by constraints depending on the satellites' environment at orbit altitude. By extensive cross-validation with other analysis centers in the frame of the European Gravity Service for Improved Emergency Management (EGSIEM) [52] project, it could be shown that mass trends and the amplitude of seasonal variations in the AIUB gravity fields are not affected by the pseudo-stochastic parameterization. Since a separate determination of sub-groups of the parameter space leads to a regularization favoring the a priori force model applied [53] orbit, stochastic and force model parameters have to be determined together in one adjustment process.

The observations used for orbit determination are either the 1D-ranges (normal points [54]) observed by SLR, the kinematic orbits of Earth observing LEOs determined by precise point positioning (PPP) [55] from high-low GPS data that are used as pseudo-observations [56], or low-low range-rates derived from the K-band inter-satellite link in case of the two satellites of the GRACE mission [57]. The sampling of these observation types is very diverse. In case of SLR, it depends on the inhomogeneous global distribution of the SLR stations [19]. Certain regions of the Earth, such as the polar regions, are not covered by observations at all due to the lack of suitably positioned stations. All information about gravity variations is derived from the orbit dynamics. Therefore, the orbit modeling of SLR satellites has to be mainly dynamical (based on physical models) and the pseudo-stochastic parameterization of the orbits has to be very limited to allow for gravity field determination. On the other hand, GPS and K-band observations are normally given at very high sampling rates of 1 s, 5 s or 10 s and in case of polar orbits the observation distribution is global and densest near the poles. Several pseudo-stochastic parameters may be set up per orbital revolution of the Swarm or GRACE satellites.

The spatial resolution of the resulting gravity fields depends on the orbital altitude and ground track pattern of the satellites. In case of sparse SLR tracking, gravity variations beyond degree 2 can only be determined by the combined evaluation of several satellites (see [10–12]). Satellites at lower inclinations are helpful to decorrelate the individual gravity field coefficients [19]. In case of the Earth observing LEOs, the achievable temporal resolution of subsequent gravity field solutions directly depends on sub-cycles of the ground track pattern, which vary depending on the orbit altitude (decreasing with time in case of the LEOs). The GRACE mission was designed to deliver monthly gravity fields [1] and since then monthly temporal resolution has become the standard, even if 10-day [58], and even daily "snapshot" solutions [59] of the gravity field are also available.

2.1. SLR

An important prerequisite for the determination of the long wavelength part of the gravity field is the exact modeling of the surface forces acting on the satellites. To simplify the modeling of the surface forces, most of the geodetic SLR satellites have a low area-to-mass ratio [9]. The high-flying SLR satellites LAGEOS 1 and 2 are mainly sensitive to the Earth's flattening C_{20} and its variations with time [60]. Higher spectral resolution of the gravity field can be achieved by combined processing of the fleet of geodetic SLR satellites at lower orbits (SLR-LEOs) and at different inclinations of the orbital plane (see [10–12,19]). We exploit the two LAGEOS satellites and the following SLR-LEOs for the determination of large scale temporal gravity field variations: Starlette, AJISAI, Stella, Larets, LARES, and the old Earth observation satellite Beacon-C that is carrying a laser retro-reflector array and is included due to its low orbit inclination. The orbit characteristics of all satellites used are compiled in Table 1.

Table 1. Orbit characteristics of satellites used and a priori observation errors, as determined by variance analysis of residuals (Beacon-C is down-weighted due to large surface forces acting on the non-spherical satellite).

Satellite	Launch Date	Orbit Altitude	Inclination	Observation Type	A Priori Error
Beacon-C	1965	940–1300 km	41.23°	SLR	50 mm
Starlette	1975	800–1100 km	49.84°	SLR	20 mm
LAGEOS-1	1976	5860 km	109.90°	SLR	8 mm
AJISAI	1986	1500 km	50.04°	SLR	25 mm
LAGEOS-2	1992	5620 km	52.67°	SLR	8 mm
Stella	1993	810 km	98.57°	SLR	20 mm
Larets	2003	690 km	97.77°	SLR	30 mm
LARES	2012	1440 km	69.56°	SLR	15 mm
GRACE A and B	2002	500 km in 03/2002	89°	GPS	2 mm
				K-band	0.3 μm/s^2
Swarm A and C	2013	460 km in 04/2014	87.4°	GPS	2–3 mm
Swarm B	2013	530 km in 04/2014	88.0°	GPS	2–3 mm

The same a priori model of gravitational forces is consistently used for the SLR, GRACE and Swarm processing. To avoid affecting the derived temporal gravity field variations by a priori information, we use a static a priori gravity field. Furthermore, the background force model consists of solid Earth tides, ocean tides, ocean pole tides, and de-aliasing of ocean and atmosphere mass variations (AOD) [61]. Specific for the SLR satellites are models of the surface forces for air drag, solar radiation pressure and albedo that take into account the properties of the individual SLR satellites. The force model constituents and the resolution of their expansion in spherical harmonics (if applicable) are listed in Table 2.

Table 2. Background force model details for processing of low Earth orbiters. Where the force model constituent is expanded in spherical harmonics, the max. degree/order is given. Where this is not the case, it is only indicated if the listed model is applied or not. ACC indicates that surface forces are observed by accelerometers.

Force	Model	SLR	GRACE	Swarm
Earth gravity	AIUB-GRACE03S (static)	degree/order 90	160	160
Ocean tides	EOT11A [62]	degree/order 30	100	100
Solid Earth tides	IERS2000 (elastic) [63]	yes	yes	yes
Ocean pole tide	Desai [63]	degree/order 100	100	100
Atmosphere and ocean de-aliasing	RL05, RL06 (since 11/2017)	degree/order 100	100	100
Air drag	NRLMSIS-00 [64]	yes	ACC	no
Solar pressure	uniform sphere approximation	yes	ACC	no
Albedo	Knocke [65,66]	yes	ACC	no

Even if SLR is sensitive to gravity field variations at higher degrees, at monthly resolution, only spherical harmonics coefficients (SHC) of degrees 2 to 5 and of degree 6 and order 1 can be determined from SLR alone, i.e., without regularization. At higher degrees SLR-only gravity field solutions suffer from the strong correlations between individual SHC [19]. A prerequisite for the separation of individual SHC, even at the low degree of 5, is that SLR LEOs orbiting at different inclinations are used [10]. In this context Beacon-C and LARES are helpful due to their rather exotic orbit inclinations (Table 1).

However, the correct localization of the mass loss signal is not possible at this low resolution [33]. Therefore, we perform a combined solution with Swarm (see Section 2.4). In combination with Swarm, the SHC are decorrelated by the Swarm observations. To exploit the sensitivity of SLR beyond degree 5 in the combination and to avoid omission or commissioning errors [67], the SHC are set up to degree and order 10. In case of SLR-only monthly solutions coefficients of degree 6 (all but order 1), and degrees 7–10 are fixed to zero. In case of combined solutions, all coefficients are estimated.

The parameter-space of the SLR solutions is complemented by monthly estimates of the SLR station coordinates, by daily piecewise-linear estimates of the geocenter coordinates and by monthly estimates of the Earth rotation parameters [68]. As indicated above, the number of pseudo-stochastic parameters is very limited due to the sparse observation coverage. Only in along-track pulses are estimated once per orbital revolution of the satellites. The stochastic orbit parameterization is completed by periodic terms on orbit revolution frequency. The complete SLR orbit parameterization and the time intervals for which the individual parameters are set up are detailed in Table 3.

Table 3. Orbit parameterization of SLR satellites and time intervals for which the individual parameters are estimated.

Parameter	LAGEOS-1/2	SLR LEOs
Station coordinates	30 days	30 days
Earth orientation parameters	piecewise-linear, daily	piecewise-linear, daily
Geocenter coordinates	30 days	30 days
Gravity field	degree/order 10	degree/order 10
Range biases	selected sites	all sites
Initial state	10 days	1 day
Const. acceleration along-track	10 days	none (air drag modelled)
Air drag scaling factor	none	1 day
Periodic along-track	10 days	1 day
Periodic cross-track	none	1 day
Pseudo-stochastic pulses	none	once per revolution along-track

2.2. GRACE

The GRACE satellites were launched in March 2002 into polar orbits at an initial orbit altitude of 500 km [1]. During 15 years, they provided information about temporal variations of the Earth gravity field at unprecedented spatial and temporal resolution. The mission ended in October 2017 due to battery failure after decay to an orbit altitude of 330 km. The satellites were equipped with GPS receivers for orbit determination [57]. The key instrument for gravity field recovery was a K-band inter-satellite link which provided range measurements with micrometer accuracy [69]. Due to long-periodic systematic errors that cancel out by differentiation [70], most analysis centers use range-rates for gravity field recovery that were derived from the original range observations by numerical differentiation. Surface forces acting on the GRACE satellites were measured with onboard accelerometers in order to separate them from the gravitational forces [49].

We estimated monthly gravity fields from K-band range-rates and kinematic orbits that were used instead of the original GPS phase observations for efficiency reasons (the equivalence of both approaches was demonstrated by Jäggi et al. [71]). Therefore, in the first step, kinematic satellite positions were determined using the Bernese GNSS Software 5.2 (AIUB, Bern, Switzerland) [72]. The kinematic orbits were introduced as pseudo-observations together with their epoch-wise covariances [56]. Normal equations (NEQs) were computed for both observation types on a daily basis, combined and summed up to monthly batches. All but the gravity field parameters were pre-eliminated from the combined daily NEQs. Note that the K-band observations only contain line-of-sight, i.e., predominantly along-track information and a K-band only orbit solution therefore would be rank-deficit [70].

The accelerometer provides biased accelerations in the instrument frame that is closely aligned to the radial, along-track, and cross-track direction of the co-rotating orbital tripod. To account for the biases, daily constant accelerations are estimated in radial and cross-track directions. In along-track, four polynomial parameters are estimated per day to also absorb temperature induced variations in the accelerometer measurements that are in conflict with the ultra-sensitive K-band observations [20]. However, the key element of the Celestial Mechanics Approach (CMA) that was developed at AIUB for orbit and gravity field determination [48] is the estimation of rather frequent pseudo-stochastic accelerations (or pulses) to compensate deficiencies in the a priori force model [51]. Accelerations are estimated at 15 min intervals in all three directions of the orbital tripod. They are constrained to zero in order to minimize absorption of gravity signal. The parameterization applied for GRACE processing is summarized in Table 4.

For the a priori force model, the static part of AIUB-GRACE03S is used, but compared to the SLR-processing with increased spherical harmonic resolution to take into account the high sensitivity of the K-band observable. In addition, the background models for tidal effects and AOD were chosen consistently with the SLR-processing (details on the force model can be found in Table 2).

2.3. Swarm

A number of Earth observation LEOs at orbit altitudes of 400–600 km that are equipped with GPS receivers may also be exploited to derive information on gravity field variations [17,56]. Eminent candidates to bridge the gap, be it at lower spatial resolution, between GRACE and the GRACE-FO mission are the three satellites of ESA's Earth's Magnetic Field and Environment Explorer (Swarm) mission [18] that were launched in November 2013. All three satellites of the constellation circle the Earth in near circular polar orbits, Swarm A and C at 460 km, Swarm B at 530 km altitude (in 04/2014). They are equipped with GPS receivers and accelerometers, but the data of the latter turned out to be disturbed by slow temperature-induced bias variations and sudden bias changes [73] and are not routinely used for orbit and gravity field determination.

At AIUB, kinematic orbits of all three Swarm satellites are determined routinely [21]. While for the first mission phase until June 2014 gravity field results are deteriorated due to high solar and consequently ionosphere activity and not optimal GPS receiver settings [74,75], for the time period critical to bridge the gap between GRACE and GRACE-FO, the data quality corresponds to the nominal. By evaluation of the kinematic Swarm orbits, monthly gravity fields can be determined that are sensitive to temporal gravity variations up to about degree 13 [47].

The background force model for the Swarm processing is defined correspondingly to SLR and GRACE (see Table 2). Again the static part of AIUB-GRACE03S is used as a priori model and AOD is applied for de-aliasing of short-term variations in the atmosphere and ocean masses. As in the case of GRACE, no models for surface forces are applied. To compensate for the not used accelerometer observations, the constraints on the pseudo-stochastic accelerations are set less strict [21]. Details on the Swarm orbit parameterization can be extracted from Table 4.

Table 4. Orbit parameterization of Earth observation satellites and time intervals for which the individual parameters are estimated.

Parameter	GRACE	Swarm
Gravity field	degree/order 60 or 90	degree/order 70
Initial state	1 day	1 day
Const. acceleration radial	1 day	1 day
Const. acceleration along-track	none	1 day
Polynomial order 3 along-track	1 day	none
Const. acceleration cross-track	1 day	1 day
Accelerometer scaling factors	1 day, in all 3 directions	none
Pseudo-stochastic accelerations	15 minutes, constrained: $1 \times 10^{-7} \frac{m}{s^2}$	15 minutes, constrained: $7 \times 10^{-6} \frac{m}{s^2}$

2.4. Combination of Swarm and SLR on the Normal Equation Level

To fill the gap between GRACE and GRACE-FO, we propose a combination of gravity fields derived by multi-SLR and Swarm analysis as detailed above. A very simple combination of SLR and GRACE, in fact the replacement of C_{20} estimates in the monthly GRACE gravity fields by values derived from SLR, is already common practice [76]. In contrast to this, we perform a combination of Swarm and SLR on the NEQ level, i.e., we solve the equation

$$(w_{SLR}\mathbf{N}_{SLR} + w_{Swarm}\mathbf{N}_{Swarm})\mathbf{dx} = w_{SLR}\mathbf{b}_{SLR} + w_{Swarm}\mathbf{b}_{Swarm} \tag{2}$$

for the vector of unknown gravity field coefficients $\mathbf{dx}$, where $\mathbf{N}_{SLR}$ and $\mathbf{N}_{Swarm}$ are the normal equation matrices of SLR and Swarm, $\mathbf{b}_{SLR}$ and $\mathbf{b}_{Swarm}$ are the corresponding right-hand side vectors of the individual normal equation systems, and w_{SLR} and w_{Swarm} are weighting factors. The solution of the combined normal equation system is superior to a combination on a solution level because correlations between gravity field coefficients and other force model, orbit, instrument and reference frame parameters are taken into account [34]. Since all satellite data have been processed consistently and all but the gravity field parameters were pre-eliminated from the individual NEQs, the combination is straightforward.

The key question of the combination is the ratio of the weights w_{Swarm} and w_{SLR} assigned to the different observation techniques [38]. We first derive monthly weights by variance component estimation (VCE) [77]. As shown in Figure 1, the ratio $w_{Swarm} : w_{SLR}$ of the weights derived by VCE ranges from 20 to 95. To assess the plausibility of these weights, we derive alternative weights based on the accuracy estimates of the SLR ranges, which typically vary around 2 cm (see Table 1), and the GPS L1 phase observations of Swarm, for which Schreiter et al. [78] provide accuracy estimates close to 3 mm in case of strong ionosphere activity and close to 2 mm in case of low ionosphere activity (as long as no advanced observation screening in the region of the geomagnetic equator is performed). Based on these accuracy assumptions, a ratio of weights in the range from 44 to 100 can be expected, which is quite close to what is determined by VCE.

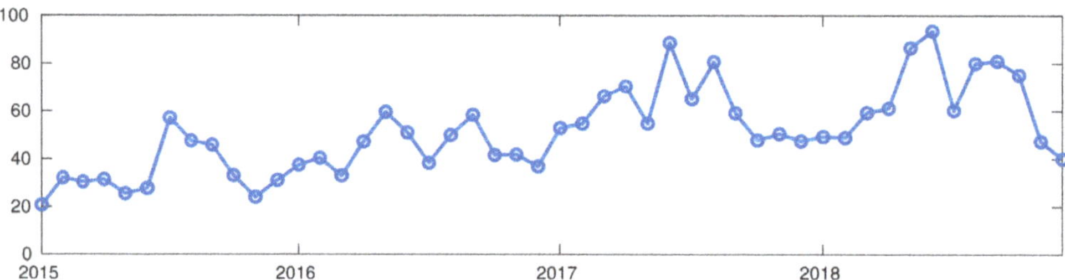

Figure 1. Ratios of monthly weights of the Swarm- with respect to the SLR-normal equations, as derived by variance component estimation.

The slightly positive trend in the relative weights visible in Figure 1 is explained by the decaying orbit altitude of the Swarm satellites and the corresponding increase in sensitivity to the gravity field. A seasonal variation probably is related to the seasonal tracking characteristics of the SLR stations and the fact that much more stations are located in the northern than in the southern hemisphere.

Test combinations were performed based on constant weighting ratios $w_{Swarm} : w_{SLR} = 100{:}1$, 10:1, 2.5:1 or 1:1. To assess the contribution of either Swarm or SLR to the combined solution, resolution matrices R_{SLR} and R_{Swarm} were determined from the individual normal equation matrices N_{SLR}, N_{Swarm} and the inverse of the combined matrix $N = w_{SLR}N_{SLR} + w_{Swarm}N_{Swarm}$ following the approach described in Sneeuw [79]:

$$R_{SLR} = N^{-1}N_{SLR}, \tag{3}$$

$$R_{Swarm} = N^{-1}N_{Swarm}. \tag{4}$$

The contribution numbers of the individual SHC are found on the main diagonals of the resolution matrices and are presented in triangle plots in Figure 2 for Swarm (left column) and SLR (right column). The contribution to the Sine-coefficients of the spherical harmonic spectrum are shown in the left hand part of the triangle plots, to the Cosine-coefficients in the right hand part, and the contribution to the zonal SHC (order 0) can be found in between. The contribution numbers vary between 0 (no influence) and 1 (determined by 100% from the corresponding observations). In the middle column of Figure 2, the mean contribution per degree is given.

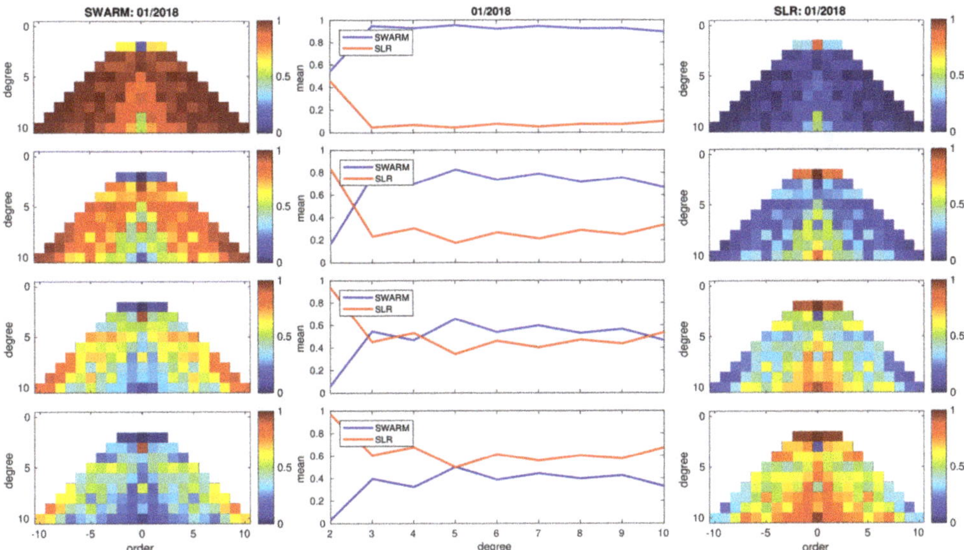

Figure 2. Contribution per spherical harmonics coefficient of Swarm (**left**) and SLR (**right**) satellites and mean contribution per degree (**middle column**) in case of relative weighting 100:1 (**top**), 10:1 (**second row**), 2.5:1 (**third row**) or equal weight (**bottom row**).

In case of a ratio of weights 100:1 (Figure 2, first row), only degree 2 SHC is significantly influenced by SLR, C_{20} in fact is dominated by SLR. Applying relative weights of 10:1 (Figure 2, second row), still mainly degree 2 SHC are impacted by SLR. Considering the strength of the SLR-derived temporal gravity variations, the suppression of the contribution of SLR to the other SHC is not justified. A combination based on the weights derived by VCE (Figure 1) therefore is pointless.

Decreasing the relative weight of the GPS observations further (Figure 2, rows three and four), the contribution of Swarm is step by step reduced to the sectorial (equal degree and order) and near sectorial SHC, where the sensitivity of GPS observations is strong [70], and to C_{30} that in case of SLR is weakly determined due to correlations with other zonal coefficients [68].

2.5. Spectral Resolution, Signal Leakage, and Filter Loss

The Earth gravity field is commonly represented by a spherical harmonic expansion (Equation (1)), truncated at a certain maximum degree $l_{max} < \infty$. The maximum degree (and consequently order) of this expansion determines the spatial scale of the represented signal. Monthly GRACE gravity fields are available to degree and order 90 (corresponding to a spatial scale of 460 km at the equator), Swarm monthly gravity fields are expanded to degree/order 70 but contain significant time-variable signal only up to about degree 13 (approx. 3100 km at the equator) [47]. SLR-derived gravity fields are even limited to degree 6 (approx. 6700 km at the equator) due to correlations between individual SHC at higher degrees that cannot be determined in unconstrained solutions given the inhomogenous observation coverage [10].

The truncation of the spherical harmonic expansion at a certain maximum degree causes signal leakage (e.g., [23–26]). To demonstrate this effect, we simulate a mass layer with uniform mass distribution within the island of Greenland, while the mass over the rest of the globe is set to zero. The simulated mass distribution is expanded in a series of spherical harmonics and reconstructed from the SHC truncating the series at various values of l_{max}. In Figure 3a, the reconstructed mass distributions are shown and the percentage of the integrated mass still contained inside the shorelines of Greenland is listed.

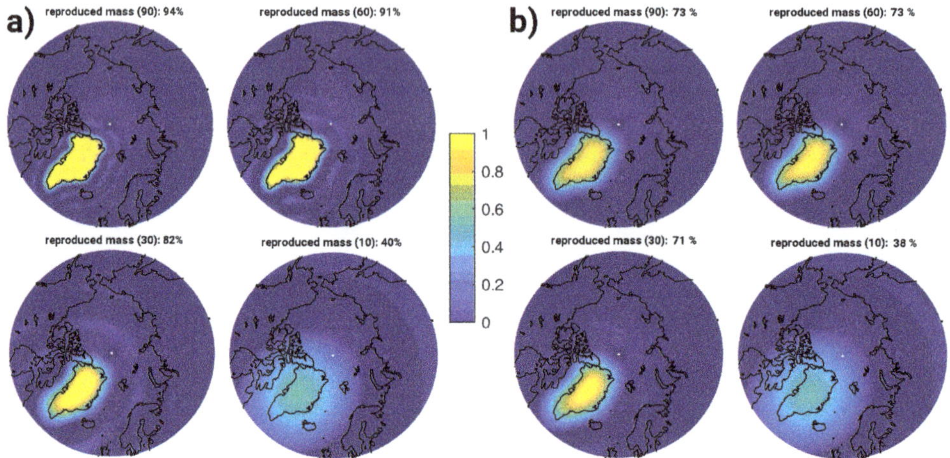

Figure 3. Integrated mass within Greenland reconstructed at different truncation degrees, (**a**) without filter or (**b**) smoothed by a 300 km Gaussian filter.

The signal content is also attenuated by filters to suppress the noise inherent to the gravity field models. A very simple filter that is commonly applied to GRACE monthly gravity fields is a Gaussian filter with 300 km filter radius [27]. In the spatial domain, the Gaussian filter represents a weighted average, the weighting function being bell-shaped with 300 km the half-width radius. In the spectral domain, the filter corresponds to degree-dependent scaling factors that quickly approach zero for higher degrees (from 1 at degree 0, the scaling factor has already dropped to 0.01 at degree 60). The experiment on the signal leakage by truncation of the spherical harmonic expansion is repeated, additionally applying a 300 km Gaussian filter for smoothing (Figure 3b). Due to the filter attenuation at medium to high degrees, the signal loss is even more dramatic than in the unfiltered case. More advanced filter types like the non-isotropic filter proposed by Han et al. [28] and the decorrelation filter DDK [29] were designed to minimize the filter loss, but the principle problem of signal attenuation persists.

Both experiments show that gravity fields truncated at different maximum degrees cannot be compared. At the very limited resolution of the SLR-derived gravity fields, only a fraction of the original mass is localized inside the borders of Greenland. The true mass localization and change can be recovered approximately applying iterative forward-modeling approaches [25,26], which require additional assumptions like mass loss being concentrated in fast flowing sections of ice streams, or at the coast. Nevertheless, leakage (in and out) and filter loss remain major limitations to the quantification of mass transport from satellite gravimetry. In the following, we truncate all gravity fields to the same low degree for the sake of comparison. Furthermore, we refrain from the use of filters.

To derive mass from SHC, first the dimensionless SHC are scaled by the factors provided by Wahr et al. [30] to transform them to EWH. Then, global 1°-grids of EWH are computed from the scaled SHC by spherical harmonic synthesis (Equation (1)). The series expansion is truncated at the maximum degree of choice (either 6 or 10). To transform the EWH grids to mass, each grid cell is multiplied by its area (dependent on latitude) and the density of water (1000 kg/m^3). By integration over the area of interest, e.g., Greenland, the final mass estimates are derived.

3. Results and Discussion

Figure 4 shows the integrated secular effect of glacial isostatic adjustment (GIA), ice mass and snow mass change in the polar regions during the period 2010 to 2014 as derived from un-smoothed degree 60 monthly gravity field models determined from GRACE GPS and K-band observations.

The mass loss is mainly related to ice loss (dynamic or by melting and run-off). GIA, i.e., the relaxation of the crust in reaction to the large scale ice melt after the last ice age, counteracts ice mass loss and therefore the actual ice mass loss is even larger (estimates for the mass change induced by GIA vary from 1 Gt/year to 20 Gt/year for Greenland and from 55 Gt/year to 110 Gt/year for Antarctica [32]) than indicated by the figures, while snow mass depends on the season and largely cancels out in a multi-year mean. The observation coverage is densest near the poles due to the polar orbits of the GRACE satellites and consequently the noise is lowest near the poles. At lower latitudes, the noisy striping typical for GRACE becomes visible.

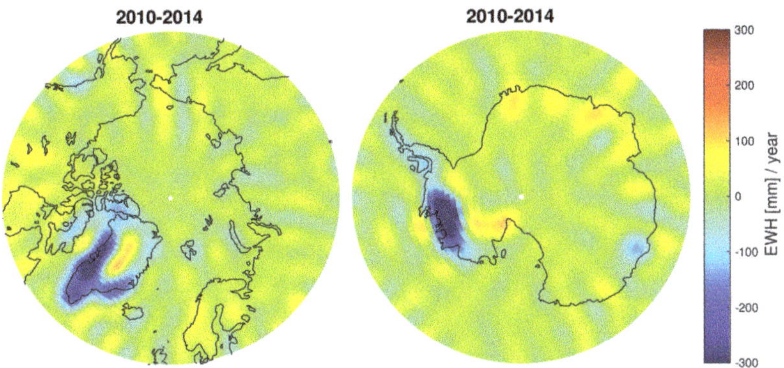

Figure 4. Trends in equivalent water height (EWH) as observed by GRACE in the Arctic (**left**) or Antarctic (**right**) region in the period 2010-2014, derived from unfiltered d/o 60 solutions.

For comparison to SLR-derived gravity field models the GRACE results are truncated at degree/order 10 and the corresponding EWH trends are shown in Figure 5. The trend signal that is well localized over the continents in Figure 4 leaks out over the oceans in Figure 5, as predicted by our simulation (Figure 3).

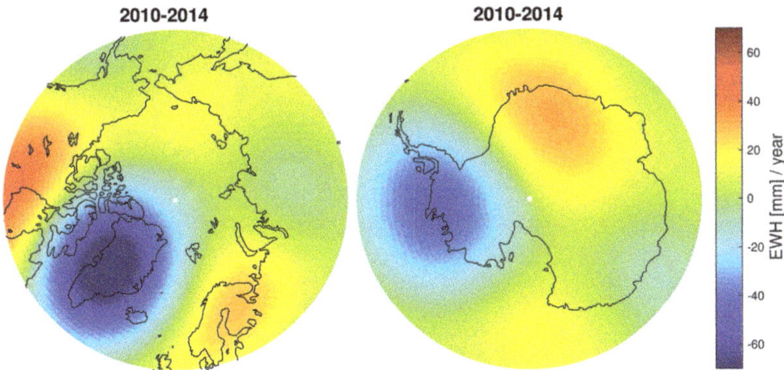

Figure 5. Trends in equivalent water height (EWH) in polar regions derived from GRACE gravity fields truncated at d/o 10.

In Figure 6, global plots of GRACE-derived trends 2010–2014 in EWH (left) and the amplitudes of seasonal variations (right) are provided, both truncated at degree/order 10. Trends are mainly visible at high latitudes, where large scale ice mass loss is the main reason for negative trends, GIA for positive trends in North America and Fennoscandia. At this low spherical harmonic resolution, it is difficult to state if negative trends near the west coast of North America are related to drought. Positive trends in the Amazon region probably are related to inter-annual variability that is not strictly seasonal.

Seasonal variations (Figure 6, right) are related to hydrology and are strongest in the tropical and subtropical regions with strong seasonal variation in rainfall.

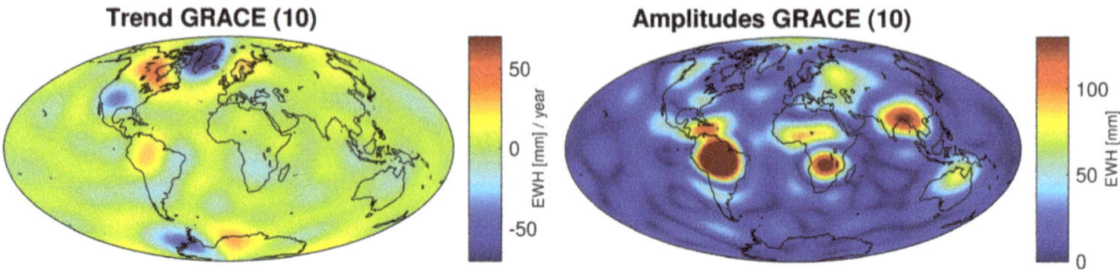

Figure 6. Global trends 2010–2014 in EWH (**left**) and amplitude of annual EWH variations (**right**) derived from GRACE gravity fields truncated at d/o 10.

Figure 7 provides the same information as Figure 6, but derived from SLR. In the top row, the unconstrained degree 6 gravity field models, and, in the bottom row, the constrained degree/order 10 solutions are evaluated. Especially in the case of the EWH trends, the improvement of the localization with a higher maximum degree is obvious. In the case of seasonal variations, the amplitudes in all but the Amazon basin are heavily attenuated compared to GRACE (Figure 6, right).

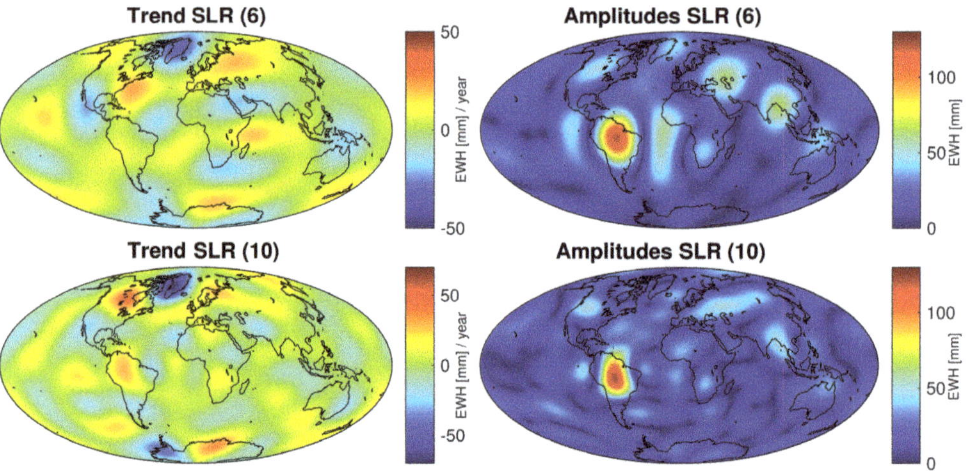

Figure 7. Global mass trends (in equivalent water height) as observed by SLR between 2010 and 2014 (**left**) and amplitude of annual variations (**right**). Top: unconstrained monthly gravity fields up to degree 6, bottom: constrained monthly gravity fields up to degree 10.

Figures 8 and 9 focus on EWH trends derived from SLR for the time periods 1995–1999, 2000–2004, 2005–2009 and 2010–2014 in polar regions. The trends are determined either from unconstrained degree 6 solutions (top row) or from regularized monthly gravity field models up to spherical harmonic degree and order 10 (bottom row). While we can confirm the findings of Bonin et al. [33] based on simulations that with degree 5 gravity fields, the correct localization of mass variations in Greenland and Antarctica is not granted; we observe rather precise localization of the SLR-derived mass trends in Greenland or Antarctica for the regularized degree 10 solutions (compare the sub-figures for the time-period 2010–2014 in Figures 8 and 9 with the corresponding results from GRACE in Figure 5). Along the coasts of Greenland and close to the coast of West Antarctica, the SLR gravity fields indicate significant mass loss over the ocean, but this has to be expected taking the effect of leakage into account (Figure 3).

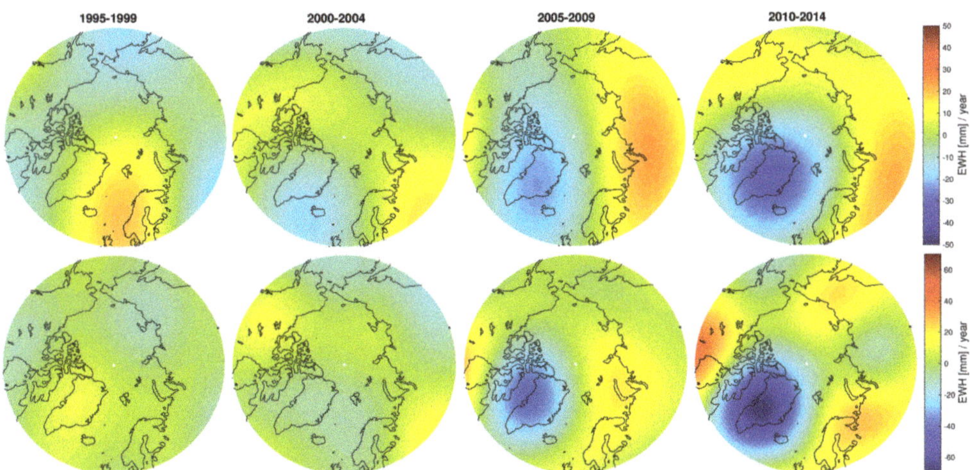

Figure 8. Trends in EWH as observed by SLR in the Arctic region during 5-year periods between 1995 and 2014. **Top**: unconstrained monthly gravity fields up to degree 6, **bottom**: constrained monthly gravity fields up to degree 10.

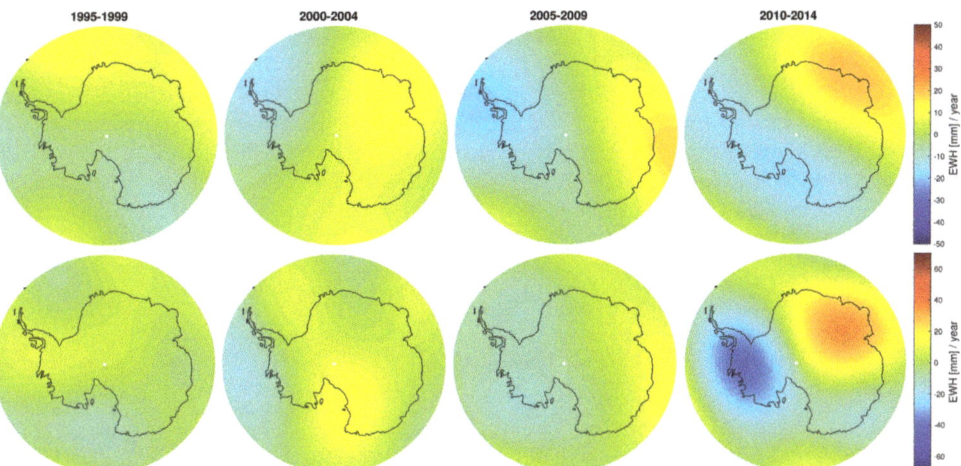

Figure 9. Trends in EWH as observed by SLR in the Antarctic region during 5-year periods between 1995 and 2014. **Top**: unconstrained monthly gravity fields up to degree 6, **bottom**: constrained monthly gravity fields up to degree 10.

In the following, time-series of mass change within certain regions are studied to demonstrate the accordance of SLR and Swarm with GRACE (both truncated at degree/order 6 corresponding to the unconstrained SLR solutions). SHC of degree 1 and C_{20} are removed, the latter because the GRACE estimates of C_{20} suffer from temporal aliasing [13] and accelerometer instrument noise [80] (the effect of C_{20} alone on mass change in Greenland sums up to about 100 Gt in the time period from 2000 to 2018). No re-scaling to compensate for signal leakage is applied and no corrections for GIA and seasonal variations due to snow mass are applied. For a complete treatment of ice mass loss as derived from GRACE data, we refer to the literature in this field (e.g., [4]).

While GRACE observations cover the period 2002–2017, earlier estimates of mass change have to be based on observations of the geodetic SLR satellites. After the end of the GRACE mission, mass estimates are based either on SLR or on Swarm observations. The good agreement of SLR- and GRACE-derived seasonal variations in the Amazon basin is obvious in Figure 10, while Swarm tends to slightly overestimate the seasonal variation.

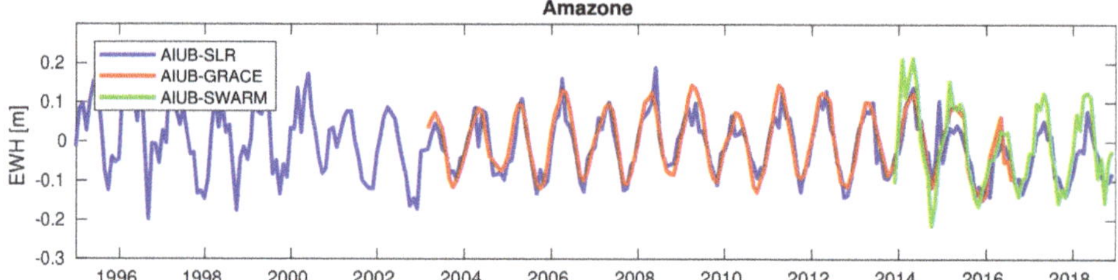

Figure 10. Monthly mass estimates for the Amazon basin. GRACE and Swarm gravity fields were truncated at degree 6 to match the resolution of SLR gravity fields. Spherical harmonic coefficient C_{20} was excluded from the comparison.

The evolution of the Greenland mass is shown in Figure 11. Again, the long-term agreement of all three observation types is very good. SLR does not reveal significant mass loss in Greenland in the pre-GRACE era. Obviously, the GRACE mission was launched just in time to observe the onset of ice loss in Greenland. After acceleration in the ice mass loss, the GRACE mass estimates almost level out in the time period from 2014 to 2016. SLR tends to slightly overestimate the Greenland mass loss between 2014 and 2017, while, after 2017, even a small mass gain is visible in both the SLR and Swarm derived mass estimates. The reduction of the SLR- and Swarm-derived mass loss therefore is time-shifted by about three years compared to GRACE. Bonin et al. [33] also observe diverging mass trends for Greenland but state that, compared to the independent time-series of mass change derived by the Ice-sheet Mass Balance Inter-comparison Exercise (IMBIE) for Greenland and Antarctica [81], in the long term, these deviations of SLR average out. The inter-annual variability is larger in the monthly SLR estimates than observed by GRACE at this low spherical harmonic degree, and even larger in the monthly Swarm solutions.

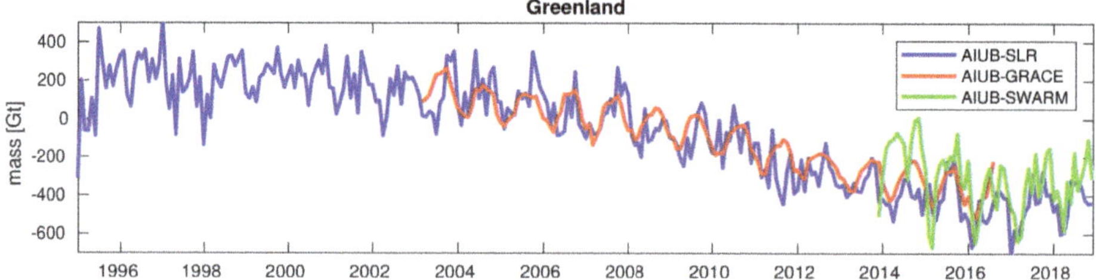

Figure 11. Monthly mass estimates for entire Greenland (see masks in Figure 6). GRACE and Swarm gravity fields were truncated at degree 6 to match the resolution of SLR gravity fields. Spherical harmonic coefficient C_{20} was excluded from the comparison.

While the accuracy of the GRACE mass estimates at this very low resolution can considered to be constant in the time-period shown, because temporal variations in noise due to the changing satellite environment and technical problems mainly are manifest in the weakly determined high degree/order SHC, the quality of the SLR-derived estimates depends on the station network that was even more sparse in the 1990s than it is today, and on the number of satellites contributing to the monthly solutions. A reduction in the scatter can be observed after the launch of LARES in 2012. The monthly Swarm solutions exhibit even greater scatter than the SLR solutions and additionally seem to over-estimate the seasonal variation compared to GRACE. After 2014, the Swarm and the SLR results match well in trend and in phase. During 2014, the Swarm results are impaired by high ionosphere activity, non-optimal GPS receiver settings, and in the first half of 2014 by reduced observation sampling [75].

Consecutively, mass trends were determined for the coastal regions of Greenland with strong mass loss, the region of the inland ice sheet, where a weak gain in mass is observed by GRACE, and the coast of West Antarctica. The masks used are provided in Figure 12. The separation of Greenland into regions of mass gain and regions of mass loss was done based on Figure 4. The mask of Antarctica and its glacial sub-basins is derived from Horwath and Dietrich [82].

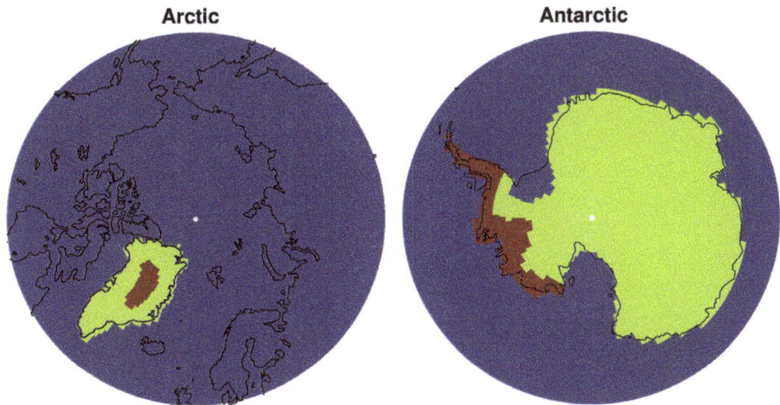

Figure 12. Masks defining inland and coastal regions of Greenland (**left**) and the coast of West Antarctica (**right**). The resolution of the masks is 1°.

Starting with the first monthly estimates from SLR in 1995, we fit deterministic models including bias, trend and seasonal variations within 5-year periods to integrated mass estimates of either SLR or GRACE monthly gravity field models summed up over the coastal areas (Figure 13, left) or the inland ice sheet (Figure 13, right) of Greenland, or the coast of West Antarctica (Figure 14).

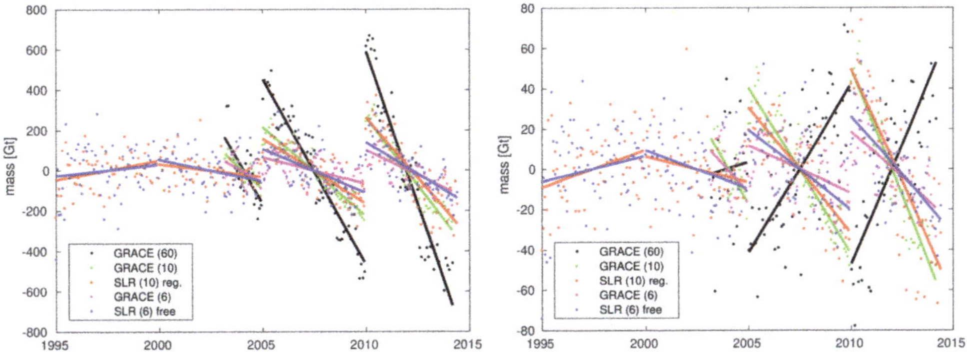

Figure 13. Monthly mass estimates in Greenland from GRACE and SLR, and best fitting trends for 5-year periods for the region of mass loss along the coast (**left**) or the inland region of mass gain (**right**). Each 5-year period is centered around zero.

In case of SLR, the trend estimation is based either on the unconstrained degree 6 monthly gravity field solutions or on the regularized degree 10 solutions (compare Figures 8 and 9 for the corresponding spatial plots). For comparison, the original degree 60 GRACE gravity fields were truncated at degrees 6 or 10 and also mass trends in 5-year intervals determined.

The SLR trend estimates reveal good agreement with GRACE (truncated at the corresponding degree) for the time periods where a direct comparison is possible. However, due to signal leakage at the very low resolution of SLR, the absolute mass loss is drastically underestimated. Truncated at degree 6 or 10, due to the very limited spatial resolution, neither GRACE nor SLR are able to trace the inland mass gain observed by GRACE at higher resolution (Figure 13, right).

Prior to the start of GRACE, no distinct mass trend within Greenland can be observed. Since the start of the highly sensitive GRACE observations, the mass loss at the coast is accelerating within the time period covered by Figure 13. This fact has already been reported, e.g., by Velicogna [83] and recently by Bevis et al. [84], and is visible in SLR and GRACE estimates alike. In case of the degree 6 SLR solutions, the five year trend estimates are 2000–2004: −21.5 Gt/year, 2005–2009: −43.6 Gt/year, 2010–2014: −63.1 Gt/year. As explained above, due to leakage, these values are drastically underestimated compared to the degree 60 GRACE solutions 2003–2004: −179.8 Gt/year, 2005–2009: −184.4 Gt/year, 2010–2014: −302.7 Gt/year (no GIA corrections applied).

Comparable results are achieved for the area of significant ice mass loss along the coast of West Antarctica (Figure 14), where, due to the bad localization of the mass loss in the degree 6 solutions, the five year trends derived from the degree 10 SLR gravity fields 2000–2004: −5.4 Gt/year, 2005–2009: −15.1 Gt/year, and 2010–2014: −50.7 Gt/year are quoted here together with the degree 60 GRACE solutions 2003–2004: −85.5 Gt/year, 2005–2009: −121.1 Gt/year, and 2010–2014: −201.0 Gt/year. Again, the trend estimates at the given low degree of SLR are underestimates due to leakage, but the increase in ice melt is captured very impressively.

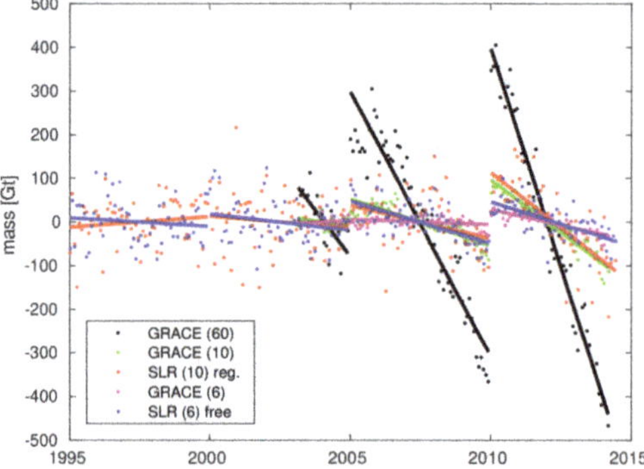

Figure 14. Monthly mass estimates at the coast of West Antarctica from GRACE and SLR, and best fitting trends for 5-year periods. Each 5-year period is centered around zero.

The individual and the combined monthly mass estimates within Greenland for the two extreme cases of 100:1 and 1:1 relative weighting of Swarm and SLR are shown in Figure 15. The combination starts with January 2015 because, in 2014, the GRACE data are still quite complete while the Swarm data is impaired by strong ionosphere activity and non-optimal GPS receiver settings (receiver settings were adapted in May 2015). During 2015 and in the first half of 2016, a direct comparison to GRACE results is possible. After August 2016, the accelerometer on GRACE B was completely switched off and the processing of GRACE data at AIUB stopped. From the direct comparison, the impression that Swarm generally slightly overestimates the amplitude of the mass variations while SLR overestimates the secular mass loss is confirmed.

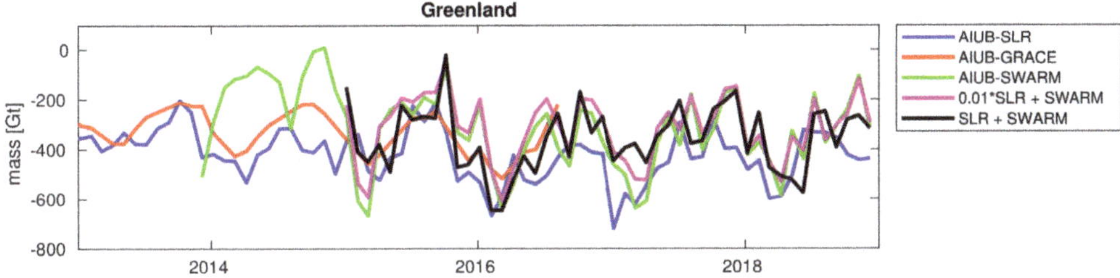

Figure 15. Monthly mass estimates within Greenland. All time-series truncated at degree 6 and with C_{20} excluded.

In case of relative weighting 100:1, the combined mass estimates closely follow the Swarm-only results. This can be expected in view of the results of the contribution analysis (Figure 2), even more so because C_{20} is excluded from the analysis. In the case of equal weights, the combined solution follows more closely the GRACE mass estimates, but still with higher inter-annual variability.

The monthly differences between GRACE and the individual or combined alternative mass estimates are shown in Figure 16. During 2015, the combination of SLR and Swarm with equal weighting has the smallest differences with respect to GRACE; in the first half of 2016, the combination 0.4*SLR+Swarm is slightly closer. The RMS of the differences with respect to GRACE over all months is 166 Gt for the SLR-only solutions, 110 Gt in case of Swarm-only, 110 Gt for the combination 0.01*SLR+Swarm, 83 Gt for 0.1*SLR+Swarm, 67 Gt for 0.4*SLR+Swarm and 68 Gt in case of equal weighting.

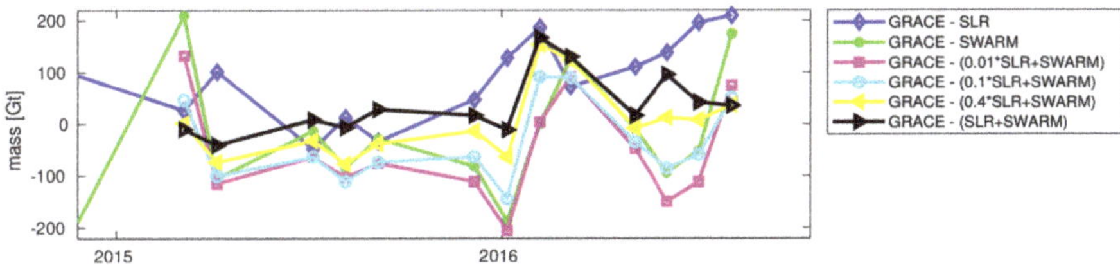

Figure 16. Monthly differences between GRACE and SLR (blue), Swarm (green) or combined mass estimates within Greenland. Missing values indicate gaps in the GRACE time-series.

These results clearly indicate that a combination with close to equal weights of Swarm (GPS) and SLR is closer to the GRACE reference than when weights are based on variance analysis of observation residuals. This can be explained by systematic errors in the kinematic orbits due to the GPS phase processing. In fact, GPS also has to be down-weighted in combination with the GRACE K-band observations (by a factor of approx. 200, see [34]). Recent experiments using one month of kinematic GRACE orbits derived with undifferenced integer-fixed GPS phase ambiguities indicate that the inconsistency between K-band and GPS is reduced. A better assessment will be possible as soon as longer time-series of kinematic orbits based on undifferenced integer-fixed ambiguities become available, but this will be the topic of a future publication.

4. Conclusions

Satellite laser ranging can contribute to the derivation of mass change estimates at the spatial scale of the Amazon basin, Greenland, or West Antarctica prior to the GRACE mission. SLR and high-low tracking of the Swarm satellites both may be used to bridge the gap between GRACE and GRACE-FO concerning the large scale mass loss in the polar regions of the Earth. When all individual gravity field results are truncated at the same spherical harmonic degree, the results can be compared directly. They match well for Greenland with a somewhat higher inter-annual variability of the SLR monthly mass estimates. Swarm can only be included in the comparison near the end of the life-time of the

GRACE satellites. The seasonal variations visible in Swarm monthly mass estimates seem to be slightly over-estimated. On the other hand, the SLR derived mass loss within Greenland is larger than what can be extracted from GRACE observations between 2014 and 2017. After 2017, both SLR and Swarm results indicate moderate mass gain within Greenland.

For a correct localization of mass change signal along the coast of West Antarctica, a spherical harmonic expansion up to degree 10 is desirable. This can be achieved by the regularization of SLR-only gravity field models or, preferably, by a combination with other LEO, e.g., Swarm data. A combination on the normal equation level, correctly taking into account all correlations between estimated and pre-eliminated parameters, is feasible as long as all satellite data is processed consistently. The best fit between a SLR/Swarm combination and GRACE for the period from January 2015 to August 2016, where both Swarm and GRACE observations of good quality are available, is achieved when SLR and Swarm normal equations are combined with almost equal weights.

It is foreseen to continue the combination of SLR data, taking into account the contributions by different analysis centers, and the combination of SLR with Swarm and possibly also GRACE-FO on the normal equation level in the frame of the Combination Service for Time-variable Gravity field models (COST-G) [34,52], the newly established product center of the International Gravity Field Service (IGFS) under the umbrella of the International Association of Geodesy (IAG).

Author Contributions: GRACE processing, derivation of mass trends, conceptualization and writing—U.M.; SLR processing and investigation—K.S.; determination of kinematic GRACE and Swarm orbits—D.A.; Swarm processing—C.D.; software development, supervision and funding—D.T.; supervision and project administration (SLR)—R.D.; supervision and project administration (GRACE, Swarm)—A.J.

Funding: This research was partly funded by BKG "Vertrag über die Weiterentwicklung der Berner GNSS Software, 21.04.2017" and by the ESA project Swarm-DISC, contract No. 4000109587/13/I-NB.

Acknowledgments: We want to thank three anonymous reviewers for their invaluable comments.

Conflicts of Interest: The authors declare no conflict of interest. The funders had no role in the design of the study; in the collection, analyses, or interpretation of data; in the writing of the manuscript, or in the decision to publish the results.

Abbreviations

The following abbreviations are used in this manuscript:

AC	Analysis Centre
AOD	Atmosphere and Ocean De-Aliasing
CMA	Celestial Mechanics Approach
GIA	Glacial Isostatic Adjustment
GPS	Global Positioning System
Gt	Gigatons (10^{12} kg)
LEO	Low Earth Orbiter
NEQ	Normal Equation
PPP	Precise Point Positioning
RMS	Root Mean Square
SHC	Spherical Harmonic Coefficients
SLR	Satellite Laser Ranging
VCE	Variance Component Estimation

References

1. Tapley, B.D.; Bettadpur, S.; Ries, J.C.; Thompson, P.F.; Watkins, M. GRACE measurements of mass variability in the Earth system. *Science* **2004**, *305*, 503–505. [CrossRef] [PubMed]
2. Sasgen, I.; Van den Broeke, M.; Bamber, J.; Rignot, E.; Sorensen, L.; Wouters, B.; Martinec, Z.; Velicogna, I.; Simonsen, S. Timing and origin of recent regional ice-mass loss in Greenland. *Earth Planet. Sci. Lett.* **2012**, *333–334*, 293–303. [CrossRef]

3.   Luthcke, S.B.; Sabaka, T.L.; Loomis, B.D.; Arendt, A.A.; McCarthy, J.J.; Camp, J. Antarctica, Greenland and Gulf of Alaska land-ice evolution from an iterated GRACE global mascon solution. *J. Glaciol.* **2013**, *59*, 613–631. [CrossRef]

4.   Velicogna, I.; Wahr, J. Time-variable gravity observations of ice sheet mass balance: Precision and limitations of the GRACE satellite data. *Geophys. Res. Lett.* **2013**, *40*, 3055–3063. [CrossRef]

5.   Flechtner, F.; Neumayer, K.-H.; Dahle, C.; Dobslaw, H.; Fagiolini, E.; Raimondo, J.-C.; Güntner, A. What Can be Expected from the GRACE-FO Laser Ranging Interferometer for Earth Science Applications. *Surv. Geophys.* **2016**, *37*, 453–470. [CrossRef]

6.   Rietbroek, R.; Fritsche, M.; Dahle, C.; Brunnabend, S.E.; Behnisch, M.; Kusche, J.; Flechtner, F.; Schröter, J.; Dietrich, R. Can GPS-Derived Surface Loading Bridge a GRACE Mission Gap? *Surv. Geophys.* **2014**, *35*, 1267–1283. [CrossRef]

7.   Peralta-Ferriz, C.; Morison, J.H.; Wallace, J.M. Proxy representation of Arctic ocean bottom pressure variability: Bridging gaps in GRACE observations. *Geophys. Res. Lett.* **2016**, *43*, 9183–9191. [CrossRef]

8.   Lück, C.; Kusche, J.; Rietbroek, R.; Löcher, A. Time-variable gravity fields and ocean mass change from 37 months of kinematic Swarm orbits. *Solid Earth* **2018**, *9*, 323–339. [CrossRef]

9.   Pearlman, M.; Arnold, D.; Davis, M.; Barlier, F.; Biancale, R.; Vasiliev, V.; Ciufolini, I.; Paolozzi, A.; Pavlis, E.; Sosnica, K.; et al. Laser geodetic satellites: A high accuracy scientific tool. *J. Geod.* **2019**, in press. [CrossRef]

10. Cheng, M.K.; Shum, C.K.; Tapley, B.D. Determination of long-term changes in the Earth's gravity field from satellite laser ranging observations. *J. Geophys. Res.* **1997**, *102*, 22377–22390. [CrossRef]

11. Bianco, G.; Devoti, R.; Fermi, M.; Luceri, V.; Rutigliano, P.; Sciaretta, C. Estimation of low degree geopotential coefficients using SLR data. *Planet. Space Sci.* **1998**, *46*, 1633–1638. [CrossRef]

12. Cheng, M.; Tapley, B.D. Seasonal variations in low degree zonal harmonics of the Earth's gravity field from satellite laser ranging observations. *J. Geophys. Res. Solid Earth* **1999**, *104*, 2667–2681. [CrossRef]

13. Cheng, M.; Ries, J. The unexpected signal in GRACE estimates of C_{20}. *J. Geod.* **2017**, *91*, 897–914. [CrossRef]

14. Cox, C.M.; Chao, B.F. Detection of a Large-Scale Mass Redistribution in the Terrestrial System Since 1998. *Science* **2002**, *297*, 831–833. [CrossRef] [PubMed]

15. Cheng, M.; Tapley, B.D. Variations in the Earth's oblateness during the past 28 years. *J. Geophys. Res.* **2004**, *109*, B09402. [CrossRef]

16. Bloßfeld, M.; Müller, H.; Gerstl, M.; Stefka, V.; Bouman, J.; Göttl, F.; Horwath, M. Second-degree Stokes coefficients from multi-satellite SLR. *J. Geod.* **2015**, *89*, 857–871. [CrossRef]

17. Zehentner, N.; Mayer-Gürr, T. Precise orbit determination based on raw GPS measurements. *J. Geod.* **2016**, *90*, 275–286. [CrossRef]

18. Friis-Christensen, E.; Lühr, H.; Hulot, G. Swarm: A constellation to study the Earth's magnetic field. *Earth Planets Space* **2006**, *58*, 351–358. [CrossRef]

19. Sosnica, K.; Jäggi, A.; Meyer, U.; Thaller, D.; Beutler, G.; Arnold, D.; Dach, R. Time variable Earth's gravity field from SLR satellites. *J. Geod.* **2015**, *89*, 945–960. [CrossRef]

20. Meyer, U.; Jäggi, A.; Jean, Y.; Beutler, G. AIUB-RL02: An improved time-series of monthly gravity fields from GRACE data. *Geophys. J. Int.* **2016**, *205*, 1196–1207. [CrossRef]

21. Jäggi, A.; Dahle, C.; Arnold, D.; Bock, H.; Meyer, U.; Beutler, G.; van den Ijssel, J. Swarm kinematic orbits and gravity fields from 18 months of GPS data. *Adv. Space Res.* **2016**, *57*, 218–233. [CrossRef]

22. Weigelt, M.; Sideris, M.G.; Sneeuw, N. On the influence of the ground track on the gravity field recovery from high–low satellite-to-satellite tracking missions: CHAMP monthly gravity field recovery using the energy balance approach revisited. *J. Geod.* **2009**, *84*, 1131–1143. [CrossRef]

23. Baur, O.; Kuhn, M.; Featherstone, W.E. GRACE-derived ice-mass variations over Greenland by accounting for leakage effects. *J. Geophys. Res.* **2009**, *114*, B06407. [CrossRef]

24. Guo, J.Y.; Duan, X.J.; Shum, C.K. Non-isotropic Gaussian smoothing and leakage reduction for determining mass changes over land and ocean using GRACE data. *Geophy. J. Int.* **2010**, *181*, 290–302. [CrossRef]

25. Jin, S.; Zou, F. Re-estimation of glacier mass loss in Greenland from GRACE with correction of land-ocean leakage effects. *Glob. Planet. Chang.* **2015**, *135*, 170–178. [CrossRef]

26. Chen, J.I.; Wilson, C.R.; Li, J.; Zhang, Z. Reducing leakage error in GRACE-observed long-term ice mass change: A case study in West Antarctica. *J. Geod.* **2015**, *89*, 925–940. [CrossRef]

27. Swenson, S.; Wahr, J. Methods for inferring regional surface-mass anomalies from Gravity Recovery and Climate Experiment (GRACE) measurements of time-variable gravity. *J. Geophys. Res.* **2002**, *107*, 2193. [CrossRef]

28. Han, S.C.; Shum, C.K.; Jekeli, C.; Kuo, C.Y.; Wilson, C.; Wilson, C.; Seo, K.W. Non-isotropic filtering of GRACE temporal gravity for geophysical signal enhancement. *Geophy. J. Int.* **2005**, *163*, 18–25. [CrossRef]

29. Kusche, J. Approximate decorrelation and non-isotropic smoothing of time-variable GRACE-type gravity field models. *J. Geod.* **2007**, *81*, 733–749. [CrossRef]

30. Wahr, J.; Molenaar, M.; Bryan, F. Time variability of the Earth's gravity field: Hydrological and oceanic effects and their possible detection using GRACE. *J. Geophys. Res.* **1998**, *103*, 30205–30230. [CrossRef]

31. Matsuo, K.; Chao, B.F.; Otsubo, T.; Heki, K. Accelerated ice mass depletion revealed by low-degree gravity fields from satellite laser ranging: Greenland, 1991-2011. *Geophys. Res. Lett.* **2013**, *40*, 4662–4667. [CrossRef]

32. Talpe, A.M.; Nerem, R.S.; Forootan, E.; Schmidt, M.; Lemoine, F.G.; Enderlin, E.M.; Landerer, F.W. Ice mass change in Greenland and Antarctica between 1993 and 2013 from satellite gravity measurements. *J. Geod.* **2017** *91*, 1283–1298. [CrossRef]

33. Bonin, J.A.; Chambers, D.P.; Cheng, M. Using satellite laser ranging to measure ice mass change in Greenland and Antarctica. *Cryosphere* **2018**, *12*, 71–79. [CrossRef]

34. Meyer, U.; Jean, Y.; Kvas, A.; Dahle, C.; Lemoine, J.M.; Jäggi, A. Combination of GRACE monthly gravity fields on the normal equation level. *J. Geod.* **2019**, submitted .

35. Moore, P.; Zhang, Q.; Alothman, A. Annual and semiannual variations of the Earth's gravitational field from satellite laser ranging and CHAMP. *J. Geophys. Res. Solid Earth* **2005**, *110*, B06401. [CrossRef]

36. Cheng, M.; Ries, C.J.; Tapley, D. Variations of the Earth's figure axis from satellite laser ranging and GRACE. *J. Geophys. Res.* **2011**, *116*, B01409. [CrossRef]

37. Maier, A.; Krauss, S.; Hausleitner, W.; Baur, O. Contribution of satellite laser ranging to combined gravity field models. *Adv. Space Res.* **2012**, *49*, 556–565. [CrossRef]

38. Haberkorn, C.; Bloßfeld, M.; Bouman, J.; Fuchs, M.; Schmidt, M. Towards a Consistent Estimation of the Earth's Gravity Field by Combining Normal Equation Matrices from GRACE and SLR. In *IAG 150 Years. International Association of Geodesy Symposia*; Rizos, C., Willis, P., Eds.; Springer: Cham, Switzerland, 2015; Volume 143, pp. 375–381. [CrossRef]

39. König, R.; Chen, Z.; Reigber, C.; Schwintzer, P. Improvement in global gravity field recovery using GFZ-1 satellite laser tracking data. *J. Geod.* **1999**, *73*, 398–406. [CrossRef]

40. Reigber, C.; Balmino, G.; Schwintzer, P.; Biancale, R.; Bode, A.; Lemoine, J.; König, R.; Loyer, S.; Neumayer, H.; Marty, J.; et al. A high quality global gravity field model from CHAMP GPS tracking data and accelerometry (EIGEN-1S). *Geophys. Res. Lett.* **2002**, *36*. [CrossRef]

41. Reigber, C.; Schmidt, R.; Flechtner, F.; König, R.; Meyer, U.; Neumayer, K.-H.; Schwintzer, P.; Zhu, S.Y. An Earth gravity field model complete to degree/order 150 from GRACE: EIGEN-GRACE02S. *J. Geodyn.* **2005**, *39*, 1–10. [CrossRef]

42. Tapley, B.; Ries, J.; Bettadpur, S.; Chambers, D.; Cheng, M.; Condi, F.; Gunter, B.; Kang, Z.; Nagel, P.; Pastor, R.; et al. GGM02—An improved Earth gravity field model from GRACE. *J. Geod.* **2005**, *79*, 467–478. [CrossRef]

43. Rummel, R.; Yi, W.; Stummer, C. GOCE gravitational gradiometry. *J. Geod.* **2011**, *85*, 777–790. [CrossRef]

44. Pail, R.; Bruinsma, S.; Migliaccio, F.; Förste Ch Goiginger, H.; Schuh, W.G.; Höck, E.; Reguzzoni, M.; Brockmann, J.M.; Abrikosov, O.; Veicherts, M.; et al. First GOCE gravity field models derived by three different approaches. *J. Geod.* **2011**, *85*, 819–843. [CrossRef]

45. Wouters, B.; Bonin, J.; Chambers, D.; Riva, R.; Sasgen, I.; Wahr, J. GRACE, time-varying gravity, Earth system dynamics and climate change. *Rep. Prog. Phys.* **2014**, *77*, 116801. [CrossRef]

46. Weigelt, M.; van Dam, T.; Jäggi, A.; Prange, L.; Tourian, M.; Keller, W.; Sneeuw, N. Time-variable gravity signal in Greenland revealed by high-low satellite-to-satellite tracking. *J. Geophy. Res. Solid Earth* **2013**, *118*, 3848–3859. [CrossRef]

47. Teixeira da Encarnacao, J.; Arnold, D.; Bezdek, A.; Dahle, C.; Doornbos, E.; van den IJssel, J.; Jäggi, A.; Mayer-Gürr, T.; Sebera, J.; Visser, J.; et al. Gravity field models derived from Swarm GPS data. *Earth Planets Space* **2016**, *68*, 127. [CrossRef]

48. Beutler, G.; Jäggi, A.; Mervart, L.; Meyer, U. The celestial mechanics approach: Theoretical foundations. *J. Geod.* **2010**, *84*, 605–624. [CrossRef]

49. Flury, J.; Bettadpur, S.; Tapley, S. Precise accelerometry onboard the GRACE gravity field satellite mission. *Adv. Space Res.* **2008**, *42*, 1414–1423. [CrossRef]
50. Heiskanen, W.H.; Moritz, H. *Physical Geodesy*; W.H. Freeman and Co.: San Francisco, CA, USA, 1967.
51. Jäggi, A.; Hubentobler, U.; Beutler, G. Pseudo-stochastic orbit modeling techniques for low-Earth orbiters. *J. Geod.* **2006**, *80*, 47–60. [CrossRef]
52. Jäggi, A.; Weigelt, M.; Flechtner, F.; Güntner, A.; Mayer-Gürr, T.; Martinis, S.; Bruinsma, S.; Flury, J.; Bourgogne, S.; Meyer, U.; et al. European Gravity Service for Improved Emergency Management (EGSIEM) - from concept to implementation. *Geophys. J. Int.* **2019**, submitted
53. Meyer, U.; Jäggi, A.; Beutler, G.; Bock, H. The impact of common versus separate estimation of orbit parameters on GRACE gravity field solutions. *J. Geod.* **2015**, *89*, 685–696. [CrossRef]
54. Kucharski, D.; Kirchner, G.; Otsubo, T.; Koidl, F. A method to calculate zero-signature satellite laser ranging normal points for millimeter geodesy—A case study with Ajisai. *Earth Planets Space* **2015**, *67*, 34. [CrossRef]
55. Zumberge, J.F.; Heflin, M.B.; Jefferson, D.C.; Watkins, M.M.; Webb, F.H. Precise point positioning for the efficient and robust analysis of GPS data from large networks. *J. Geophy. Res.* **1997**, *102*, 5005–5017. [CrossRef]
56. Jäggi, A.; Bock, H.; Prange, L.; Meyer u Beutler, G. GPS-only gravity field recovery with GOCE, CHAMP, and GRACE. *Adv. Space Res.* **2011**, *47*, 1020–1028. [CrossRef]
57. Dunn, C.E.; Bertiger, W.; Bar-Sever, Y.; Desai, S.; Haines, B.; Kuang, D.; Franklin, G.; Harris, I.; Kruizinga, G.; Meehan, T.; et al. Instrument of GRACE: GPS augments gravity measurements. *GPS World* **2003**, *14*, 16–28.
58. Bruinsma, S.; Lemoine, J.M.; Biancale, R.; Vales, N. CNES/GRGS 10-day gravity field models (release 2) and their evaluation. *Adv. Space Res.* **2010**, *45*, 587–601. [CrossRef]
59. Kurtenbach, E.; Mayer-Gürr, T.; Eicker, A. Deriving daily snapshots of the Earth's gravity field from GRACE L1B data using Kalman filtering. *Geophys. Res. Lett.* **2009**, *36*, L17102. [CrossRef]
60. Yoder, C.; Williams, J.; Dickey, J.; Schutz, B.; Eanes, R.; Tapley, B. Secular variations of Earth's gravitational harmonic J2 coefficient from Lageos and nontidal acceleration of Earth rotation. *Nature* **1983**, *303*, 757–762. [CrossRef]
61. Dobslaw, H.; Bergmann-Wolf, I.; Dill, R.; Poropat, L.; Thomas, M.; Dahle, C.; Esselborn, S.; König, R.; Flechtner, F. A new high-resolution model of non-tidal atmosphere and ocean mass variability for de-aliasing of satellite gravity observations: AOD1B RL06. *Geophys. J. Int.* **2017**, *211*, 263–269. [CrossRef]
62. Savcenko, R.; Bosch, W. EOT11a—A new tide model from Multi-Mission Altimetry. In Proceedings of the OSTST Meeting, San Diego, CA, USA, 16–21 October 2011.
63. Petit, G.; Luzum, B. (Eds.) *IERS Conventions*; IERS Technical Note No. 36, International Earth Rotation and Reference Systems Service; Verlag des Bundesamtes für Karthographie und Gedäsie: Frankfurt am Main, Germany, 2010.
64. Picone, J.; Hedin, A.; Drob, D.; Aikin, A. NRLMSISE-00 empirical model of the atmosphere: Statistical comparisons and scientific issues. *J. Geophys. Res.* **2002**, *107*, 1468. [CrossRef]
65. Knocke, P.; Ries, J.; Tapley, B. Earth radiation pressure effects on satellites. In Proceedings of the AIAA/AAS Astrodynamics Conference, Minneapolis, MN, USA, 15–17 August 1988; pp. 577–587.
66. Rodriguez-Solano, C.; Hugentobler, U.; Steigenberger, P.; Lutz, S. Impact of Earth radiation pressure on GPS position estimates. *J. Geod.* **2012**, *86*, 309–317. [CrossRef]
67. Gunter, B.; Ries, J.; Bettadput, S.; Tapley, B. A simulation study of the errors of omission and comission for GRACE RL01 gravity fields. *J. Geod.* **2006**, *80*, 341–351. [CrossRef]
68. Sosnica, K. Determination of precise satellite orbits and geodetic parameters using satellite laser ranging. In *Geodätisch-Geophysikalische Arbeiten in der Schweiz*; Swiss Geodetic Commission: Zürich, Switzerland, 2015; Volume 93, ISBN 978-3-908440-38-3.
69. Thomas, J.B. *An Analysis of Gravity-Field Estimation Based on Intersatellite Dual-1-Way Biased Ranging*; JPL Publication; Jet Propulsion Laboratory: Pasadena, CA, USA, 1999.
70. Beutler, G.; Jäggi, A.; Mervart, L.; Meyer, U. The celestial mechanics approach: Application to data of the GRACE mission. *J. Geod.* **2010**, *84*, 661–681. [CrossRef]
71. Jäggi, A.; Prange, L.; Hugentobler, U. Impact of covariance information of kinematic positions on orbit reconstruction and gravity field recovery. *Adv. Space Res.* **2011**, *47*, 1472–1479. [CrossRef]
72. Dach, R.; Lutz, S.; Walser, P.; Fridez, P. (Eds.) *Bernese GNSS Software*, Version 5.2.; University of Bern, Bern Open Publishing: Bern, Switzerland, 2015; ISBN 978-3-906813-05-9. [CrossRef]

73. Siemes, C.; Teixeira da Encarnacao, J.; Doornbos, E.; van den IJssel, J.; Kraus, J.; Peresty, R.; Grunwaldt, L.; Apelbaum, G.; Flury, J.; Olsen, P.E.H. Swarm accelerometer data processing from raw accelerations to thermospheric neutral densities. *Earth Planets Space* **2016**, *8*, 92. [CrossRef]

74. van den IJssel, J.; Forte, B.; Montenbruck, O. Impact of Swarm GPS Receiver Updates on POD Performance. *Earth Planets Space* **2016**, *68*, 85. [CrossRef]

75. Dahle, C.; Arnold, D.; Jäggi, A. Impact of tracking loop settings of the Swarm GPS receiver on gravity field recovery. *Adv. Space Res.* **2017**, *59*, 2843–2854. [CrossRef]

76. Cheng, M.K.; Ries, J.C. GRACE Technical Note 05. Center for Space Research, The University of Texas at Austin, USA, 2012. Available online: ftp://podaac.jpl.nasa.gov/allData/grace/docs/TN-05_C20_SLR.txt (accessed on 21 April 2019).

77. Koch, K.R.; Kusche, J. Regularization of geopotential determination from satellite data by variance components. *J. Geod.* **2002**, *76*, 259–268. [CrossRef]

78. Schreiter, L.; Arnold, D.; Sterken, V.; Jäggi, A. Mitigation of ionospheric signatures in Swarm GPS gravity field estimation using weighting strategies. *Ann. Geophys.* **2019**, *37*, 111–127. [CrossRef]

79. Sneeuw, N. *A Semi-Analytical Approach to Gravity Field Analysis From Satellite Observations*; Deutsche Geodätische Kommission, C 527: München, Germany, 2000; ISBN 3-7696-9566-6.

80. Klinger, B.; Mayer-Gürr, T. The role of accelerometer data calibration within GRACE gravity field recovery: Results from ITSG-Grace2016. *Adv. Space Res.* **2016**, *58*, 1597–1609. [CrossRef]

81. Shepherd, A.; Ivins, E.R.; Geruo, A.; Barletta, V.R.; Bentley, M.J.; Bettadpur, S.; Briggs, K.H.; Bromwich, D.H.; Forsberg, R.; Galin, N.; et al. A reconciled estimate of ice-sheet mass balance. *Science* **2012**, *338*, 1183–1189. [CrossRef] [PubMed]

82. Horwath, M.; Dietrich, R. Signal and error in mass change inferences from GRACE: The case of Antarctica. *Geophys. J. Int.* **2009**, *177*, 849–864. [CrossRef]

83. Velicogna, I. Increasing rates of ice mass loss from the Greenland and Antarctic ice sheets revealed by GRACE. *Geophys. Res. Lett.* **2009**, *36*, L19503. [CrossRef]

84. Bevis, M.; Harig, C.; Khan, S.A.; Brown, A.; Simons, F.J.; Willis, M.; Fettweis, X.; van den Broeke, M.R.; Bo Madsen, F.; Kendrick, E.; et al. Accelerating changes in ice mass within Greenland, and the ice sheet's sensitivity to atmospheric forcing. *Proc. Natl. Acad. Sci. USA* **2019**, *116*, 1934–1939. [CrossRef] [PubMed]

Article

A New Approach to Earth's Gravity Field Modeling Using GPS-Derived Kinematic Orbits and Baselines

Xiang Guo * and Qile Zhao

GNSS Research Center, Wuhan University, Wuhan 430079, China
* Correspondence: xiangguo@whu.edu.cn

Received: 13 June 2019; Accepted: 19 July 2019; Published: 21 July 2019

Abstract: Earth's gravity field recovery from GPS observations collected by low earth orbiting (LEO) satellites is a well-established technique, and kinematic orbits are commonly used for that purpose. Nowadays, more and more satellites are flying in close formations. The GPS-derived kinematic baselines between them can reach millimeter precision, which is more precise than the centimeter-level kinematic orbits. Thus, it has long been expected that the more precise kinematic baselines can deliver better gravity field solutions. However, this expectation has not been met yet in practice. In this study, we propose a new approach to gravity field modeling, in which kinematic orbits of the reference satellite and baseline vectors between the reference satellite and its accompanying satellite are jointly inverted. To validate the added value, data from the Gravity Recovery and Climate Experiment (GRACE) satellite mission are used. We derive kinematic orbits and inter-satellite baselines of the twin GRACE satellites from the GPS data collected in the year of 2010. Then two sets of monthly gravity field solutions up to degree and order 60 are produced. One is derived from kinematic orbits of the twin GRACE satellites ('orbit approach'). The other is derived from kinematic orbits of GRACE A and baseline vectors between GRACE A and B ('baseline approach'). Analysis of observation postfit residuals shows that noise in the kinematic baselines is notably lower than the kinematic orbits by 50, 47 and 43% for the along-track, cross-track and radial components, respectively. Regarding the gravity field solutions, analysis in the spectral domain shows that noise of the gravity field solutions beyond degree 10 can be significantly reduced when the baseline approach is applied, with cumulative errors up to degree 60 being reduced by 34%, when compared to the orbit approach. In the spatial domain, the recovered mass changes with the baseline approach are more consistent with those inferred from the K-Band Ranging based solutions. Our results demonstrate that the proposed baseline approach is able to provide better gravity field solutions than the orbit approach. The findings may facilitate, among others, bridging the gap between GRACE and GRACE Follow-On satellite mission.

Keywords: Earth's gravity field; kinematic orbit; kinematic baseline; GRACE

---

1. Introduction

Knowledge of temporal variations of the Earth's gravity field is of importance to understand large-scale mass transport at and below the Earth's surface. It has been mostly observed with the ultra-precise K-Band ranging (KBR) measurements from the Gravity Recovery and Climate Experiment (GRACE) mission [1] and its successor, i.e., the GRACE Follow-On (GFO) mission. However, there exists a gap between GRACE and GFO. Nowadays, there is a consensus that GPS-based high-low satellite-to-satellite tracking (hl-SST) can play an important role in bridging the gap [2–7]. For that purpose, GPS-derived kinematic orbits are usually taken as pseudo-observations in gravity field modeling. Since GPS data are about three orders of magnitude less precise than the KBR data, they are only sensitive to temporal variations at low spherical harmonic degrees (<20) of the Earth's gravity field. Furthermore, it is well known that kinematic orbits are very sensitive to observation geometry and various systematic errors, e.g., mismodeling of GPS antenna phase center, high-order ionosphere-induced errors, near-field multipath, and uncalibrated hardware delays [8–10]. As a result, the recovered gravity field solutions usually suffer from systematic errors, particularly at low degrees [11]. Therefore, efforts are still ongoing to further improve the GPS-based gravity field solutions. Among others, GPS-derived kinematic baselines between formation flying satellites have drawn much attention, due to their higher precision than the kinematic orbits.

To exploit the more precise space baselines, two approaches have been proposed in the context of the celestial mechanics approach to gravity field modeling [12]: The 'observation equation approach' and the 'GRACE-type approach' [13]. In the former, individual observation equations for both satellites are first established based on their positions; then the two individual observation equations are differenced to form the baseline observation equations; finally, the normal equations (NEQs) are established from the baseline observation equations. Since only the baselines are used, orbit parameters for one of the two satellites have to be fixed to the a priori values in this approach. Unfortunately, this operation will seriously degrade the gravity field solutions at very low degrees [13,14]. In the GRACE-type approach, individual observation and normal equations for both satellites are established based on their kinematic orbits in a first step; in a second step, the observation and normal equations for baseline vectors or baseline lengths are established based on kinematic baselines; the NEQs from the first and second step are combined to form the final NEQs in the third step. Unlike the observation equation approach, orbit parameters for both satellites are estimated in the GRACE-type approach. Recent studies based on this approach all use baseline lengths to perform gravity field recovery, the improvements are demonstrated to be insignificant [13] or negligible [15], when compared to the gravity field solutions based solely on kinematic orbits.

To make full use of the precise kinematic baseline vectors, we propose a new approach in this study, where kinematic orbits of the reference satellite and kinematic baseline vectors between the reference satellite and its accompanying satellite are taken as observations during gravity field recovery. Hereafter, we denote the new approach as the 'baseline approach'. Accordingly, the approach based solely on kinematic orbits is denoted as the 'orbit approach'. To demonstrate the added value of the new approach, data collected by the GRACE mission in 2010 are used. We first compute the kinematic orbits and baselines of the twin GRACE satellites. Then two sets of monthly gravity field solutions are produced: One is based on kinematic orbits of the twin satellites with the orbit approach (hereafter denoted as the 'orbit solution' where applicable), the other is based on kinematic orbits of GRACE A and kinematic baseline vectors between GRACE A and B with the baseline approach (hereafter denoted as the 'baseline solution' where applicable). To evaluate the quality of the GPS-based solutions, we use as the reference the WHU RL01 monthly gravity field solutions produced at the GNSS Research Center of Wuhan University [16,17]. Those solutions are mainly based on the GRACE KBR measurements and therefore are much more accurate than any GPS-based solutions.

This paper is organized as follows: Section 2 presents in detail the data and methods adopted for gravity field modeling. Results are shown in Section 3, and discussed in Section 4. Finally, Section 5 is left for conclusions.

2. Data and Methods

2.1. Data and Models

Data processing in this study is performed with the Position And Navigation Data Analyst (PANDA) software, which is developed at the GNSS Research Center of Wuhan University and has been widely used in precise orbit determination for both GNSS satellites and low Earth orbiters [18,19]. Recently, the dynamic approach to gravity field modeling has been implemented in PANDA and successfully applied to produce GRACE monthly gravity field solutions [16,17,20,21]. In this approach, data processing consists of two steps when the gravity field is estimated from kinematic orbits. In the first step, a priori dynamic orbits are computed by fitting to the kinematic orbits through numerically integrating the equation of motion defined by the a priori force models (cf. Table 1). During this process, orbits are expressed as truncated Taylor series with respect to the unknown parameters (cf. Table 1), about the computed a priori orbits. The partials with respect to those parameters are obtained by resolving the so-called variational equations. Only non-gravity parameters are adjusted at this step. In the second step, the gravity parameters are included. Based on the partial derivatives with respect to all unknown parameters, daily NEQs are set up for the purpose of the subsequent least-squares adjustment. Arc-specific parameters are then pre-eliminated, and the daily NEQs are accumulated into monthly NEQs. The monthly NEQ matrix is eventually inverted in order to obtain the corrections of the spherical harmonic coefficients (SHCs) with respect to the a priori gravity values.

Details about the data and models used in this study are also described in Table 1. The kinematic orbits of the twin GRACE satellites used in this study are produced in house based on a zero-difference data processing scheme. To achieve the best accuracy, both the single-difference ambiguities between GPS satellites at each GRACE satellite and double-difference ambiguities between the GRACE satellites and GPS satellites are resolved. Accuracy of the kinematic orbits obtained in this way is about 1.5 cm as inferred from satellite laser ranging residuals. More details can be found in Reference [22]. As to the kinematic baselines, they are produced based on a single-difference scheme data processing as described in Reference [23]. In this scheme, the positions of the reference satellite (which is GRACE A in this study), are not estimable, and have to be fixed to the a priori values. This requires accuracy of the a priori orbits to be better than 10 cm for millimeter precision baseline determination [24]. For that purpose, we exploit the abovementioned ambiguity-fixed kinematic orbits, which are fully qualified with an accuracy better than 2 cm. Accuracy of the kinematic baselines obtained in this way is demonstrated to be slightly better than 4 mm, as inferred from KBR residuals, which is consistent with other studies [13,25]. During the solution process, the accelerometer observations are used to model the non-gravitational forces. For each component in the accelerometer frame, the scale factors are estimated on a monthly basis, and the biases are modeled with a three-order polynomial and the polynomial coefficients are estimated on a daily basis to account for thermal variations. Finally, the arc length is set equal in our data processing scheme to 24 h.

Table 1. Data and models used for gravity field modeling.

Background Force Models	Description
Mean gravity field model	EIGEN-6C4 [26] (120 × 120)
Solid Earth and pole tides	IERS Conventions 2010 [27]
Ocean tides	EOT11a [28] (120 × 120)
Ocean pole tides	Desai [29] (30 × 30)
Atmosphere and ocean de-aliasing	AOD1B RL06 [30]
Third-body perturbations	DE421 [31]
General relativistic effects	IERS Conventions 2010 [27]
Reference frames	
Conventional inertial reference frame	IERS Conventions 2010 [27]
Precession/nutation	IAU 2006/2000A [27]
Earth orientation parameters	IERS EOP 08 C04
Input data	
Kinematic orbits and baselines	30 s sampling
GRACE attitudes	Level 1B RL03, 5 s sampling
Non-gravitational accelerations	Level 1B RL02, 5 s sampling
Estimated parameters	
Initial state vector	Position and velocity per satellite and per arc
Accelerometer bias	Three order polynomial for each component per arc
Accelerometer scale factor	One per month per component
Spherical harmonic coefficients	Between degrees 2 and 60

2.2. The Orbit and Baseline Approach

In the orbit approach, only kinematic orbits of the twin GRACE satellites are taken as observations. The linearized observation equations are expressed as follows:

$$
\begin{aligned}
\mathbf{y}_A &= \mathbf{y}_A^0 + \mathbf{d}_A^{ng}\mathbf{x}_A^{ng} + \mathbf{d}_A^{g}\mathbf{x}^g + \varepsilon_{y_A} \\
\mathbf{y}_B &= \mathbf{y}_B^0 + \mathbf{d}_B^{ng}\mathbf{x}_B^{ng} + \mathbf{d}_B^{g}\mathbf{x}^g + \varepsilon_{y_B}
\end{aligned}
\tag{1}
$$

where $\mathbf{y}$ is the satellite position vector, which is extracted from the kinematic orbits, and $\mathbf{y}^0$ is its computed counterpart; $\mathbf{x}^{ng}$ denotes the non-gravity parameter vector, which consists of satellite initial state vector, accelerometer scale and bias parameters (cf. Table 1); $\mathbf{x}^g$ denotes the gravity parameters in the form of SHCs up to degree and order 60; $\mathbf{d}^{ng}$ and $\mathbf{d}^g$ denote the design matrices with respect to the non-gravity and gravity parameters, respectively; Finally, ε denotes the noise which includes noise in both the position and its computed counterpart. Data processing consists of three steps. First, individual observation and normal equations for the two satellites are established based on kinematic orbits. Second, the two individual NEQs are combined, and arc-specific parameters are pre-eliminated from the daily NEQs. Third, the daily NEQs are accumulated into monthly NEQs, which are eventually solved to obtain monthly gravity field solutions.

In the baseline approach, the observations consist of kinematic orbits of GRACE A and kinematic baseline vectors between GRACE A and B. The linearized observation equations are expressed as follows:

$$
\begin{aligned}
\mathbf{y}_A &= \mathbf{y}_A^0 + \mathbf{d}_A^{ng}\mathbf{x}_A^{ng} + \mathbf{d}_A^{g}\mathbf{x}^g + \varepsilon_{y_A} \\
\mathbf{y}_{AB} &= \mathbf{y}_{AB}^0 + \mathbf{d}_B^{ng}\mathbf{x}_B^{ng} - \mathbf{d}_A^{ng}\mathbf{x}_A^{ng} + \left(\mathbf{d}_B^{g} - \mathbf{d}_A^{g}\right)\mathbf{x}^g + \varepsilon_{y_{AB}}
\end{aligned}
\tag{2}
$$

where $\mathbf{y}_{AB}$ and $\mathbf{y}_{AB}^0$ are the baseline vector and its computed counterpart, respectively; $\varepsilon_{y_{AB}}$ denote the noise which consists of noise in both the baseline and its counterpart; other symbols are same to those in Equation (1). Data processing includes the following steps: First, kinematic orbits of GRACE B are derived from the kinematic orbits of GRACE A and the kinematic baseline vectors between GRACE A and B, i.e., $\mathbf{y}_B = \mathbf{y}_A + \mathbf{y}_{AB}$. Second, individual observation equations for both satellites are established based on the kinematic orbits. Third, the baseline observation equations are formed by differencing the two orbit observation equations, i.e., $\mathbf{y}_{AB} = \mathbf{y}_B - \mathbf{y}_A$. Fourth, individual NEQs are established from observation equations of kinematic orbits of GRACE A and observation equations of kinematic baseline vectors between GRACE A and B, respectively. Fifth, the individual NEQs are combined and arc-parameters are pre-eliminated from the daily NEQs. Finally, the daily NEQs are accumulated into monthly NEQs, which are eventually inverted to obtain monthly gravity field solutions.

As described above, six observations per epoch are used in both the orbit and baseline approach. While observations consist of the position vectors of GRACE A and B in the orbit approach, they are composed of the position vector of GRACE A and baseline vector between GRACE A and B in the baseline approach. It should be noted that the same kinematic orbits of GRACE A are used, and the same parameters are set up in both approaches (cf. Table 1). In the following calculations, we further replace the kinematic orbits of GRACE B with those constructed from kinematic orbits of GRACE A and kinematic baseline vectors between GRACE A and B in the context of the orbit approach. Our test experiments show that impacts of this replacement on the gravity field solutions are minor. Now, one may consider that the baseline approach is equivalent to the orbit approach, since the baseline vectors are just linear combinations of the two kinematic orbits used in the orbit approach and this will not change the final solutions in frame of the weighted least square adjustment if the stochastic models for different observations are properly accounted for [32]. However, this is not the case when the observations are contaminated by systematic errors, which has long been a problem for time-varying gravity field recovery from spaceborne GPS data. In that case, the baseline approach has two advantages at least over the orbit approach: First, the baseline vectors are free of common-mode errors from the GPS satellites, due to the adopted single-difference data processing scheme when deriving them from the GPS measurements. Thus, the baselines are more precise than kinematic orbits calculated with a zero-difference scheme. Second, according to the Hill's equation [33], errors in the initial state vectors and background force models would lead to constant or slowly varying perturbations in the computed a priori orbits which are used to establish the observation equations of kinematic orbits. While these systematic errors will enter into the **O−C** (observed minus computed) vectors and contaminate the final gravity field solutions in the orbit approach, they can be mitigated when forming the baseline observation equations through differencing the two orbit observation equations. This is particularly true for the GRACE mission, because errors for the two satellites are largely common, since the two spacecraft are identical and not far separated (about 220 km) in a coplanar orbit. Therefore, one may expect that the baseline approach will deliver better gravity field solutions than the orbit approach, which will also be demonstrated later in this article.

2.3. Data Weighting Scheme

Generally speaking, errors in kinematic orbits are often temporally correlated, although ambiguity-fixing can largely reduce the correlations [34]. On the other hand, deficiencies in the background force models also lead to correlated errors in the computed dynamic orbits [35]. As a result, noise in the residuals is usually frequency-dependent (i.e., colored). In this study, we adopt the frequency-dependent data weighting (FDDW) concept proposed by Reference [36] to account for the colored noise during the solution process. Recently, the FDDW concept has successfully been applied to GRACE KBR data processing with the classical dynamic approach and has notably reduced noise in the WHU RL01 monthly gravity solutions [17]. To represent the dependence of noise on frequency, we consider noise power spectral density (PSD), which is estimated from postfit observation residuals. For that purpose, we start from a simple assumption of white noise in the observations during the first computation. Then the noise PSDs are estimated from the postfit residuals and are used for applying the FDDW scheme in the second computation. No further iteration is needed, due to the accurate background models. The reader is referred to Reference [36] for more details.

3. Results

3.1. Analysis in the Spectral Domain

Figure 1 shows the square-root PSD, hereafter denoted as $PSD^{1/2}$, estimated from the postfit residuals of kinematic orbits of GRACE A and kinematic baselines between GRACE A and B for a typical month of January 2010. Since the $PSD^{1/2}$s for kinematic orbits of both satellites and in both approaches show similar patterns, only those for GRACE A are displayed here. It can be observed that the $PSD^{1/2}$s for both kinematic orbits and baselines exhibit clear frequency-dependent behaviors, which necessitates the FDDW scheme adopted in this study. A comparison between the two sets of $PSD^{1/2}$s reveals that $PSD^{1/2}$s for the kinematic baselines are generally lower than those for the kinematic orbits. On average, $PSD^{1/2}$s for the kinematic baselines are reduced by 50, 47 and 44% at the along-track, cross-track and radial components, respectively, when compared to those for the kinematic orbits. Obviously, the kinematic baselines are of higher precision than the kinematic orbits.

As mentioned above, the KBR-based WHU RL01 monthly gravity field solutions are chosen as the ground truth. In fact, we also tried the official GRACE monthly solutions for that purpose. The results revealed that the differences were negligible. Therefore, to assess the computed GPS-based gravity field solutions, we subtract from them the WHU RL01 solutions and the obtained residual SHCs are interpreted as the 'true' errors of the GPS-based solutions. To facilitate analysis in the spectral domain, we calculate the degree errors (DEG) and cumulative errors (CUM) of different solutions in terms of geoid heights with the following formulas:

$$DEG_l = R \sqrt{\sum_{m=0}^{l} \left(\Delta C_{lm}^2 + \Delta S_{lm}^2 \right)}$$

$$CUM_l = R \sqrt{\sum_{i=2}^{l} DEG_i^2} \tag{3}$$

where R is the reference radius (6,378,137.46 m), $\Delta C_{lm}, \Delta S_{lm}$ are the residual SHCs with respect to WHU RL01. Figure 2 displays the degree errors of different GPS-based solutions and degree signals represented by WHU RL01 in January 2010. It can be seen that the degree errors of the baseline solutions are smaller than those of the orbit solutions, except for relatively low degree terms. To make the discussion more comprehensive, we further calculate the RMS (root mean square) for each SHC time series, as well as for the associated formal errors in 2010. The results are displayed in Figure 3, which in general are similar to those displayed in Figure 2. One can see that degree errors beyond degree 10 can be significantly reduced when the baseline approach is applied. One can also observe that the true errors clearly depart from the formal errors below degree 15 in both cases. This may likely be attributed to the presence of systematic errors in the observations, e.g., mismodeling of GPS antenna phase center, high-order ionosphere-induced errors as mentioned before. Table 2 further lists the cumulative geoid height errors up to degree 10, 20 and 60. It reveals that the cumulative errors up to degree 20 and 60 are reduced by 15 and 34%, respectively, when the baseline approach is applied. From these results, we can conclude that the proposed baseline approach can improve the gravity field beyond degree 10.

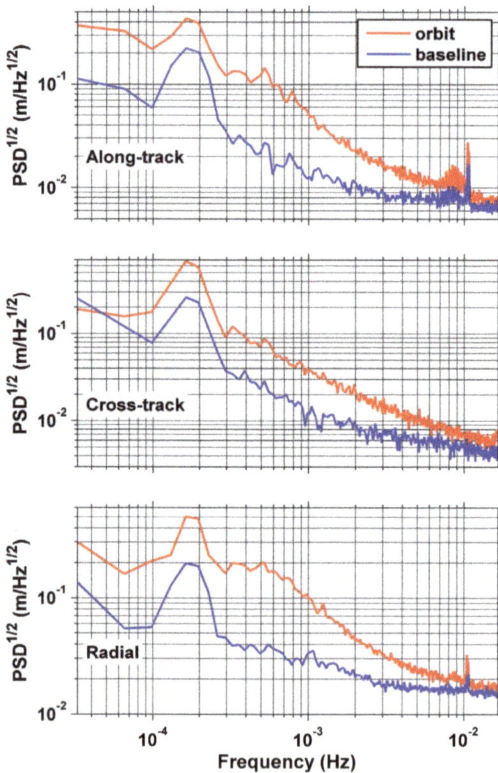

Figure 1. PSD$^{1/2}$s estimated from postfit residuals of kinematic orbits of GRACE A and kinematic baselines between GRACE A and B.

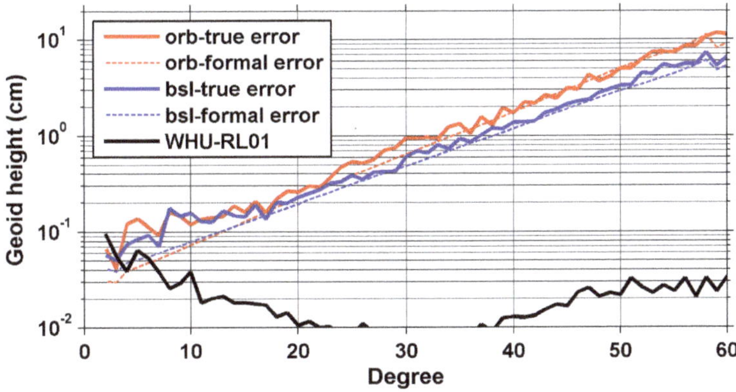

Figure 2. Geoid height errors per degree for different gravity field solutions in January 2010.

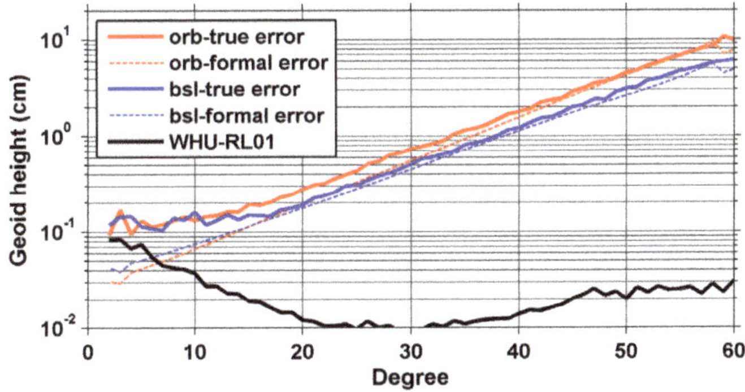

Figure 3. Geoid height errors per degree for different gravity field solutions (RMS values in 2010).

Table 2. Cumulative geoid height errors (cm) up to degree 10, 20 and 60 of different solutions (RMS values in 2010).

Degree	True Error		
	Orbit	Baseline	$\frac{\text{baseline} - \text{orbit}}{\text{orbit}}$
10	0.38	0.38	0%
20	0.73	0.62	−15%
60	26.2	17.2	−34%

3.2. Analysis in the Spatial Domain

In this section, we compare the ability of the orbit and baseline approaches to recover temporal gravity field variations and mass anomalies in the spatial domain. Since the unfiltered solutions are dominated by high-frequency noise, post-processing is required. In this study, we use the Gaussian filter [37] for that purpose. To account for the impact of the filter width on the results, we consider two radii—750 and 1000 km. Hereafter, we denote the corresponding solutions as the 'G750' and 'G1000' filtered solutions, respectively. As it was done in the spectral domain, we also compare the GPS-based solutions with the KBR-based WHU RL01 solutions. To keep consistency with the GPS-based solutions, the latter is also post-processed with the same Gaussian filters. In addition, the C_{20} terms, to which GRACE is less sensitive, due to the polar orbits [38], are replaced in all gravity field solutions with the SLR-derived values [39]. As regards the degree 1 terms, which cannot be derived from GRACE data alone, we use the estimates obtained by combining GRACE data and geophysical models as described in References [40,41].

To perform the analysis in the spatial domain, we first transform the monthly SHCs to mass anomalies in terms of equivalent water heights (EWHs) on a $1° \times 1°$ grid, as explained in Reference [37]. The correction for the Earth's oblateness has been applied as proposed by Ditmar [42]. As mass variations (if exist) are primarily linear or/and seasonal, a deterministic model composed of an offset, trend, annual, and semi-annual terms is fitted to the time-series of mass anomalies per grid node. Then, a comparison with the WHU RL01 solutions allows us to assess the signals and noise of the mass anomalies inferred from the GPS-based solutions.

3.2.1. Gridded Mass Anomalies

Figure 4 displays geographical maps of the derived periodic annual signals in mass anomalies inferred from the G750 filtered solutions. It can be seen that after applying the G750 filter, the estimates are still dominated by strong high-frequency noise. In the case of the G1000 filtered solutions (Figure 5), high-frequency noise is largely suppressed, and the signal patterns inferred from the GPS-based solutions become similar to WHU RL01. Among others, annual signals over the Amazon river basin in South America can be clearly seen in the GPS-based solutions.

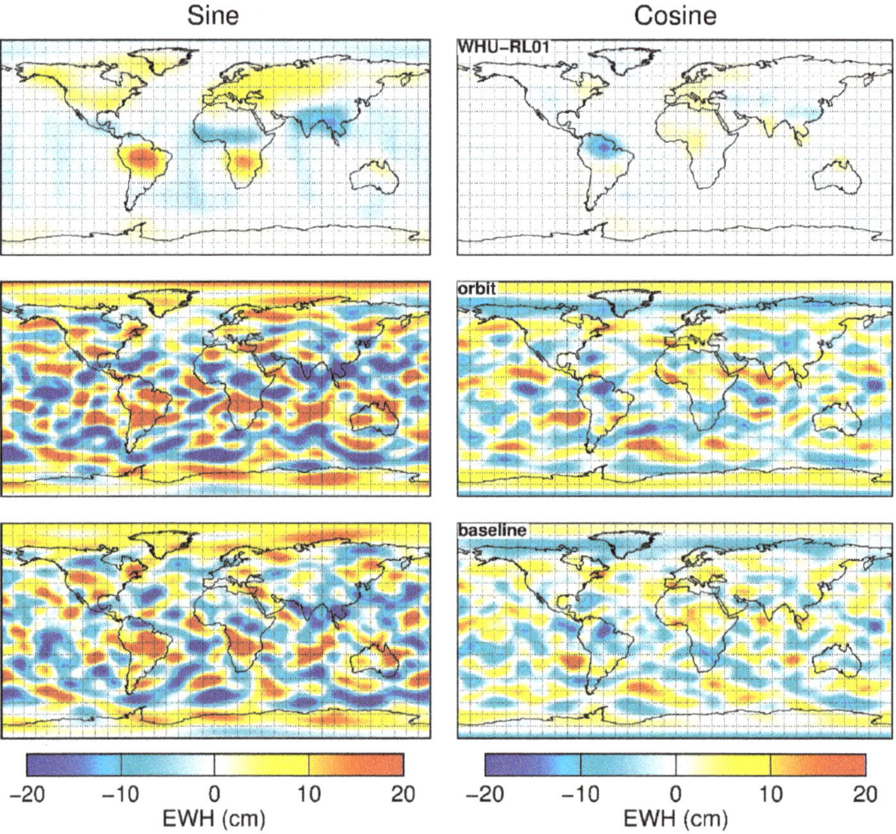

Figure 4. Periodic annual signals in mass anomalies in terms of equivalent water heights inferred from different G750 filtered solutions. From top to bottom, the plots display the results based on the WHU-RL01, orbit and baseline solutions, respectively.

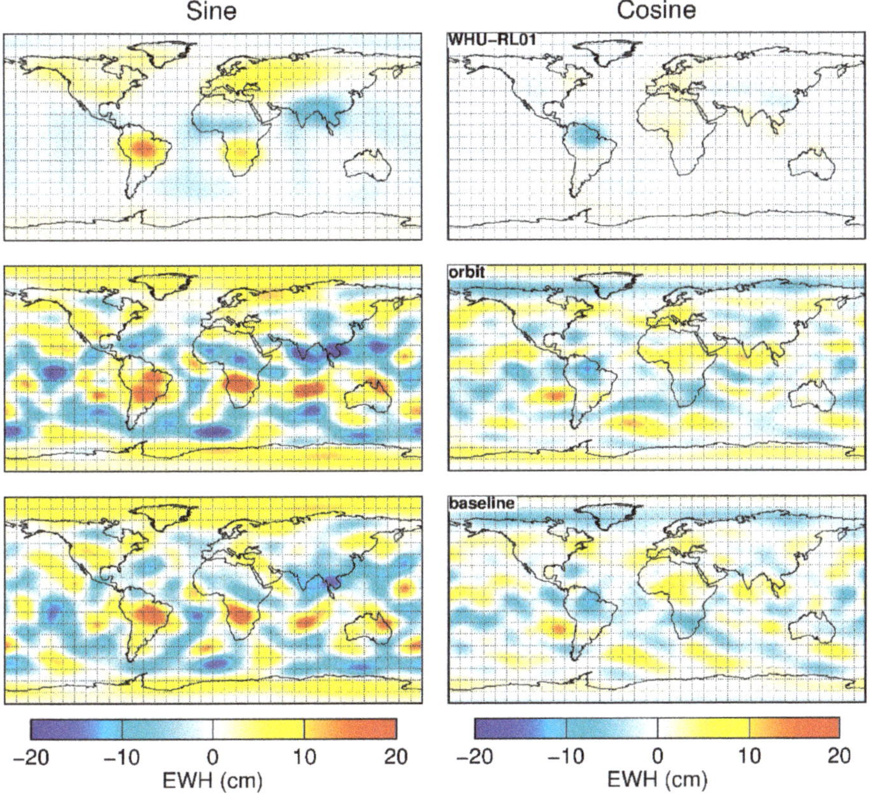

Figure 5. Same as in Figure 4, but for the G1000 filtered solutions.

We have also computed the RMS differences of gridded mass anomalies between the GPS-based solutions and WHU RL01. The geographical distribution of the RMS differences is displayed in Figure 6. It can be observed that the RMS differences in the case of the baseline solutions are generally smaller than those in case of the orbit solutions when the G750 filter is applied. A further calculation reveals that the weighted mean of the gridded RMS differences (weighted by the cosine of latitude) in the case of the baseline solutions are reduced by 14 and 6% for the G750 and G1000 filtered solutions, respectively, when compared to those in case of the orbit solutions (cf. Table 3). These results demonstrate that the mass anomalies inferred from the baseline solutions are more consistent with WHU RL01 and, therefore, are more accurate, as compared to the orbit solutions.

Table 3. Weighted mean of the gridded RMS differences of mass anomalies (cm) with respect to consistently filtered WHU RL01 solutions, for different GPS-based solutions.

	Orbit	Baseline	$\frac{\text{Baseline-Orbit}}{\text{Orbit}}$
G750	13.3	11.5	−14%
G1000	7.2	6.8	−6%

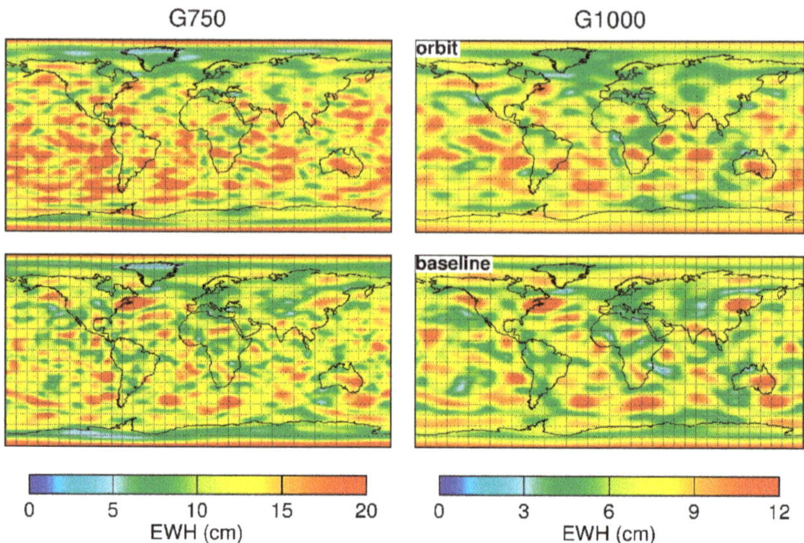

Figure 6. Geographical distribution of the gridded RMS differences with respect to WHU RL01 for different GPS-based solutions. The top and bottom rows show the results for the orbit and baseline solutions, respectively.

3.2.2. Regional Mass Anomalies

To compare the performance of the orbit and baseline approaches in the context of mean regional signals, we select the Amazon river basin as the target regions. As shown in Figure 5, the Amazon river basin exhibits significant annual variations, which are mainly of hydrological origin [43,44].

Figure 7 displays the time series of mean mass anomalies over the Amazon river basin inferred from different solutions in 2010. To evaluate the consistency between the GPS-based solutions and WHU RL01, we calculate the correlation coefficients and RMS differences between the mass anomaly time-series. The results confirm that the GPS-based solutions obtained with the baseline approach are more consistent with WHU RL01, as evidenced by larger correlation coefficients and smaller RMS differences (cf. Table 4).

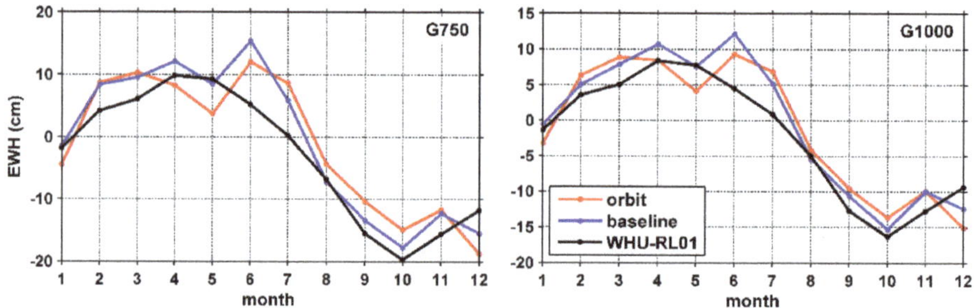

Figure 7. Time series of mean mass anomalies over the Amazon river basin inferred from the G750 (left) and G1000 (right) filtered solutions in 2010.

Table 4. Correlation coefficients and RMS differences (cm) between mass anomaly time series over the Amazon river basin inferred from the GPS-based solutions and WHU RL01.

	Correlation Coefficient			RMS Differences		
	Orbit	Baseline	$\frac{\text{Baseline–Orbit}}{\text{Orbit}}$	Orbit	Baseline	$\frac{\text{Baseline–Orbit}}{\text{orbit}}$
G750	0.90	0.96	7%	5.1	4.1	−20%
G1000	0.93	0.97	4%	3.6	3.1	−14%

Table 5 further lists the annual amplitudes and phases and the associated formal errors over the Amazon river basin inferred from different solutions. One can see that the GPS-based solutions (after applying a Gaussian filter) can reproduce the signals as WHU RL01 does, particularly when the baseline approach is applied. The signals inferred from the baseline solutions are closer to those inferred from the WHU RL01 solutions when compared to the orbit solutions.

Table 5. Annual amplitudes (cm) and phases (day) and the associated formal errors of mass anomalies over the Amazon river basin derived from different solutions.

	Amplitude (cm)			Phase (Day)		
	WHU RL01	Orbit	Baseline	WHU RL01	Orbit	Baseline
G750	13.4 ± 0.88	11.8 ± 3.62	13.2 ± 2.95	−21.1 ± 0.07	−38.2 ± 0.31	−31.1 ± 0.23
G1000	11.2 ± 0.82	10.5 ± 2.76	10.9 ± 2.05	−21.4 ± 0.07	−34.7 ± 0.27	−32.3 ± 0.19

All these results lead us to conclude that the proposed baseline approach is able to provide better regional mass anomaly estimates, when compared to the orbit approach.

4. Discussion

As shown in the text, with the proposed baseline approach we have achieved more promising results than previous studies which are based on the GRACE-type approach [13,15], when exploiting GPS-derived kinematic baselines for Earth's gravity field recovery. While improvement of the gravity field solutions was insignificant or even negligible in previous investigations, it is shown to be significant in our case with cumulative errors up to degree 60 being reduced by 34%. We attribute this improvement mainly to the fact that we have made full use of the precise three-dimensional baseline vectors in our approach, whereas only the baseline lengths were used in previous studies.

Our results also show that the gravity field solutions still cannot benefit from the proposed baseline approach at the very low degree terms (below degree 10) at present. Because these low degree terms are very prone to systematic errors in the observations, further efforts are still needed to identify and reduce the possible systematic errors.

Another issue is that we have made use of the high precision accelerometer observations to account for the non-gravitational forces acting on the GRACE satellites, which may not be available for other formation flying satellites. Indeed, high precision accelerometer observations are indispensable to achieve the best possible gravity field solutions from hl-SST data. In fact, it is hardly possible to obtain realistic gravity field solutions if the non-gravitational forces cannot be adequately modeled either with accelerometer observations or with physical models. Thus, this is the prerequisite for both the orbit and baseline approach. Because the kinematic baselines are more precise than kinematic orbits (Figure 1), it is reasonable that the baseline approach can deliver better gravity field solutions, at least for formation flying satellites equipped with high precision accelerometers.

5. Conclusions

In this study, we have proposed a new approach to Earth's gravity field modeling based on GPS-derived kinematic orbits and baselines from formation flying satellites. In this approach, kinematic orbits of the reference satellite and kinematic baselines between the reference satellite and its accompanying satellite are taken as observations to perform gravity field recovery (denoted as the 'baseline approach' in the text). Compared to the orbit approach, which is based solely on kinematic orbits, systematic errors in observations and background force models can be suppressed in the proposed approach. To demonstrate the added value of the proposed approach, two sets of monthly gravity field solutions have been produced based on one year of kinematic orbits and baseline vectors of the GRACE satellites using both the baseline and orbit approaches. A comparison in the spectral domain shows that the gravity field solutions beyond degree 10 can be notably improved when the baseline approach is applied, with cumulative errors up to degree 20 and 60 being reduced by 15% and 34%, respectively, as compared to the orbit approach. Analysis in the spatial domain shows that the GPS-based gravity field solutions obtained with the baseline approach are more consistent with the KBR-based solutions. This is evidenced by smaller RMS differences and larger correlation coefficients with respect to the KBR-based solutions. These results, therefore, demonstrate that the proposed baseline approach can provide better time-varying gravity field solutions than the orbit approach. Our findings may facilitate, among others, bridging the gap between GRACE and GFO missions by using hl-SST GPS data from formation flying satellites.

Author Contributions: X.G. conceived and performed the experiments, analyzed the data and wrote the paper. Q.Z. supported the research and helped with the text.

Funding: This research was funded by the National Natural Science Foundation of China, grant number 41574030, 41374083 and the APC was funded by 41574030.

Acknowledgments: The GRACE level-1B data are publicly available from the ftp site: ftp://isdcftp.gfz-potsdam.de/grace/Level-1B. The numerical calculations in this research have been done on the supercomputing system in the Supercomputing Center of Wuhan University. Finally, we thank the three anonymous reviewers for their helpful comments, which helped us improve the manuscript.

Conflicts of Interest: The authors declare no conflict of interest. The funders had no role in the design of the study; in the collection, analyses, or interpretation of data; in the writing of the manuscript, or in the decision to publish the results.

References

1. Tapley, B.D.; Bettadpur, S.; Ries, J.C.; Thompson, P.F.; Watkins, M.M. GRACE measurements of mass variability in the Earth system. *Science* **2004**, *305*, 503–505. [CrossRef] [PubMed]
2. Weigelt, M.; van Dam, T.; Jäggi, A.; Prange, L.; Tourian, M.J.; Keller, W.; Sneeuw, N. Time-variable gravity signal in Greenland revealed by high-low satellite-to-satellite tracking. *J. Geophys. Res. Solid Earth* **2013**, *118*, 3848–3859. [CrossRef]
3. Bezděk, A.; Sebera, J.; Teixeira da Encarnação, J.; Klokočník, J. Time-variable gravity fields derived from GPS tracking of Swarm. *Geophys. J. Int.* **2016**, *205*, 1665–1669. [CrossRef]
4. Jäggi, A.; Dahle, C.; Arnold, D.; Bock, H.; Meyer, U.; Beutler, G.; van den Ijssel, J. Swarm kinematic orbits and gravity fields from 18 months of GPS data. *Adv. Space Res.* **2016**, *57*, 218–233. [CrossRef]
5. Visser, P.N.A.M.; van der Wal, W.; Schrama, E.J.O.; van den Ijssel, J.; Bouman, J. Assessment of observing time-variable gravity from GOCE GPS and accelerometer observations. *J. Geod.* **2014**, *88*, 1029–1046. [CrossRef]
6. Guo, X.; Ditmar, P.; Zhao, Q.; Klees, R.; Farahani, H.H. Earth's gravity field modelling based on satellite accelerations derived from onboard GPS phase measurements. *J. Geod.* **2017**, *91*, 1049–1068. [CrossRef]
7. Jäggi, A.; Bock, H.; Meyer, U.; Beutler, G.; van den Ijssel, J. GOCE: Assessment of GPS-only gravity field determination. *J. Geod.* **2014**, *89*, 33–48. [CrossRef]
8. Jäggi, A.; Dach, R.; Montenbruck, O.; Hugentobler, U.; Bock, H.; Beutler, G. Phase center modeling for LEO GPS receiver antennas and its impact on precise orbit determination. *J. Geod.* **2009**, *83*, 1145–1162. [CrossRef]
9. Bock, H.; Jäggi, A.; Beutler, G.; Meyer, U. GOCE: Precise orbit determination for the entire mission. *J. Geod.* **2014**, *88*, 1047–1060. [CrossRef]
10. Montenbruck, O.; Hackel, S.; Jäggi, A. Precise orbit determination of the Sentinel-3A altimetry satellite using ambiguity-fixed GPS carrier phase observations. *J. Geod.* **2017**, *92*, 711–726. [CrossRef]

11. Jäggi, A.; Bock, H.; Prange, L.; Meyer, U.; Beutler, G. GPS-only gravity field recovery with GOCE, CHAMP, and GRACE. *Adv. Space Res.* **2011**, *47*, 1020–1028. [CrossRef]

12. Beutler, G.; Jäggi, A.; Mervart, L.; Meyer, U. The celestial mechanics approach: Theoretical foundations. *J. Geod.* **2010**, *84*, 605–624. [CrossRef]

13. Jäggi, A.; Dahle, C.; Arnold, D.; Meyer, U.; Bock, H. Kinematic space-baselines and their use for gravity field recovery. In Proceedings of the 40th COSPAR Scientific Assembly, Moscow, Russia, 2–10 August 2014.

14. Jäggi, A.; Beutler, G.; Prange, L.; Dach, R.; Mervart, L. Assessment of GPS-only Observables for Gravity Field Recovery from GRACE. In *Observing Our Changing Earth*; Sideris, M.G., Ed.; Springer: Berlin, Germany, 2009; pp. 113–123.

15. Zehentner, N.; Mayer-Gürr, T.; Ellmer, M.; Teixeira da Encarnacao, J.; Visser, P.; Doornbos, E.; Van den IJssel, J.; Mao, X.; Iorfida, E.; Arnold, D. Investigations of GNSS-derived baselines for gravity field recovery. In Proceedings of the EGU General Assembly, Vienna, Austria, 4–13 April 2018; p. 11920.

16. Guo, X.; Zhao, Q.; Ditmar, P.; Liu, J. A new time-series of GRACE monthly gravity field solutions obtained by accounting for the colored noise in the K-Band range-rate measurements. *GFZ Data Serv.* **2017**. [CrossRef]

17. Guo, X.; Zhao, Q.; Ditmar, P.; Sun, Y.; Liu, J. Improvements in the monthly gravity field solutions through modeling the colored noise in the GRACE data. *J. Geophys. Res. Solid Earth* **2018**, *123*, 7040–7054. [CrossRef]

18. Liu, J.; Ge, M. PANDA software and its preliminary result of positioning and orbit determination. *Wuhan Univ. J. Nat. Sci.* **2003**, *8*, 603–609.

19. Shi, C.; Zhao, Q.; Geng, J.; Lou, Y.; Ge, M.; Liu, J. Recent development of PANDA software in GNSS data processing. *Proceedings of the Society of Photographic Instrumentation Engineers.* Li, D., Gong, J., Wu, H., Eds.; 2008, Volume 7285, p. 72851S. Available online: https://www.spiedigitallibrary.org/conference-proceedings-of-spie/7285/72851S/Recent-development-of-PANDA-software-in-GNSS-data-processing/10.1117/12.816261.short?SSO=1 (accessed on 13 June 2019).

20. Guo, X.; Zhao, Q. GRACE time-varying gravity field solutions based on PANDA software. *Geod. Geodyn.* **2018**, *9*, 162–168. [CrossRef]

21. Zhao, Q.; Guo, J.; Hu, Z.; Shi, C.; Liu, J.; Cai, H.; Liu, X. GRACE gravity field modeling with an investigation on correlation between nuisance parameters and gravity field coefficients. *Adv. Space Res.* **2011**, *47*, 1833–1850. [CrossRef]

22. Guo, X.; Geng, J.; Chen, X.; Zhao, Q. Enhanced Orbit Determination for Formation-Flying Satellites through Integrated Single- and Double-Difference GPS Ambiguity Resolution. under review.

23. Kroes, R.; Montenbruck, O.; Bertiger, W.; Visser, P. Precise GRACE baseline determination using GPS. *GPS Solut.* **2005**, *9*, 21–31. [CrossRef]

24. Kroes, R. Precise Relative Positioning of Formation Flying Spacecraft Using GPS. Ph.D. Thesis, Delft University of Technology, Delft, The Netherlands, 2006.

25. Allende-Alba, G.; Montenbruck, O.; Jäggi, A.; Arnold, D.; Zangerl, F. Reduced-dynamic and kinematic baseline determination for the Swarm mission. *GPS Solut.* **2017**, *21*, 1275–1284. [CrossRef]

26. Förste, C.; Bruinsma, S.L.; Abrikosov, O.; Lemoine, J.-M.; Marty, J.C.; Flechtner, F.; Balmino, G.; Barthelmes, F.; Biancale, R. EIGEN-6C4 The latest combined global gravity field model including GOCE data up to degree and order 2190 of GFZ Potsdam and GRGS Toulouse. In Proceedings of the EGU General Assembly, Vienna, Austria, 27 April–2 May 2014.

27. Petit, G.; Luzum, B. *IERS Conventions (2010)*; Verlag des Bundesamts für Kartographie und Geodäsie: Frankfurt am Main, Germany, 2010.

28. Rieser, D.; Mayer-Gürr, T.; Savcenko, R.; Bosch, W.; Wünsch, J.; Dahle, C.; Flechtner, F. *The Ocean Tide Model EOT11a in Spherical Harmonics Representation*; Institute of Theoretical Geodesy and Satellite Geodesy (ITSG): TU Graz, Austria; Deutsches Geodätisches Forschungsinstitut (DGFI): Munich, Germany; GFZ German Research Centre for Geosciences: Potsdam, Germany, 2012.

29. Desai, S.D. Observing the pole tide with satellite altimetry. *J. Geophys. Res.* **2002**, *107*, 1–13. [CrossRef]

30. Flechtner, F.; Dobslaw, H.; Fagiolini, E. *AOD1B Product Description Document for Product Release 05*; GFZ German Research Centre for Geosciences: Potsdam, Germany, 2015.

31. Folkner, W.M.; Williams, J.G.; Boggs, D.H. *The Planetary and Lunar Ephemeris DE 421*; Jet Propulsion Laboratory, California Institute of Technology: Pasadena, CA, USA, 2009; pp. 1–34.

32. Ditmar, P.; van Eck van der Sluijs, A.A. A technique for modeling the Earth's gravity field on the basis of satellite accelerations. *J. Geod.* **2004**, *78*, 12–33. [CrossRef]

33. Kaplan, M.H. *Modern Spacecraft Dynamics and Control*; John Wiley and Sons, Inc.: New York, NY, USA, 1976.

34. Montenbruck, O.; Hackel, S.; van den Ijssel, J.; Arnold, D. Reduced dynamic and kinematic precise orbit determination for the Swarm mission from 4 years of GPS tracking. *GPS Solut.* **2018**, *22*, 79. [CrossRef]

35. Ditmar, P.; Teixeira da Encarnação, J.; Hashemi Farahani, H. Understanding data noise in gravity field recovery on the basis of inter-satellite ranging measurements acquired by the satellite gravimetry mission GRACE. *J. Geod.* **2012**, *86*, 441–465. [CrossRef]

36. Ditmar, P.; Klees, R.; Liu, X. Frequency-dependent data weighting in global gravity field modeling from satellite data contaminated by non-stationary noise. *J. Geod.* **2007**, *81*, 81–96. [CrossRef]

37. Wahr, J.; Molenaar, M.; Bryan, F. Time variability of the Earth's gravity field: Hydrological and oceanic effects and their possible detection using GRACE. *J. Geophys. Res. Solid Earth* **1998**, *103*, 30205–30229. [CrossRef]

38. Cheng, M.; Ries, J. The unexpected signal in GRACE estimates of C_{20}. *J. Geod.* **2017**, *91*, 897–914. [CrossRef]

39. Cheng, M.; Tapley, B.D.; Ries, J.C. Deceleration in the Earth's oblateness. *J. Geophys. Res. Solid Earth* **2013**, *118*, 740–747. [CrossRef]

40. Sun, Y.; Riva, R.; Ditmar, P. Optimizing estimates of annual variations and trends in geocenter motion and J_2 from a combination of GRACE data and geophysical models. *J. Geophys. Res. Solid Earth* **2016**, *121*, 8352–8370. [CrossRef]

41. Sun, Y.; Ditmar, P.; Riva, R. Statistically optimal estimation of degree-1 and C_{20} coefficients based on GRACE data and an ocean bottom pressure model. *Geophys. J. Int.* **2017**, *210*, 1305–1322. [CrossRef]

42. Ditmar, P. Conversion of time-varying Stokes coefficients into mass anomalies at the Earth's surface considering the Earth's oblateness. *J. Geod.* **2018**, *92*, 1401–1412. [CrossRef]

43. Xavier, L.; Becker, M.; Cazenave, A.; Longuevergne, L.; Llovel, W.; Filho, O.C.R. Interannual variability in water storage over 2003–2008 in the Amazon Basin from GRACE space gravimetry, in situ river level and precipitation data. *Remote Sens. Environ.* **2010**, *114*, 1629–1637. [CrossRef]

44. Chen, J.L.; Wilson, C.R.; Tapley, B.D. The 2009 exceptional Amazon flood and interannual terrestrial water storage change observed by GRACE. *Water Resour. Res.* **2010**, *46*. [CrossRef]

Article

Improved Estimates of Geocenter Variability from Time-Variable Gravity and Ocean Model Outputs

Tyler C. Sutterley [1,2,*,†] **and Isabella Velicogna** [2,3]

1   NASA Goddard Space Flight Center, Cryospheric Sciences Laboratory, Greenbelt, MD 20771, USA
2   Department of Earth System Science, University of California, Irvine, CA 92697, USA; isabella@uci.edu
3   Jet Propulsion Laboratory, California Institute of Technology, Pasadena, CA 91109, USA
*   Correspondence: tsutterl@uw.edu; Tel.: +1-206-616-0361
†   Current address: Polar Science Center, Applied Physics Laboratory, University of Washington, Seattle, WA 98105, USA.

Received: 7 August 2019; Accepted: 4 September 2019; Published: 10 September 2019

Abstract: Geocenter variations relate the motion of the Earth's center of mass with respect to its center of figure, and represent global-scale redistributions of the Earth's mass. We investigate different techniques for estimating of geocenter motion from combinations of time-variable gravity measurements from the Gravity Recovery and Climate Experiment (GRACE) and GRACE Follow-On missions, and bottom pressure outputs from ocean models. Here, we provide self-consistent estimates of geocenter variability incorporating the effects of self-attraction and loading, and investigate the effect of uncertainties in atmospheric and oceanic variation. The effects of self-attraction and loading from changes in land water storage and ice mass change affect both the seasonality and long-term trend in geocenter position. Omitting the redistribution of sea level affects the average annual amplitudes of the x, y, and z components by 0.2, 0.1, and 0.3 mm, respectively, and affects geocenter trend estimates by 0.02, 0.04 and 0.05 mm/yr for the the x, y, and z components, respectively. Geocenter estimates from the GRACE Follow-On mission are consistent with estimates from the original GRACE mission.

Keywords: GRACE; GRACE-FO; time-variable gravity; geocenter; reference frames; self-attraction and loading

---

1. Introduction

Variations in the Earth's geocenter reflect the largest scale variability of mass within the Earth system, and are essential inclusions for the complete recovery of surface mass change from time-variable gravity [1,2]. The Earth's geocenter is the difference between the Earth's center of mass and center of figure, which is represented by the degree one spherical harmonic terms [3,4]. Estimates of geocenter position have important applications in the determination of terrestrial reference frame variations, satellite altimeter orbit fluctuations, and mass transport from time-variable gravity [1,5]. Geocenter positions are not stationary. Variations in geocenter have been attributed to changes in terrestrial water storage, glacier and ice sheet mass, atmospheric and oceanic circulation, geodynamic processes and other mass transport processes, Figure 1 [1,6–8].

Measurements of time-variable gravity from the Gravity Recovery and Climate Experiment (GRACE) and the GRACE Follow-On (GRACE-FO) missions are set in a center of mass (CM) reference frame, in which the total degree one variations are inherently zero [9–11]. However, the individual contributions to degree one variations in the CM reference frame, such as from oceanic processes or terrestrial water storage change, are not necessarily zero [4]. Applications set in a center of figure (CF) reference frame, such as the recovery of mass variations of the oceans, hydrosphere and cryosphere, also require the inclusion of degree one terms to be fully accurate [2,4]. The exclusion of degree one terms can have a significant impact on estimates of ocean mass [12], ice sheet mass change [13] and terrestrial hydrology [14] due to far-field signals leaking into each regional estimate.

There are presently two methods that regularly provide estimates of geocenter variability: (1) measurements from satellite laser ranging (SLR) [5,15] and (2) calculations from time-variable gravity and ocean model outputs [2,16]. Global inversions of GRACE data, GPS data, and modeled ocean bottom pressure can also provide long-term or detrended estimates of geocenter variability [17,18]. Trends in SLR-derived solutions can be contaminated by network effects, such as station drift [19,20]. They are not necessarily related to long-term changes in geocenter position. Geocenter estimates from inversions of time-variable gravity and ocean model outputs can provide trend estimates. However, they require information about the oceanic redistribution of mass from changes in terrestrial water storage and ice mass change [2]. Here, we improve geocenter estimates from time-variable gravity and ocean model output combinations by including the effects of self-attraction and loading. We test the overall sensitivity of the estimates to uncertainties in atmospheric pressure, ocean bottom pressure and glacial isostatic adjustment. In the following sections, we discuss (1) the data utilized for this research, (2) how we implement and test our method, (3) the results from our method and (4) the overall implications of the research.

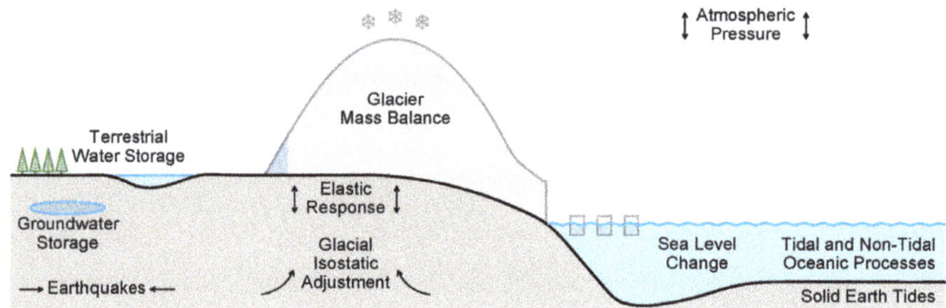

Figure 1. Schematic of the major geophysical processes contributing to mass variability sensed by time-variable gravity.

2. Data

2.1. Time-Variable Gravity

We use monthly GRACE/GRACE-FO Release-6 (RL06) gravity solutions provided by the University of Texas Center for Space Research (UTCSR), the German Research Centre for Geosciences (GeoForschungsZentrum, GFZ) and the Jet Propulsion Laboratory (JPL) for the period April 2002–May 2019 [9–11,21]. Each Level-2 gravity field solution consists of fully-normalized spherical harmonic coefficients ($\tilde{C}_{lm}$, $\tilde{S}_{lm}$) of degree, l, and order, m. We substitute the $\tilde{C}_{20}$ coefficients derived from GRACE/GRACE-FO data with estimates from satellite laser ranging (SLR) from NASA Goddard Space Flight Center due to anomalous variability in the GRACE-derived coefficients [22]. For periods when the satellite pairs had a single fully-operating accelerometer, we replace the $\tilde{C}_{30}$ coefficients derived from GRACE/GRACE-FO data with SLR-derived estimates [22]. The monthly GRACE/GRACE-FO RL06 products have non-tidal atmospheric mass variability removed using outputs from the European Centre for Medium-Range Weather Forecasts (ECMWF) and ERA-Interim [23]. Non-tidal oceanic mass variability is removed from the Release-6 GRACE/GRACE-FO data using outputs from the Max Planck Institute ocean model (MPIOM) [24]. We correct for the effects of Glacial Isostatic Adjustment (GIA), which is the viscoelastic response to changes in ice mass since the last glacial maximum, using outputs from the A et al. [25] compressible Earth model using ICE-6G ice history. The impact of different ice histories and mantle rheologies are tested using expected GIA rates from Caron et al. [26]. We account for the elastic deformation of the solid Earth induced by variations in surface mass loading using load Love numbers of gravitational potential, k_l, calculated by Wahr et al. [27]. The degree one load Love number of gravitational potential, k_1, is updated to reflect the use of a CF reference frame [28]. We smooth the GRACE/GRACE-FO coefficients using a 300 km radius Gaussian averaging function to reduce the impact of random spherical harmonic errors [4,29]. We filter the coefficients to reduce the impact of correlated north/south "striping" errors [30].

As an independent assessment of geocenter variability, we use monthly estimates derived from satellite laser ranging (SLR) [15]. The traditional SLR-derived geocenter solutions reflect the best fit to the center of the SLR network (CN) [31]. However, SLR-derived solutions can be affected by network-effects when individual stations drift relative to each other due to tectonics, surface loading or other processes [19]. Here, we use both the traditional CN-CM solutions [15] and the new UTCSR CF-CM solutions that try to account for these network-effect range biases [5]. We correct both sets of SLR solutions for the gravitational effects of non-tidal atmospheric and oceanic variability using the 6-h Release-6 de-aliasing product provided by GFZ [23].

2.2. Atmospheric Reanalyses and Ocean Models

We estimate the effect of uncertainty in atmospheric pressure by comparing the geocenter outputs from the GRACE/GRACE-FO atmospheric de-aliasing products (GAA) with estimates from ERA-Interim [32], MERRA-2 [33], NCEP-DOE-2 [34] and JRA-55 [35] reanalysis outputs. ERA-Interim is computed by ECMWF and is available starting from 1979 [32]. MERRA-2 is computed by the NASA Global Modeling and Assimilation Office (GMAO) and is available starting from 1980 [33]. NCEP-DOE-2 is computed by the National Centers for Environmental Prediction (NCEP) and is available starting from 1979 [34]. JRA-55 is computed by the Japan Meteorological Agency (JMA) and is available starting from 1958 [35]. Atmospheric pressure anomalies for each reanalysis are calculated relative to the mean over 2001–2002.

We estimate the effect of uncertainty in oceanic circulation by substituting the GRACE/GRACE-FO ocean bottom pressure (OBP) product (GAD) derived from MPIOM with estimates from ECCO-JPL near real-time Kalman-filtered (kf080i) simulations [36,37] and ECCO Version 4 Release 3 (V4r3) simulations [38,39]. Each ocean model incorporates different model physics, data assimilation schemes, and atmospheric forcings. MPIOM is a global ocean model forced with atmospheric outputs from ECMWF medium-range forecasts, and is coupled with a prognostic sea ice model [24]. ECCO-JPL is a regional ocean model configured to resolve tropical ocean circulation (79.5°S–78.5°N latitudinal limits) that is forced with atmospheric outputs from NCEP Reanalysis, and the model does not include sea ice [36]. ECCO V4r3 is a global ocean model that is forced with atmospheric outputs from ERA-Interim, and is coupled with a sea ice model [38,40,41]. The OBP data from ECCO V4r3 were converted from anomalies relative to depth, ϕ_{bot}, into a time series of absolute OBP in pascals using the model standard density, ρ_0, and the average gravitational acceleration of the Earth, g. As Boussinesq-type models conserve volume rather than mass, OBP anomalies for each ECCO model were calculated relative to the global average OBP at each time step [42]. Temporal anomalies in OBP are calculated relative to the mean over 2003–2007, following the JPL Tellus OBP product documentation [36].

3. Methods

A change in surface mass density, $\sigma(\theta, \phi, t)$, of a thin layer at the Earth's surface at time, t, colatitude, θ, and longitude, ϕ, can be decomposed into a series of fully-normalized spherical harmonic coefficients, $\tilde{C}_{lm}(t)$ and $\tilde{S}_{lm}(t)$, after allowing for the Earth's elastic response [4].

$$\begin{Bmatrix} \tilde{C}_{lm}(t) \\ \tilde{S}_{lm}(t) \end{Bmatrix} = \frac{3}{4\pi a \rho_e} \frac{1 + k_l}{2l + 1} \int \sigma(\theta, \phi, t) \, P_{lm}(\cos\theta) \begin{Bmatrix} \cos m\phi \\ \sin m\phi \end{Bmatrix} d\Omega \tag{1}$$

where a is average radius of the Earth, ρ_e is the average density of the Earth, k_l is the gravitational load Love number of degree l, P_{lm} is the associated Legendre Polynomial of degree l and order m, and $d\Omega$ is the element of solid angle $\sin\theta \, d\theta \, d\phi$. The change in the surface mass density field, $\sigma(\theta, \phi, t)$, can be calculated through a summation of the fully-normalized spherical harmonics as shown in Wahr et al. [4].

$$\sigma(\theta, \phi, t) = \frac{a\rho_e}{3} \sum_{l=0}^{\infty} \sum_{m=0}^{l} \frac{2l + 1}{1 + k_l} P_{lm}(\cos\theta) \left\{ \tilde{C}_{lm}(t) \cos m\phi + \tilde{S}_{lm}(t) \sin m\phi \right\} \tag{2}$$

For this work, we use surface mass spherical harmonic coefficients, denoted here as $C_{lm}(t)$ and $S_{lm}(t)$, which are calculated from the fully-normalized spherical harmonic coefficients, $\tilde{C}_{lm}(t)$ and $\tilde{S}_{lm}(t)$.

$$\begin{Bmatrix} C_{lm}(t) \\ S_{lm}(t) \end{Bmatrix} = \frac{a\rho_e}{3} \frac{2l + 1}{1 + k_l} \begin{Bmatrix} \tilde{C}_{lm}(t) \\ \tilde{S}_{lm}(t) \end{Bmatrix} \tag{3}$$

Time-variable gravity measurements from GRACE and GRACE-FO are set in a reference frame of instantaneous center of mass (CM) [9–11]. In this reference frame, spherical harmonic coefficients of degree one, C_{10}, C_{11}, and S_{11}, are inherently zero [4]. However, applications set in a center of figure (CF) reference frame require estimates of degree one variability to be accurate [2,4]. Here, we estimate the degree one variations by combining time-variable gravity measurements with ocean model outputs following Swenson et al. [2]. We start by partitioning the surface mass density changes into the individual land and ocean components (Equation (4)). For this, we use an ocean function, $\vartheta(\theta, \phi)$, with coastlines buffered by 300 km in order to limit the leakage of mass from the land into the ocean estimate [4].

$$\sigma(\theta, \phi, t) = \sigma_{land}(\theta, \phi, t) + \sigma_{ocean}(\theta, \phi, t)$$
$$\sigma_{ocean}(\theta, \phi, t) = \vartheta(\theta, \phi) \, \sigma(\theta, \phi, t) \tag{4}$$

The oceanic components of the degree one spherical harmonics (C_{10}^{ocean}, C_{11}^{ocean}, and S_{11}^{ocean}) can be calculated from the changes in ocean mass, $\sigma_{ocean}(\theta, \phi, t)$ [2,4].

$$C_{10}^{ocean}(t) = \frac{1}{4\pi} \int P_{10}(\cos\theta) \, \sigma_{ocean}(\theta, \phi, t) \, d\Omega$$
$$C_{11}^{ocean}(t) = \frac{1}{4\pi} \int P_{11}(\cos\theta) \, \sigma_{ocean}(\theta, \phi, t) \, \cos\phi \, d\Omega \tag{5}$$
$$S_{11}^{ocean}(t) = \frac{1}{4\pi} \int P_{11}(\cos\theta) \, \sigma_{ocean}(\theta, \phi, t) \, \sin\phi \, d\Omega$$

Assuming that the changes in oceanic mass can be determined from the global surface mass density field using an ocean function (Equation (4)), the oceanic contributions to degree one can also be estimated from the global set of spherical harmonics [2,4]. Here, we separate the degree one terms in the spherical harmonic summation from the higher degree terms.

$$
\begin{aligned}
C_{10}^{ocean}(t) = &\frac{C_{10}(t)}{4\pi} \int P_{10}(\cos\theta)\ \vartheta(\theta,\phi)\ P_{10}(\cos\theta)\ d\Omega\ + \\
&\frac{C_{11}(t)}{4\pi} \int P_{10}(\cos\theta)\ \vartheta(\theta,\phi)\ P_{11}(\cos\theta)\ \cos\phi\ d\Omega\ + \\
&\frac{S_{11}(t)}{4\pi} \int P_{10}(\cos\theta)\ \vartheta(\theta,\phi)\ P_{11}(\cos\theta)\ \sin\phi\ d\Omega\ + \\
&\frac{1}{4\pi} \int P_{10}(\cos\theta)\ \vartheta(\theta,\phi) \sum_{l=2}^{\infty}\sum_{m=0}^{l} P_{lm}(\cos\theta)\ \{C_{lm}(t)\cos m\phi + S_{lm}(t)\sin m\phi\}\ d\Omega
\end{aligned}
\tag{6}
$$

$$
\begin{aligned}
C_{11}^{ocean}(t) = &\frac{C_{10}(t)}{4\pi} \int P_{11}(\cos\theta)\ \cos\phi\ \vartheta(\theta,\phi)\ P_{10}(\cos\theta)\ d\Omega\ + \\
&\frac{C_{11}(t)}{4\pi} \int P_{11}(\cos\theta)\ \cos\phi\ \vartheta(\theta,\phi)\ P_{11}(\cos\theta)\ \cos\phi\ d\Omega\ + \\
&\frac{S_{11}(t)}{4\pi} \int P_{11}(\cos\theta)\ \cos\phi\ \vartheta(\theta,\phi)\ P_{11}(\cos\theta)\ \sin\phi\ d\Omega\ + \\
&\frac{1}{4\pi} \int P_{11}(\cos\theta)\ \cos\phi\ \vartheta(\theta,\phi) \sum_{l=2}^{\infty}\sum_{m=0}^{l} P_{lm}(\cos\theta)\ \{C_{lm}(t)\cos m\phi + S_{lm}(t)\sin m\phi\}\ d\Omega
\end{aligned}
\tag{7}
$$

$$
\begin{aligned}
S_{11}^{ocean}(t) = &\frac{C_{10}(t)}{4\pi} \int P_{11}(\cos\theta)\ \sin\phi\ \vartheta(\theta,\phi)\ P_{10}(\cos\theta)\ d\Omega\ + \\
&\frac{C_{11}(t)}{4\pi} \int P_{11}(\cos\theta)\ \sin\phi\ \vartheta(\theta,\phi)\ P_{11}(\cos\theta)\ \cos\phi\ d\Omega\ + \\
&\frac{S_{11}(t)}{4\pi} \int P_{11}(\cos\theta)\ \sin\phi\ \vartheta(\theta,\phi)\ P_{11}(\cos\theta)\ \sin\phi\ d\Omega\ + \\
&\frac{1}{4\pi} \int P_{11}(\cos\theta)\ \sin\phi\ \vartheta(\theta,\phi) \sum_{l=2}^{\infty}\sum_{m=0}^{l} P_{lm}(\cos\theta)\ \{C_{lm}(t)\cos m\phi + S_{lm}(t)\sin m\phi\}\ d\Omega
\end{aligned}
\tag{8}
$$

If the oceanic contributions to degree one variability (C_{10}^{ocean}, C_{11}^{ocean}, and S_{11}^{ocean}) can be estimated from an ocean model, then the unknown complete degree one terms (C_{10}, C_{11}, and S_{11}) in Equations (6)–(8) can be calculated from the residual between the oceanic degree one terms and the measured mass change over the ocean calculated using all other degrees of the global spherical harmonics [2].

$$
\begin{bmatrix} C_{10}(t) \\ C_{11}(t) \\ S_{11}(t) \end{bmatrix} =
\begin{bmatrix}
I_{10C}^{10C} & I_{11C}^{10C} & I_{11S}^{10C} \\
I_{10C}^{11C} & I_{11C}^{11C} & I_{11S}^{11C} \\
I_{10C}^{11S} & I_{11C}^{11S} & I_{11S}^{11S}
\end{bmatrix}^{-1}
\begin{bmatrix}
C_{10}^{ocean}(t) - G_{10C}(t) \\
C_{11}^{ocean}(t) - G_{11C}(t) \\
S_{11}^{ocean}(t) - G_{11S}(t)
\end{bmatrix}
\tag{9}
$$

The *I*-matrix in Equation (9) is comprised of the degree one terms in the spherical harmonic summations from Equations (6)–(8) [2,4]:

$$
\begin{aligned}
I_{10C}^{10C} &= \frac{1}{4\pi} \int P_{10}(\cos\theta)\ \vartheta(\theta,\phi)\ P_{10}(\cos\theta)\ d\Omega \\
I_{11C}^{10C} &= \frac{1}{4\pi} \int P_{10}(\cos\theta)\ \vartheta(\theta,\phi)\ P_{11}(\cos\theta)\ \cos\phi\ d\Omega \\
I_{11S}^{10C} &= \frac{1}{4\pi} \int P_{10}(\cos\theta)\ \vartheta(\theta,\phi)\ P_{11}(\cos\theta)\ \sin\phi\ d\Omega \\
I_{10C}^{11C} &= \frac{1}{4\pi} \int P_{11}(\cos\theta)\ \cos\phi\ \vartheta(\theta,\phi)\ P_{10}(\cos\theta)\ d\Omega \\
I_{11C}^{11C} &= \frac{1}{4\pi} \int P_{11}(\cos\theta)\ \cos\phi\ \vartheta(\theta,\phi)\ P_{11}(\cos\theta)\ \cos\phi\ d\Omega \\
I_{11S}^{11C} &= \frac{1}{4\pi} \int P_{11}(\cos\theta)\ \cos\phi\ \vartheta(\theta,\phi)\ P_{11}(\cos\theta)\ \sin\phi\ d\Omega \\
I_{10C}^{11S} &= \frac{1}{4\pi} \int P_{11}(\cos\theta)\ \sin\phi\ \vartheta(\theta,\phi)\ P_{10}(\cos\theta)\ d\Omega \\
I_{11C}^{11S} &= \frac{1}{4\pi} \int P_{11}(\cos\theta)\ \sin\phi\ \vartheta(\theta,\phi)\ P_{11}(\cos\theta)\ \cos\phi\ d\Omega \\
I_{11S}^{11S} &= \frac{1}{4\pi} \int P_{11}(\cos\theta)\ \sin\phi\ \vartheta(\theta,\phi)\ P_{11}(\cos\theta)\ \sin\phi\ d\Omega
\end{aligned}
\tag{10}
$$

The measured mass changes over the ocean, G_{10C}, G_{11C}, and G_{11S}, are estimated from time-variable gravity measurements of degree 2 and greater as listed in Equations (6)–(8) [2].

$$G_{10C}(t) = \frac{1}{4\pi} \int P_{10}(\cos\theta)\, \vartheta(\theta,\phi) \sum_{l=2}^{\infty} \sum_{m=0}^{l} P_{lm}(\cos\theta)\, \{C_{lm}(t)\cos m\phi + S_{lm}(t)\sin m\phi\}\; d\Omega$$

$$G_{11C}(t) = \frac{1}{4\pi} \int P_{11}(\cos\theta)\cos\phi\, \vartheta(\theta,\phi) \sum_{l=2}^{\infty} \sum_{m=0}^{l} P_{lm}(\cos\theta)\, \{C_{lm}(t)\cos m\phi + S_{lm}(t)\sin m\phi\}\; d\Omega \qquad (11)$$

$$G_{11S}(t) = \frac{1}{4\pi} \int P_{11}(\cos\theta)\sin\phi\, \vartheta(\theta,\phi) \sum_{l=2}^{\infty} \sum_{m=0}^{l} P_{lm}(\cos\theta)\, \{C_{lm}(t)\cos m\phi + S_{lm}(t)\sin m\phi\}\; d\Omega$$

3.1. Eustatic Sea Level from Land Surface Fluxes

The MPIOM model used to correct GRACE/GRACE-FO data for ocean circulation does not include the effects of land-sea exchange and the corresponding sea level response [24]. As the land-sea flux signal would be present in the ocean mass variations, we estimate the monthly land-sea flux using GRACE/GRACE-FO time-variable gravity measurements when calculating the oceanic contributions to degree one (C_{10}^{ocean}, C_{11}^{ocean}, and S_{11}^{ocean}) [12]. We use the same 300 km buffered coastline mask to calculate our land function ($\mathcal{L} = 1 - \vartheta$). At each time step, we use solutions to the sea level equation to calculate the spatial pattern of sea level variation induced by the change in land mass [43–45]. These changes in sea level are often referred to as the effects of self-attraction and loading (SAL) or as the sea level fingerprints (SLF) of the land mass change [46]. As the SLFs can differ significantly from global ocean averages, geocenter estimates that assume a uniform redistribution of terrestrial water fluxes can be negatively impacted [16]. Here, we use a pseudo-spectral approach for solving the sea level equation in which we assume the Earth deforms elastically from the modern-day mass change without the manifestation of viscoelastic effects [47,48]. When calculating each SLF, we use a static ocean function with the same 300 km buffered coastlines to verify that mass is being conserved.

3.2. Iterated Solutions

In order to estimate the full component of land-sea mass transport, an initial estimate of the geocenter variability is needed to calculate the total land mass [2]. Initially, we use annual and semi-annual geocenter estimates from Chen et al. [7], which are calculated from land-surface model estimates of soil moisture and snow water equivalent. We then calculate the eustatic sea level change induced from the land-sea flux and use it to generate a full geocenter estimate with Equation (9). The initial geocenter estimate from Chen et al. [7] is then replaced with the newly calculated estimate, and the sequence is repeated until the difference between successive iterations falls below a threshold value (see flowchart in Figure 2).

3.3. Spherical Harmonics of Atmospheric and Oceanic Variability

Changes in atmospheric pressure, p, and ocean bottom pressure, p_{bot}, at colatitude, θ, and longitude, ϕ, will impact the Earth's gravitational field. These changes can be decomposed into a series of fully-normalized spherical harmonic coefficients representing the induced gravitational change after allowing for the Earth's elastic response [49,50]:

$$\left\{ \begin{array}{c} \tilde{C}_{lm}(t) \\ \tilde{S}_{lm}(t) \end{array} \right\} = \frac{3}{4\pi\rho_e} \frac{1+k_l}{(2l+1)} \int \xi_l(\theta,\phi,t)\, P_{lm}(\cos\theta) \left\{ \begin{array}{c} \cos m\phi \\ \sin m\phi \end{array} \right\} d\Omega \qquad (12)$$

The vertical integral, $\xi_l(\theta,\phi,t)$, in Equation (12) is determined based on assumptions of the Earth's geometry and the vertical structure of the atmosphere or ocean [49].

Here, spherical harmonics from NCEP-DOE-2 and JRA-55 reanalyses are calculated assuming a thin layer two-dimensional atmosphere with a realistic Earth geometry incorporating the model orography, $h(\theta,\phi)$, and estimates of geoid height, $N(\theta,\phi)$ (Equations (12) and (13)), while harmonics from ERA-Interim and MERRA-2 reanalyses are calculated assuming a three-dimensional atmospheric geometry integrating over the model layers (Equations (12) and (14)) [49].

$$\xi_l(\theta,\phi,t) = \left(\frac{a + h(\theta,\phi) + N(\theta,\phi)}{a}\right)^{l+2} \frac{p_0(\theta,\phi,t)}{g(\theta,\phi)} \tag{13}$$

$$\xi_l(\theta,\phi,t) = -\int_{p_0}^{0} \left(\frac{a + z(\theta,\phi) + N(\theta,\phi)}{a}\right)^{l+2} \frac{dp}{g(\theta,\phi,z)} \tag{14}$$

Similarly, spherical harmonics from ECCO kf080i and ECCO V4r3 ocean bottom pressure are calculated assuming a thin layer two-dimensional ocean with a realistic Earth geometry incorporating estimates of geoid height and ocean bathymetry, $d(\theta,\phi)$ (Equations (12) and (15)).

$$\xi_l(\theta,\phi,t) = \left(\frac{a + N(\theta,\phi) - d(\theta,\phi)}{a}\right)^{l+2} \frac{p_{bot}(\theta,\phi,t)}{g(\theta,\phi)} \tag{15}$$

3.4. Time Series Analysis

We calculate the average geocenter change by simultaneously fitting a least-squares model with constant, linear and quadratic terms with annual, semiannual and 161-day oscillating terms to the estimate geocenter time series [51]. The 161-day oscillating terms account for aliasing of the S2 tidal constituent in the monthly GRACE/GRACE-FO time-variable gravity fields [52].

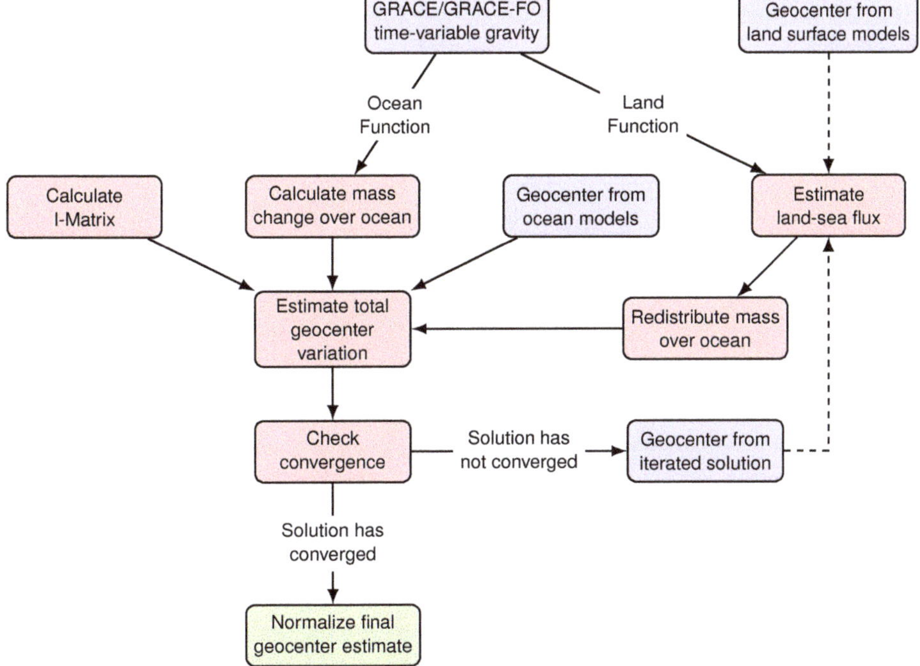

Figure 2. Flowchart of our processing scheme for estimating geocenter variations from time-variable gravity data and ocean model outputs. Blue nodes denote datasets, red nodes denote calculations, and the green node denotes the final converged solution.

4. Results

4.1. Simulated Geocenter Estimates

We test the efficacy of our methodology for estimating geocenter variations by running experiments using fields of simulated global mass variability [2]. Sets of coefficients are constructed using estimates of terrestrial water storage (TWS) calculated from the Global Land Data Assimilation Systems (GLDAS) NOAH land surface model [53], estimates of surface mass balance (SMB) change for glaciers and ice caps calculated from the Regional Atmospheric and Climate Model (RACMO2.3) [54,55] and estimates of Greenland and Antarctic ice sheet mass balance from the mass budget method (MBM) [56]. Monthly GLDAS TWS estimates were calculated by combining the snow water equivalent, canopy water storage and soil moisture variables for non-glaciated regions [57]. We include the mass changes from glaciers and ice sheets to incorporate more processes that affect inter-annual geocenter variability [8]. The SMB of a glacier represents the sum of mass accumulation from snow and rain minus the surface ablation from meltwater runoff, sublimation, and snow drift erosion [58,59]. Cumulative anomalies in SMB were calculated for glaciers and ice caps in reference to a 1961–1990 baseline [60]. MBM estimates of ice sheet mass balance were calculated combining SMB outputs from RACMO2 with estimates of total ice discharge [56,61]. The sea level fingerprints of the synthetic data were calculated as an estimate of oceanic variability [48]. The ocean function used when calculating the sea level fingerprints in the synthetic is an update of Hall et al. [62] that incorporates Antarctic grounded ice delineations [63] and more accurate Greenland coastlines [64].

We test three different scenarios: (1) a uniform ocean redistribution of the land mass change with a static seasonal geocenter estimate similar to Swenson et al. [2], (2) a uniform ocean distribution of the land mass change that iteratively solves for the geocenter and (3) an oceanic redistribution taking into account self-attraction and loading effects that iteratively solves for the geocenter. Each model run uses spherical harmonic coefficients of degree two and greater that are calculated from the synthetic data. The harmonics are truncated to degree and order 60 and processed using the same 300 km Gaussian averaging function and decorrelation filter used with the GRACE/GRACE-FO data [4,29,30]. In order to calculate the full component of the land mass change, we include seasonal estimates of degree one variation from Chen et al. [7]. In the two iteration scenarios, we replace the degree one coefficients for each run with the calculated results of the previous run. The process is repeated until the difference between estimates on successive iterations falls below a threshold value.

We compare the results of each scenario against the exact time series of degree one variations calculated from the synthetic dataset (Figure 3). The RMS differences between the scenarios and the original time series are 0.32 mm, 0.21 mm, and 0.11 mm for the static, iterated, and iterated SLF scenarios, respectively. This represents differences of 21%, 14%, and 7%, respectively, for each of the three scenarios compared to the standard deviation of the original time series. Calculating the land mass change using a static seasonal estimate of geocenter variability underestimates both the short-term and long-term variability in geocenter motion. Self-attraction and loading effects limit the recovery of geocenter variations in the two scenarios that assume a uniform redistribution of land-sea fluxes. The recovery of the seasonal variability of each coefficient is significantly improved when self-attraction and loading effects are incorporated (Figure 4).

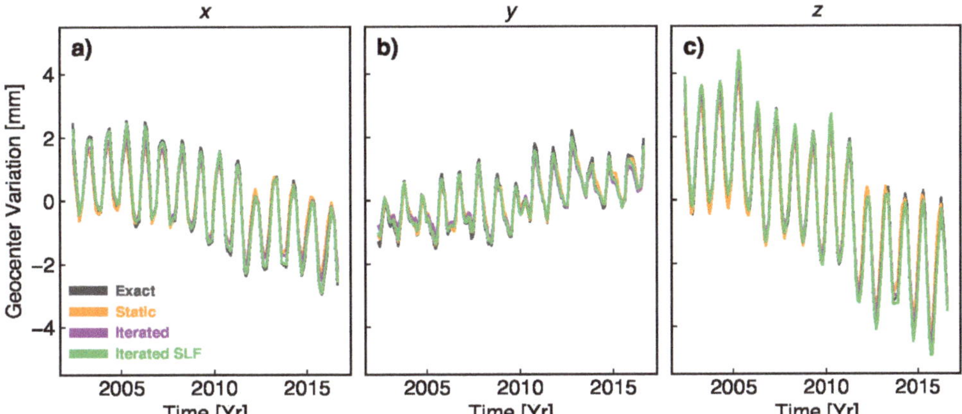

Figure 3. Time series of actual and recovered geocenter variations, (**a**) x, (**b**) y and (**c**) z, in mm from synthetic fields of mass variability derived from GLDAS NOAH v2.1 land surface model outputs [53], RACMO2.3 surface mass balance outputs [54,55] and ice sheet mass balance estimates [56]. The gray line is the original degree one time series calculated directly from the synthetic dataset. The orange, purple and green lines are the derived degree one time series using different scenarios to calculate the land-sea fluxes. The orange line uses a static seasonal geocenter from Chen et al. [7], the purple line uses an iterated self-consistent geocenter, and the green line uses an iterated self-consistent geocenter taking into account the effects of self-attraction and loading.

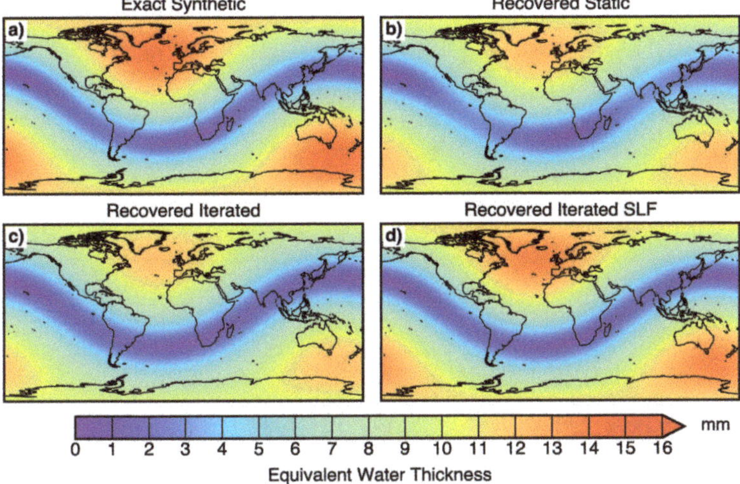

Figure 4. Annual amplitudes of (**a**) actual and (**b**–**d**) recovered degree one surface mass variations in mm water equivalent calculated from synthetic fields of mass variability derived from GLDAS NOAH v2.1 land surface model outputs [53], RACMO2.3 surface mass balance outputs [54,55] and ice sheet mass balance estimates [56]. In the recovered solutions, the land-sea exchange is estimated using (**b**) a static seasonal geocenter from Chen et al. [7], (**c**) an iterated self-consistent geocenter, and (**d**) an iterated self-consistent geocenter taking into account the effects of self-attraction and loading.

4.2. Recovered Geocenter Estimates

We estimate sets of degree one coefficients that are corrected for the effects of non-tidal atmospheric and oceanic variation [2]. These coefficients are directly applicable to time-variable gravity applications for estimating surface mass change [4]. We calculate geocenter variability estimates for each of the processing centers (CSR, GFZ, JPL) of the GRACE/GRACE-FO Science Data System (SDS) using the iterated self-consistent geocenter method that incorporates self-attraction and loading effects (Figure 5). Geocenter results from all three processing centers largely agree in terms of the amplitude and phase of the seasonal signal, along with the long-term trend in each coefficient. Self-attraction and loading effects impact the trend of the estimated geocenter solution for all three processing centers (Figure 6). Using a static seasonal geocenter to estimate the total land-sea flux results in weaker trends for all three processing centers. The trend and annual amplitudes of each coefficient calculated using time-variable gravity fields from each processing center are listed in Table 1. Annual amplitudes for solutions derived from satellite laser ranging (SLR) after correcting for non-tidal atmospheric and oceanic variability are listed for comparison [5,15]. The listed uncertainties for each coefficient are 95% confidence intervals, but do not take into account spherical harmonic errors or uncertainties in the geophysical corrections.

Table 1. Geocenter motion annual amplitudes, annual phase, and trends for 2002–2017 derived using satellite laser ranging (SLR) and time-variable gravity fields from the Center for Space Research (CSR), the German Research Centre for Geosciences (GFZ) and the Jet Propulsion Laboratory (JPL) corrected for the effects of non-tidal atmospheric and oceanic variation. Errors denote the 95% confidence level.

	x	y	z
Annual Amplitude [mm]			
CSR	1.34 ± 0.11	1.54 ± 0.12	2.25 ± 0.16
GFZ	1.38 ± 0.14	1.56 ± 0.13	2.30 ± 0.16
JPL	1.31 ± 0.11	1.52 ± 0.12	2.20 ± 0.17
SLR CN-CM	1.93 ± 0.38	1.17 ± 0.38	4.25 ± 0.57
SLR CF-CM	1.29 ± 0.29	1.48 ± 0.23	2.97 ± 0.46
Annual Phase [day]			
CSR	356.5 ± 5.0	151.9 ± 4.6	9.6 ± 4.2
GFZ	352.7 ± 5.7	150.7 ± 4.9	4.5 ± 4.1
JPL	355.0 ± 4.8	151.4 ± 4.7	7.9 ± 4.4
SLR CN-CM	5.5 ± 11.7	194.7 ± 18.9	52.4 ± 7.7
SLR CF-CM	347.9 ± 13.3	169.6 ± 9.0	46.3 ± 9.1
Trend [mm/yr]			
CSR	−0.15 ± 0.02	0.10 ± 0.02	−0.62 ± 0.03
GFZ	−0.19 ± 0.03	0.21 ± 0.03	−0.66 ± 0.03
JPL	−0.15 ± 0.02	0.11 ± 0.03	−0.63 ± 0.03

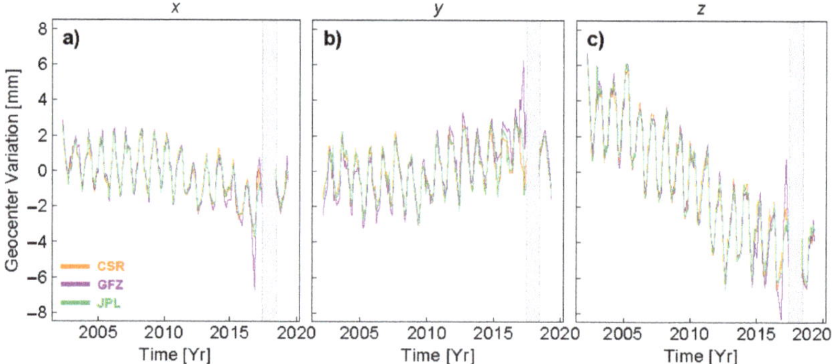

Figure 5. Time series of recovered geocenter variations, (**a**) x, (**b**) y and (**c**) z, in mm calculated using an iterated self-consistent geocenter with self-attraction and loading effects from time-variable gravity fields provided by the Center for Space Research (orange), the German Research Centre for Geosciences (purple) and the Jet Propulsion Laboratory (green). The gray shading denotes the period between the GRACE and GRACE-FO missions.

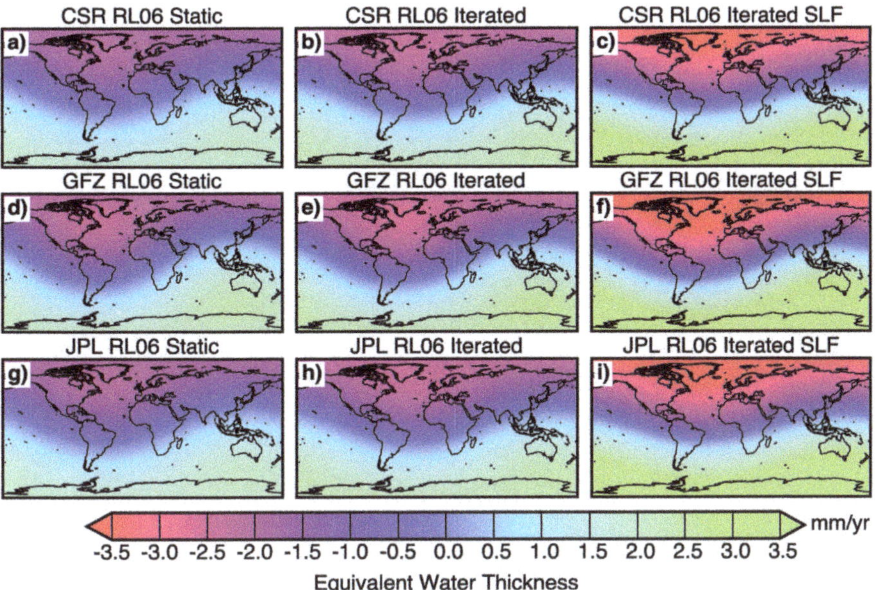

Figure 6. Trends in recovered degree one surface mass variations in mm water equivalent calculated from time-variable gravity fields provided by the Center for Space Research (**a–c**), the German Research Centre for Geosciences (**d–f**) and the Jet Propulsion Laboratory (**g–i**). The land-sea exchange is calculated in the first column (**a,d,g**) with a static seasonal geocenter from Chen et al. [7], in the second column (**b,e,h**) with an iterated self-consistent geocenter, and in the third column (**c,f,i**) with an iterated self-consistent geocenter taking into account the effects of self-attraction and loading.

4.3. Uncertainty Estimates

The total uncertainty in the estimated geocenter time series represents a combination of GRACE/GRACE-FO measurement error, signal leakage between ocean and land, and uncertainty in the geophysical corrections. We assess the impact of these uncertainties, which we assume are uncorrelated, on the trend, annual amplitude and annual phase to the 95% confidence level. Monthly errors in the GRACE.GRACE-FO Level-2 spherical harmonics are estimated by first smoothing each coefficient with a 13-month Loess-type algorithm [13] and then calculating the residuals between the original and smoothed coefficients [65]. Spherical harmonic uncertainties contribute 0.11, 0.09 and 0.19 mm to the x, y, and z components for coefficients derived from CSR time-variable gravity fields.

We estimate the contribution from uncertainty in ocean circulation using ocean bottom pressure outputs from ECCO-JPL near real-time Kalman-filtered simulations and ECCO V4r3 simulations [36,38]. The time series of recovered degree one coefficients derived from CSR time-variable gravity fields using ocean bottom pressure outputs from ECCO-JPL kf080i [36], ECCO V4r3 [38] and MPIOM [23] are shown in Figure 7. The trend and annual amplitudes of each coefficient derived using the different ocean bottom pressure estimates are listed in Table 2. Solutions from ECCO-JPL agree with ECCO V4r3 and MPIOM derived solutions for the y and z components, but differs in terms of trend in the x-component [2]. Solutions from ECCO V4r3 and MPIOM agree for all coefficients between 2002 and 2014.

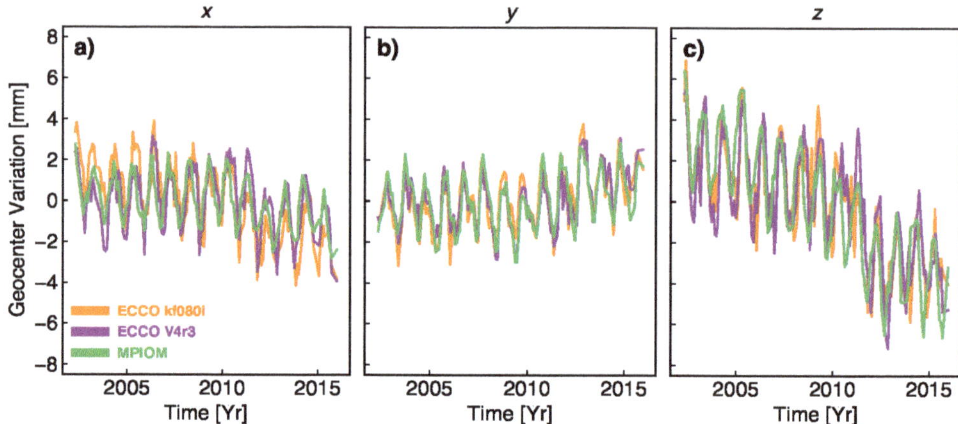

Figure 7. Time series of recovered geocenter variations, (**a**) x, (**b**) y and (**c**) z, in mm calculated using an iterated self-consistent geocenter with self-attraction and loading effects from time-variable gravity fields provided by the Center for Space Research using ocean bottom pressure outputs from ECCO-JPL real-time Kalman-filtered simulations (orange) [36], ECCO Version 4 Release 3 simulations (purple) [38], and Max Planck Institute ocean model (MPIOM) [24].

Table 2. Geocenter motion annual amplitudes, annual phase, and trends for 2002–2015 derived using time-variable gravity fields from the Center for Space Research (CSR) using ocean bottom pressure outputs from ECCO-JPL real-time Kalman-filtered simulations (kf080i) [36] and ECCO Version 4 Release 3 simulations (V4r3) [38]. Errors denote the 95% confidence level.

	x	y	z
Annual Amplitude [mm]			
ECCO-JPL kf080i	1.46 ± 0.20	1.28 ± 0.17	1.80 ± 0.31
ECCO V4r3	1.63 ± 0.18	1.21 ± 0.16	2.31 ± 0.27
MPIOM	1.34 ± 0.11	1.55 ± 0.13	2.24 ± 0.16
Annual Phase [day]			
ECCO-JPL kf080i	307.3 ± 7.9	165.5 ± 7.9	327.8 ± 9.9
ECCO V4r3	323.0 ± 6.5	150.8 ± 7.7	326.2 ± 6.9
MPIOM	358.3 ± 5.0	150.3 ± 4.8	9.6 ± 4.3
Trend [mm/yr]			
ECCO-JPL kf080i	−0.32 ± 0.04	0.16 ± 0.03	−0.48 ± 0.06
ECCO V4r3	−0.10 ± 0.04	0.12 ± 0.04	−0.44 ± 0.06
MPIOM	−0.12 ± 0.02	0.08 ± 0.03	−0.62 ± 0.03

We estimate the uncertainty in atmospheric circulation by using an ensemble of reanalysis outputs [32–35]. Differences between the GRACE/GRACE-FO atmospheric de-aliasing product (GAA) and outputs from ERA-Interim [32], MERRA-2 [33], NCEP-DOE-2 [34], and JRA-55 [35] reanalyses will affect the estimates of geocenter motion. The average monthly uncertainty in geocenter variation due to uncertainties in atmospheric circulation is 0.07, 0.10, and 0.24 mm for the x, y, and z components, respectively.

Glacial Isostatic Adjustment (GIA) affects both ocean mass and land-sea flux calculations used to reconstruct the geocenter variation. The contribution of GIA uncertainty is calculated by comparing our best case estimate against the expected GIA rate from Caron et al. [26]. Using coefficients from Caron et al. [26] affects trend estimates by 0.04, 0.003, and 0.20 mm/yr for the x, y, and z components, respectively. This represents trend differences in the x, y, and z components of 21%, 3%, and 28%, respectively, compared with the values derived using GIA outputs from A et al. [25] with ICE-6G ice history. Uncertainty in GIA is a limiting factor for determining rates of geocenter change from reconstructions with time-variable gravity and ocean models.

5. Discussion

There is better agreement between the estimates derived from GRACE/GRACE-FO and the new CF-CM SLR-derived solutions compared with SLR-derived CN-CM estimates Figure 8, Table 1 [5]. Self-attraction and loading effects improve the correspondence with SLR-derived solutions, particularly in terms of average annual amplitude (Figure 9). However, there are still differences in the seasonal amplitudes of the z-component of geocenter motion (C_{10}). Using a different ocean model to derive the oceanic component of geocenter variability cannot fully explain the difference between the solutions (Figure 8). Uncertainties in the GRACE spherical harmonics and uncertainties in atmospheric variation can partially explain the differences in the solutions. In addition, uncertainties in the ability to correct for network-effects in the SLR-derived CF-CM solutions can also help explain some discrepancies with the solutions derived from GRACE/GRACE-FO.

There is strong agreement between the three processing centers for each of the degree one coefficients during both the GRACE and GRACE-FO periods. However, between the end of 2015 and the end of the GRACE mission, there is disagreement between centers in estimates of the x and y components of geocenter motion (C_{11} and S_{11}). Operational procedures enacted to maintain the battery life of the GRACE satellites during the latter stages of the mission affected the quality of the time-variable gravity fields. The accelerometer onboard GRACE-B was turned off in September 2016 to reduce the battery load and maintain the operation of the microwave ranging instrument. Independent methods have been developed by the GRACE processing centers to spatiotemporally transplant the accelerometer data retrieved from GRACE-A for GRACE-B [66]. The increased uncertainty in the gravitational fields likely affects our ability to estimate the geocenter variability, particularly for the months with a single operating accelerometer.

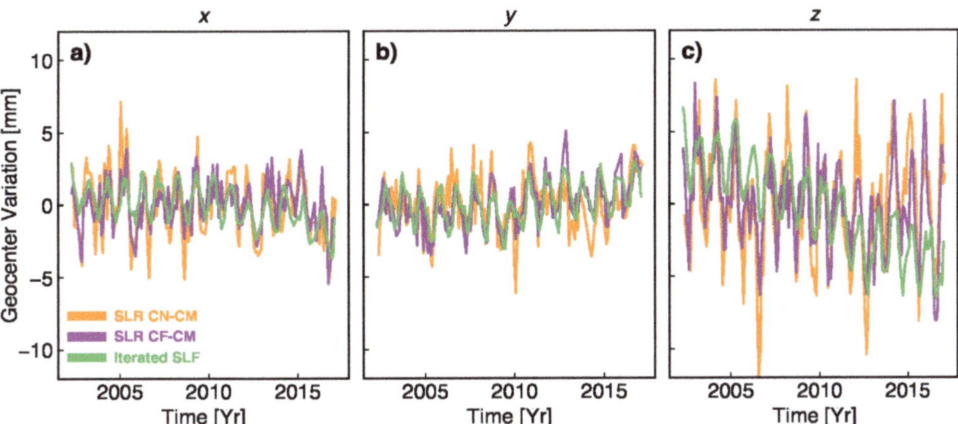

Figure 8. Time series of measured and recovered geocenter variations, (**a**) x, (**b**) y and (**c**) z, in mm from satellite laser ranging (orange and purple) and time-variable gravity fields provided by the Center for Space Research (green). The SLR-derived solutions in orange (CN-CM) are the traditional solutions that center the SLR network [15], and the SLR-derived solutions in purple (CF-CM) include the effects of local site displacements [5]. The solutions derived from GRACE/GRACE-FO in green use an iterated self-consistent geocenter that takes into account the effects of self-attraction and loading.

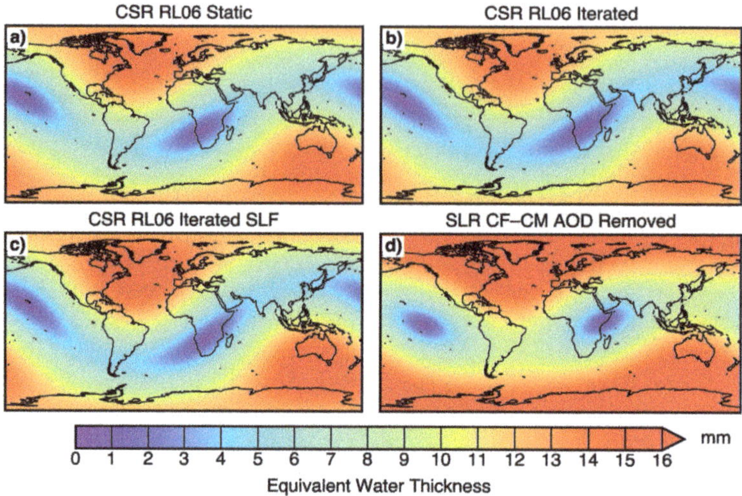

Figure 9. Annual amplitudes of measured and recovered degree one surface mass variations in mm water equivalent calculated from (**a–c**) time-variable gravity fields provided by the Center for Space Research, and (**d**) satellite laser ranging (SLR) CF-CM solutions. In the solutions derived from GRACE/GRACE-FO, the land-sea exchange is estimated using (**a**) a static seasonal geocenter from Chen et al. [7], (**b**) an iterated self-consistent geocenter, and (**c**) an iterated self-consistent geocenter taking into account the effects of self-attraction and loading.

Swenson et al. [2] provides a method for deriving geocenter motion from time-variable gravity measurements and estimates of ocean bottom pressure. The full component of oceanic mass is estimated in Swenson et al. [2] by calculating the land-sea fluxes using time-variable gravity measurements. We expand upon this method by using an iterated, self-consistent geocenter when calculating the full component of the land mass change, and by including self-attraction and loading effects when redistributing the mass over the ocean. The necessity of these inclusions only became evident with a longer record of time-variable gravity measurements. Using a static, seasonal geocenter omits the contribution of inter-annual fluctuations in degree one to the total land mass change. Self-attraction and loading effects impact both the seasonal amplitudes and the long-term trend of the recovered geocenter estimate (Figures 6 and 9). This is due to the spatial patterns of sea level fingerprints that can deviate strongly from the uniform average, particularly over the long-term from changes in glaciers and ice sheets [48,67].

Sun et al. [16] expanded on Swenson et al. [2] by simultaneously estimating oblateness, C_{20}, variations along with geocenter variations. They test the sensitivity of geocenter estimation methods to geophysical processes, such as glacial isostatic adjustment and ocean redistributions due to self-attraction and loading. In Sun et al. [16] the GRACE spherical harmonics are truncated to degree and order 45 and not processed with a decorrelation filter. Here, we use the full expansion of spherical harmonics for each processing center and filter for correlated errors in the GRACE/GRACE-FO harmonics [30]. The test the efficacy of the two methods using our synthetic reconstruction of global mass change. While statistically similar between the two techniques, we find that expanding to higher degree and orders and filtering produces more consistent estimates compared with synthetic estimate. The Sun et al. [16] method is used when calculating the JPL Tellus geocenter product. The differences between the geocenter estimates computed here and the estimates provided by JPL Tellus are largely during the final months of the GRACE mission and initial months of the GRACE-FO mission. Estimates of Antarctic ice sheet mass balance from time-variable gravity are sensitive to uncertainties in geocenter variation [68]. We find that differences between the two geocenter estimates affect Antarctic ice sheet mass balance estimates by 8–10 Gt/yr between 2002 and 2019.

Here, we use a buffered 300 km ocean function to calculate the sea level fingerprints in order to conserve mass and to reduce the leakage between land and ocean [4]. A full-resolution ocean function with realistic coastlines could be used if sets of scaling factors could be derived for both the land and ocean mass change. Typically, the use of scaling factors is applicable only for deriving seasonal fluctuations in land mass [69]. A more-complete scaling factor, such as from Hsu and Velicogna [67] for the land mass change, could possibly improve estimates of geocenter variation. The effects of self-attraction and loading would likely become more evident with a full-resolution ocean function as the effects can be more pronounced in coastal areas [48].

6. Conclusions

Geocenter variations are an important representation of global mass change, with applications in satellite gravimetry and in satellite orbit determination. Here, we investigate the effects of different calculations of the eustatic sea level caused by changes in land mass for estimating geocenter variations from combinations of ocean model outputs and time-variable gravity measurements from the GRACE and GRACE Follow-On missions. We find that using an iterated, self-consistent geocenter incorporating self-attraction and loading effects provides the best estimate of geocenter variation. Uncertainties from glacial isostatic adjustment, ocean circulation and atmospheric circulation limit the determination of geocenter position from this technique. The annual amplitudes of the z-component of geocenter variation differs between estimates from this technique and estimates derived from satellite laser ranging (SLR). The effects of self-attraction and loading improve the correspondence with SLR-derived coefficients but there are still discrepancies worth further investigation. Estimates of geocenter variations using data from the GRACE Follow-On mission are consistent with estimates from the GRACE mission, which enables the extension of the geocenter time series going forward.

Author Contributions: T.C.S. performed the analysis and wrote the manuscript. I.V. supervised the project, contributed to the analysis, helped analyze the data and provided comments/feedback.

Funding: Research was supported by an appointment to the NASA Postdoctoral Program at NASA Goddard Space Flight Center, administered by Universities Space Research Association under contract with NASA.

Acknowledgments: This work was performed at NASA Goddard Space Flight Center, the Jet Propulsion Laboratory, and University of California, Irvine. The authors thank our editor Thomas Gruber, and four anonymous reviewers for their comments and suggestions on improving this manuscript. The authors would like to thank John Ries (UTCSR) for his work to produce the geocenter solutions from satellite laser ranging (SLR) and for his helpful discussions on geocenter variability. The authors wish to thank Byron Tapley, Frank Flechtner, Michael Watkins, Srinivas Bettadpur and the GRACE/GRACE-FO Science Data System (SDS) teams for their work to produce the GRACE and GRACE-FO gravity solutions. The authors also wish to thank the German Space Operations Center (GSOC) of the German Aerospace Center (DLR) for providing the raw GRACE and GRACE-FO telemetry data to the processing centers. GRACE and GRACE-FO data are available from the NASA Physical Oceanography Distributed Active Archive Center (PO.DAAC) at https://podaac.jpl.nasa.gov/grace and the GFZ Information System and Data Center (ISDC) at http://isdc.gfz-potsdam.de/grace-isdc/. Reanalysis outputs are available from ERA-Interim at https://www.ecmwf.int/en/forecasts/datasets/reanalysis-datasets/era-interim, from MERRA-2 at https://gmao.gsfc.nasa.gov/reanalysis/MERRA-2/, from NCEP-DOE-2 at https://rda.ucar.edu/datasets/ds091.0/, and from JRA-55 at http://jra.kishou.go.jp/JRA-55/index_en.html. Documentation for the JPL Tellus ocean bottom pressure product is available at https://grace.jpl.nasa.gov/data/get-data/ocean-bottom-pressure/. Geocenter data from this project are available on Figshare under a CC BY 4.0 license at https://doi.org/10.6084/m9.figshare.7388540. The following programs are provided by this project for processing the GRACE/GRACE-FO data: https://github.com/tsutterley/read-GRACE-harmonics reads the Level-2 spherical harmonic data, and https://github.com/tsutterley/read-GRACE-geocenter reads the geocenter data provided by this project.

Conflicts of Interest: The authors declare no conflict of interest.

References

1. Dong, D.; Dickey, J.O.; Chao, Y.; Cheng, M.K. Geocenter variations caused by atmosphere, ocean and surface ground water. *Geophys. Res. Lett.* **1997**, *24*, 1867–1870. [CrossRef]
2. Swenson, S.C.; Chambers, D.P.; Wahr, J. Estimating geocenter variations from a combination of GRACE and ocean model output. *J. Geophys. Res. Solid Earth* **2008**, *113*, B08410. [CrossRef]
3. Petit, G.; Luzum, B. *IERS Conventions (2010)*; Technical Report 36, International Earth Rotation and Reference Systems Service (IERS); Verlag des Bundesamts für Kartographie und Geodäsie: Frankfurt am Main, Germany, 2010.
4. Wahr, J.; Molenaar, M.; Bryan, F. Time variability of the Earth's gravity field: Hydrological and oceanic effects and their possible detection using GRACE. *J. Geophys. Res. Solid Earth* **1998**, *103*, 30205–30229. [CrossRef]
5. Ries, J. Reconciling Estimates of Annual Geocenter Motion from Space Geodesy. In Proceedings of the 20th International Workshop on Laser Ranging, Potsdam, Germany, 10–14 October 2016.
6. Stolz, A. Changes in the Position of the Geocentre due to Seasonal Variations in Air Mass and Ground Water. *Geophys. J. Int.* **1976**, *44*, 19–26. [CrossRef]
7. Chen, J.L.; Wilson, C.R.; Eanes, R.J.; Nerem, R.S. Geophysical interpretation of observed geocenter variations. *J. Geophys. Res. Solid Earth* **1999**, *104*, 2683–2690. [CrossRef]
8. Wu, X.; Ray, J.; van Dam, T.M. Geocenter motion and its geodetic and geophysical implications. *J. Geodyn.* **2012**, *58*, 44–61. [CrossRef]
9. Bettadpur, S. *UTCSR Level-2 Processing Standards Document*; Technical Report GRACE 327-742; Center for Space Research, University of Texas: Austin, TX, USA, 2018.
10. Dahle, C.; Flechtner, F.; Murböck, M.; Michalak, G.; Neumayer, H.; Abrykosov, O.; Reinhold, A.; König, R. *GFZ Processing Standards Document for Level-2 Product Release 06*; Technical Report GRACE 327-743; GFZ German Research Centre for Geosciences: Potsdam, Germany, 2018.
11. Yuan, D.N. *JPL Level-2 Processing Standards Document for Level-2 Product Release 06*; Technical Report GRACE 327-744; Jet Propulsion Laboratory: Pasadena, CA, USA, 2018.
12. Chambers, D.P.; Wahr, J.; Nerem, R.S. Preliminary observations of global ocean mass variations with GRACE. *Geophys. Res. Lett.* **2004**, *31*, L13310. [CrossRef]
13. Velicogna, I. Increasing rates of ice mass loss from the Greenland and Antarctic ice sheets revealed by GRACE. *Geophys. Res. Lett.* **2009**, *36*, L19503. [CrossRef]
14. Chen, J.L.; Rodell, M.; Wilson, C.R.; Famiglietti, J.S. Low degree spherical harmonic influences on Gravity Recovery and Climate Experiment (GRACE) water storage estimates. *Geophys. Res. Lett.* **2005**, *32*, L14405. [CrossRef]
15. Cheng, M.K. *Geocenter Variations from Analysis of SLR Data*; Reference Frames for Applications in Geosciences; Springer: Berlin/Heidelberg, Germany, 2013; pp. 19–25. [CrossRef]
16. Sun, Y.; Riva, R.; Ditmar, P. Optimizing estimates of annual variations and trends in geocenter motion and J2 from a combination of GRACE data and geophysical models. *J. Geophys. Res. Solid Earth* **2016**, *121*, 8352–8370. [CrossRef]
17. Rietbroek, R.; Fritsche, M.; Brunnabend, S.E.; Daras, I.; Kusche, J.; Schröter, J.; Flechtner, F.M.; Dietrich, R. Global surface mass from a new combination of GRACE, modelled OBP and reprocessed GPS data. *J. Geodyn.* **2012**, *59–60*, 64–71. [CrossRef]
18. Wu, X.; Heflin, M.B.; Schotman, H.; Vermeersen, B.L.A.; Dong, D.; Gross, R.S.; Ivins, E.R.; Moore, A.W.; Owen, S.E. Simultaneous estimation of global present-day water transport and glacial isostatic adjustment. *Nat. Geosci.* **2010**, *3*, 642–646. [CrossRef]
19. Collilieux, X.; Altamimi, Z.; Ray, J.; van Dam, T.M.; Wu, X. Effect of the satellite laser ranging network distribution on geocenter motion estimation. *J. Geophys. Res. Solid Earth* **2009**, *114*, B04402. [CrossRef]
20. Zannat, U.J.; Tregoning, P. Estimating network effect in geocenter motion: Theory. *J. Geophys. Res. Solid Earth* **2017**, *122*, 8360–8375. [CrossRef]
21. Tapley, B.D. GRACE Measurements of Mass Variability in the Earth System. *Science* **2004**, *305*, 503–505. [CrossRef] [PubMed]
22. Loomis, B.D.; Rachlin, K.E.; Luthcke, S.B. Improved Earth Oblateness Rate Reveals Increased Ice Sheet Losses and Mass-Driven Sea Level Rise. *Geophys. Res. Lett.* **2019**, *46*, 6910–6917. [CrossRef]

23. Flechtner, F.; Dobslaw, H.; Fagiolini, E. *AOD1B Product Description Document for Product Release 05*; Technical Report GRACE 327-750; GFZ German Research Centre for Geosciences: Potsdam, Germany, 2015.

24. Dobslaw, H.; Bergmann-Wolf, I.; Dill, R.; Poropat, L.; Thomas, M.; Dahle, C.; Esselborn, S.; König, R.; Flechtner, F. A new high-resolution model of non-tidal atmosphere and ocean mass variability for de-aliasing of satellite gravity observations: AOD1B RL06. *Geophys. J. Int.* **2017**, *211*, 263–269. [CrossRef]

25. Geruo, A.; Wahr, J.; Zhong, S. Computations of the viscoelastic response of a 3-D compressible Earth to surface loading: An application to Glacial Isostatic Adjustment in Antarctica and Canada. *Geophys. J. Int.* **2013**, *192*, 557–572. [CrossRef]

26. Caron, L.; Ivins, E.R.; Larour, E.; Adhikari, S.; Nilsson, J.; Blewitt, G. GIA Model Statistics for GRACE Hydrology, Cryosphere, and Ocean Science. *Geophys. Res. Lett.* **2018**, *45*, 2203–2212. [CrossRef]

27. Wahr, J.; DaZhong, H.; Trupin, A.S. Predictions of vertical uplift caused by changing polar ice volumes on a viscoelastic Earth. *Geophys. Res. Lett.* **1995**, *22*, 977–980. [CrossRef]

28. Trupin, A.S.; Meier, M.F.; Wahr, J. Effect of melting glaciers on the Earth's rotation and gravitational field: 1965–1984. *Geophys. J. Int.* **1992**, *108*, 1–15. [CrossRef]

29. Jekeli, C. *Alternative Methods to Smooth the Earth's Gravity Field*; Technical Report 327; Ohio State University, Department of Geodetic Science and Surveying, 1958 Neil Avenue: Columbus, OH, USA, 1981.

30. Swenson, S.C.; Wahr, J. Post-processing removal of correlated errors in GRACE data. *Geophys. Res. Lett.* **2006**, *33*, L08402. [CrossRef]

31. Cheng, M. *Geocenter Variations from Analysis of TOPEX/Poseidon SLR Data*; Technical Report 25; International Earth Rotation and Reference Systems Service (IERS): Frankfurt am Main, Germany, 1998.

32. Berrisford, P.; Dee, D.P.; Poli, P.; Brugge, R.; Fielding, K.; Fuentes, M.; Kållberg, P.W.; Kobayashi, S.; Uppala, S.; Simmons, A. *The ERA-Interim Archive Version 2.0*; ERA Report Series; ECMWF: Reading, UK, 2011; p. 23.

33. Bosilovich, M.G.; Lucchesi, R.; Suarez, M. MERRA-2: File Specification. *GMAO Office Note* **2016**, *9*, 1–73.

34. Kanamitsu, M.; Ebisuzaki, W.; Woollen, J.; Yang, S.K.; Hnilo, J.J.; Fiorino, M.; Potter, G.L. NCEP–DOE AMIP-II Reanalysis (R-2). *Bull. Am. Meteorol. Soc.* **2002**, *83*, 1631–1644. [CrossRef]

35. Kobayashi, S.; Ota, Y.; Harada, Y.; Ebita, A.; Moriya, M.; Onoda, H.; Onogi, K.; Kamahori, H.; Kobayashi, C.; Endo, H.; et al. The JRA-55 Reanalysis: General Specifications and Basic Characteristics. *J. Meteorol. Soc. Jpn. Ser. II* **2015**, *93*, 5–48. [CrossRef]

36. Fukumori, I. A Partitioned Kalman Filter and Smoother. *Mon. Weather Rev.* **2002**, *130*, 1370–1383. [CrossRef]

37. Kim, S.B.; Lee, T.; Fukumori, I. Mechanisms Controlling the Interannual Variation of Mixed Layer Temperature Averaged over the Niño-3 Region. *J. Clim.* **2007**, *20*, 3822–3843. [CrossRef]

38. Fukumori, I.; Wang, O.; Fenty, I.; Forget, G.; Heimbach, P.; Ponte, R.M. *ECCO Version 4 Release 3*; Technical report; JPL/Caltech and NASA Physical Oceanography: Pasadena, CA, USA, 2017,

39. Forget, G.; Campin, J.M.; Heimbach, P.; Hill, C.N.; Ponte, R.M.; Wunsch, C. ECCO version 4: An integrated framework for non-linear inverse modeling and global ocean state estimation. *Geosci. Model Dev.* **2015**, *8*, 3071–3104. [CrossRef]

40. Forget, G.; Ponte, R.M. The partition of regional sea level variability. *Prog. Oceanogr.* **2015**, *137*, 173–195. [CrossRef]

41. Losch, M.; Menemenlis, D.; Campin, J.M.; Heimbach, P.; Hill, C. On the formulation of sea-ice models. Part 1: Effects of different solver implementations and parameterizations. *Ocean Model.* **2010**, *33*, 129–144. [CrossRef]

42. Greatbatch, R.J. A note on the representation of steric sea level in models that conserve volume rather than mass. *J. Geophys. Res. Oceans* **1994**, *99*, 12767–12771. [CrossRef]

43. Farrell, W.E.; Clark, J.A. On Postglacial Sea Level. *Geophys. J. R. Astron. Soc.* **1976**, *46*, 647–667. [CrossRef]

44. Clark, J.A.; Lingle, C.S. Future sea-level changes due to West Antarctic ice sheet fluctuations. *Nature* **1977**, *269*, 206–209. [CrossRef]

45. Mitrovica, J.X.; Milne, G.A. On post-glacial sea level: I. General theory. *Geophys. J. Int.* **2003**, *154*, 253–267. [CrossRef]

46. Tamisiea, M.E.; Mitrovica, J.X. The Moving Boundaries of Sea Level Change: Understanding the Origins of Geographic Variability. *Oceanography* **2011**, *24*, 24–39. [CrossRef]

47. Kendall, R.A.; Mitrovica, J.X.; Milne, G.A. On post-glacial sea level – II. Numerical formulation and comparative results on spherically symmetric models. *Geophys. J. Int.* **2005**, *161*, 679–706. [CrossRef]

48. Tamisiea, M.E.; Hill, E.M.; Ponte, R.M.; Davis, J.L.; Velicogna, I.; Vinogradova, N.T. Impact of self-attraction and loading on the annual cycle in sea level. *J. Geophys. Res. Oceans* **2010**, *115*, C07004. [CrossRef]
49. Boy, J.P.; Chao, B.F. Precise evaluation of atmospheric loading effects on Earth's time-variable gravity field. *J. Geophys. Res. Solid Earth* **2005**, *110*, B08412. [CrossRef]
50. Swenson, S.; Wahr, J. Estimated effects of the vertical structure of atmospheric mass on the time-variable geoid. *J. Geophys. Res. Solid Earth* **2002**, *107*, 2194. [CrossRef]
51. Velicogna, I.; Sutterley, T.C.; van den Broeke, M.R. Regional acceleration in ice mass loss from Greenland and Antarctica using GRACE time-variable gravity data. *Geophys. Res. Lett.* **2014**, *41*, 8130–8137. [CrossRef]
52. Chen, J.L.; Wilson, C.R.; Seo, K.W. S2 tide aliasing in GRACE time-variable gravity solutions. *J. Geod.* **2009**, *83*, 679–687. [CrossRef]
53. Rodell, M.; Houser, P.R.; Jambor, U.; Gottschalck, J.; Mitchell, K.; Meng, C.J.; Arsenault, K.; Cosgrove, B.; Radakovich, J.; Bosilovich, M.; et al. The Global Land Data Assimilation System. *Bull. Am. Meteorol. Soc.* **2004**, *85*, 381–394. [CrossRef]
54. Noël, B.; van de Berg, W.J.; Lhermitte, S.; Wouters, B.; Schaffer, N.; van den Broeke, M.R. Six decades of glacial mass loss in the Canadian Arctic Archipelago. *J. Geophys. Res. Earth Surf.* **2018**, *123*, 1430–1449. [CrossRef]
55. Noël, B.; van de Berg, W.J.; van Wessem, J.M.; van Meijgaard, E.; van As, D.; Lenaerts, J.T.M.; Lhermitte, S.; Kuipers Munneke, P.; Smeets, C.J.P.P.; van Ulft, L.H.; et al. Modelling the climate and surface mass balance of polar ice sheets using RACMO2 – Part 1: Greenland (1958–2016). *Cryosphere* **2018**, *12*, 811–831. [CrossRef]
56. Rignot, E.J.; Velicogna, I.; van den Broeke, M.R.; Monaghan, A.J.; Lenaerts, J.T.M. Acceleration of the contribution of the Greenland and Antarctic ice sheets to sea level rise. *Geophys. Res. Lett.* **2011**, *38*, L05503. [CrossRef]
57. Syed, T.H.; Famiglietti, J.S.; Rodell, M.; Chen, J.; Wilson, C.R. Analysis of terrestrial water storage changes from GRACE and GLDAS. *Water Resour. Res.* **2008**, *44*. [CrossRef]
58. Ettema, J.; van den Broeke, M.R.; van Meijgaard, E.; van de Berg, W.J.; Bamber, J.L.; Box, J.E.; Bales, R.C. Higher surface mass balance of the Greenland ice sheet revealed by high-resolution climate modeling. *Geophys. Res. Lett.* **2009**, *36*, L12501. [CrossRef]
59. Lenaerts, J.T.M.; van den Broeke, M.R.; van de Berg, W.J.; van Meijgaard, E.; Kuipers Munneke, P. A new, high-resolution surface mass balance map of Antarctica (1979–2010) based on regional atmospheric climate modeling. *Geophys. Res. Lett.* **2012**, *39*, L04501. [CrossRef]
60. Van den Broeke, M.R.; Bamber, J.L.; Ettema, J.; Rignot, E.J.; Schrama, E.; van de Berg, W.J.; van Meijgaard, E.; Velicogna, I.; Wouters, B. Partitioning Recent Greenland Mass Loss. *Science* **2009**, *326*, 984–986. [CrossRef]
61. Rignot, E.J.; Bamber, J.L.; van den Broeke, M.R.; Davis, C.H.; Li, Y.; van de Berg, W.J.; van Meijgaard, E. Recent Antarctic ice mass loss from radar interferometry and regional climate modelling. *Nat. Geosci.* **2008**, *1*, 106–110. [CrossRef]
62. Hall, F.G.; Brown de Colstoun, E.; Collatz, G.J.; Landis, D.; Dirmeyer, P.; Betts, A.; Huffman, G.J.; Bounoua, L.; Meeson, B. ISLSCP Initiative II global data sets: Surface boundary conditions and atmospheric forcings for land-atmosphere studies. *J. Geophys. Res. Atmos.* **2006**, *111*. [CrossRef]
63. Mouginot, J.; Scheuchl, B.; Rignot, E. *MEaSUREs Antarctic Boundaries for IPY 2007–2009 from Satellite Radar, Version 2*; NASA National Snow and Ice Data Center Distributed Active Archive Center: Boulder, CO, USA, 2017. [CrossRef]
64. Henriksen, N.; Higgins, A.; Kalsbeek, F.; Pulvertaft, T.C.R. Greenland from Archaean to Quaternary, Descriptive text to the 1995 Geological Map of Greenland 1:2,500,000. *Geol. Surv. Den. Greenl. Bull.* **2009**, *18*, 1–126.
65. Wahr, J.; Swenson, S.C.; Velicogna, I. Accuracy of GRACE mass estimates. *Geophys. Res. Lett.* **2006**, *33*, L06401. [CrossRef]
66. Bandikova, T.; McCullough, C.; Kruizinga, G.L.; Save, H.; Christophe, B. GRACE accelerometer data transplant. *Adv. Space Res.* **2019**, *64*, 623–644. [CrossRef]
67. Hsu, C.W.; Velicogna, I. Detection of Sea Level Fingerprints derived from GRACE gravity data. *Geophys. Res. Lett.* **2017**, 2017GL074070. [CrossRef]

68. Velicogna, I.; Wahr, J. Time-variable gravity observations of ice sheet mass balance: Precision and limitations of the GRACE satellite data. *Geophys. Res. Lett.* **2013**, *40*, 3055–3063. [CrossRef]
69. Landerer, F.W.; Swenson, S.C. Accuracy of scaled GRACE terrestrial water storage estimates. *Water Resour. Res.* **2012**, *48*. [CrossRef]

 *remote sensing*

Article

An Assessment of the GOCE High-Level Processing Facility (HPF) Released Global Geopotential Models with Regional Test Results in Turkey

Bihter Erol *, Mustafa Serkan Işık and Serdar Erol

Istanbul Technical University, Civil Engineering Faculty, Geomatics Engineering Department, Maslak 34469, Istanbul, Turkey; isikm@itu.edu.tr (M.S.I.); erol@itu.edu.tr (S.E.)
* Correspondence: bihter@itu.edu.tr; Tel.: +90-212-285-3821

Received: 23 December 2019; Accepted: 6 February 2020; Published: 10 February 2020

Abstract: The launch of dedicated satellite missions at the beginning of the 2000s led to significant improvement in the determination of Earth gravity field models. As a consequence of this progress, both the accuracies and the spatial resolutions of the global geopotential models increased. However, the spectral behaviors and the accuracies of the released models vary mainly depending on their computation strategies. These strategies are briefly explained in this article. Comprehensive quality assessment of the gravity field models by means of spectral and statistical analyses provides a comparison of the gravity field mapping accuracies of these models, as well as providing an understanding of their progress. The practical benefit of these assessments by means of choosing an optimal model with the highest accuracy and best resolution for a specific application is obvious for a broad range of geoscience applications, including geodesy and geophysics, that employ Earth gravity field parameters in their studies. From this perspective, this study aims to evaluate the GOCE High-Level Processing Facility geopotential models including recently published sixth releases using different validation methods recommended in the literature, and investigate their performances comparatively and in addition to some other models, such as GOCO05S, GOGRA04S and EGM2008. In addition to the validation statistics from various countries, the study specifically emphasizes the numerical test results in Turkey. It is concluded that the performance improves from the first generation RL01 models toward the final RL05 models, which were based on the entire mission data. This outcome was confirmed when the releases of different computation approaches were considered. The accuracies of the RL05 models were found to be similar to GOCO05S, GOGRA04S and even to RL06 versions but better than EGM2008, in their maximum expansion degrees. Regarding the results obtained from these tests using the GPS/leveling observations in Turkey, the contribution of the GOCE data to the models was significant, especially between the expansion degrees of 100 and 250. In the study, the tested geopotential models were also considered for detailed geoid modeling using the remove-compute-restore method. It was found that the best-fitting geopotential model with its optimal expansion degree (please see the definition of optimal degree in the article) improved the high-frequency regional geoid model accuracy by almost 15%.

Keywords: gravity field satellite missions; GOCE; GRACE; GOCE High-Level Processing Facility (HPF), earth gravity field; geoid; spectral enhancement method (SEM), GPS/leveling

---

1. Introduction

The low-earth orbiting (LEO) satellite missions initiated a new stage in Earth gravity field studies and led to unprecedented progress in determining global gravitational field models and related parameters. In addition to the satellite laser ranging (SLR) tracking data, as a fundamental technique for gravity field determination since the earlier research [1], the recent contribution of the LEO satellites significantly improved global geopotential models (GGMs). The mentioned improvements are both in terms of precision and spatial resolution, and hence they can be assumed as a breakthrough with a stunning impact on geodetic science and related fields [2,3]. The observations by LEO satellites primarily rely on satellite-to-satellite tracking (SST), which includes high-to-low (hl-SST) as well as low-to-low SST (ll-SST) [4–8]. The fundamental principle of SST techniques is based on the precise measurements of distance variations between the satellites, which can be used to determine the orbit perturbations and hence the gravitational force acting on the satellite. In addition to SST, dedicated satellite gravity gradiometry (SGG), which is a sensitive detection technique of the space gravitational gradients, is also used in the LEO satellite technology and the gravitational field of the Earth is determined globally with high precision and resolution by means of combined SGG and SST observations.

As a chronological overview of the dedicated Earth gravity field satellite missions, the Challenging Minisatellite Payload (CHAMP), launched first, in July 2000 as a German Research Center for Geosciences (GFZ) project, and pioneered the LEO satellite missions for mapping the gravity field in a global scale. High-low SST with an equipped space-borne GPS-receiver on CHAMP enabled the determination of the long-wavelength static gravity field of Earth [9]. Thereafter, the Gravity Recovery and Climate Experiment (GRACE) twin satellites were launched in March 2002 as a US National Space Agency (NASA) and National Aeronautics and Space Research Center of Germany (DLR) joint project, aiming to track the changes in the gravity field initially for a five-year duration. The GRACE mission employed the dual-frequency one-way K-band phase measurements transmitted and received by both satellites [7,10]. Hence, precise mapping of the Earth's gravity field with a 200–300 km half-wavelength resolution every 30 days was managed with the contribution of the mission data. Thus, many studies and applications in Earth science disciplines have benefited from the information relating to mass transport and redistribution in the Earth system by GRACE-based observations [11]. The science mission of GRACE ended in October 2017. As its successor, the Gravity Recovery and Climate Experiment Follow-On (GRACE-FO) mission was launched in May 2018 to extend GRACE's observations and hence to continue its contribution to research [11,12].

Other than the GRACE and GRACE-FO missions, the Gravity field and steady-state Ocean Circulation Explorer (GOCE) was launched in March 2009 as a European Space Agency (ESA) project. The GOCE satellite mission utilized the satellite gravity gradiometer (SGG) for the first time in combination with the hl-SST technique. In addition to a novel SGG technique (see, e.g., [13,14]), with its low orbit altitude, the GOCE mission was dedicated to producing a near-global map of Earth's static gravity field with high accuracy and resolution [5,8,15,16]. However, it should also be noted that considering the noise characteristics of the gradiometer, the SGG data alone are not sufficient to determine the low-degree harmonic coefficients of the gravity field with high accuracy. Therefore, the GPS SST data are essential to determine the precise orbit, and thus geo-locate the tensor observables, and the long-wavelength part of the gravity field. With the GOCE mission, an approximate 1–2 cm accuracy of geoid undulations and 1–2 mGal accuracy of gravity anomalies were targeted. An unprecedented spatial resolution of 100 km, which approximately corresponds to 200 degrees of spherical harmonic expansion, was accomplished in mapping the gravity field with the contribution of the GOCE satellite mission data [17].

A number of gravity field solutions were produced by the High-Level Processing Facility for GOCE (HPF) using the GOCE data. The official level 2 products of HPF can be obtained through datacenters such as the International Center of Global Earth Models (ICGEM) for GOCE users (geodesists, solid Earth scientists, oceanographers, etc.) [18–21].

In the released products, when the spherical harmonic expansion models (GGMs) are considered, it is seen that the maximum degrees of these models and hence their spectral contents vary in a rather large spectrum. Regarding the spectral bands, expressed by the maximum expansion degree, the GGMs can be classified as a low degree (having a degree/order (d/o) of harmonic expansion between 0 and 100), medium degree (100 < d/o ≤ 300), high-degree (300 < d/o ≤ 3000) and ultra-high degree

(3000 < d/o) models (see, e.g., [15]). This described categorization of the GGMs is adopted in the present article.

This article aims to provide an overview of the geopotential models, particularly those released after the GOCE mission, in terms of the results of comprehensive quality validations using independent terrestrial data. The numerical tests include the GGM validations and their use in Turkey with the high-accuracy GPS/leveling data. These data include high order (first, second and third order) network quality GNSS coordinates (approximately 1–2 cm three-dimensional position accuracy) and orthometric heights (with ~2–3 cm accuracy) in the benchmarks. Turkey is one of the restricted areas in terms of terrestrial gravity field data sharing for global geopotential model calculations. As a consequence, the results of this study may be assumed to be a new statistical contribution to the literature on the model performance assessments of gravity field mapping using the recent HPF models in this territory. However, it should be emphasized that rather than suggesting a new methodology, it was aimed to demonstrate the performance gain of the geopotential models and to highlight the progress of HPF GOCE models from release to release, with the numerical assessment results using terrestrial data from Turkey. The impact of the improvement in geopotential model accuracies on the detailed geoid modeling in the area with rather sparse terrestrial gravity data was investigated as a second aim of this article. In summary, this article was written as a case study for providing a well-coordinated reference to users for geopotential model assessments, addressing the appropriate numerical methods that were suggested in the notable literature (i.e., spectral enhancement method or filtering the terrestrial data), rather than providing an innovative contribution to the theory.

The numerical validations of the GOCE- and GRACE-based geopotential models in the study area are accompanied by the high degree EGM2008 model. EGM2008, with 2190 d/o of spherical harmonic expansion, has been developed as a combination of the ITG-GRACE03S (complete to spherical harmonic d/o 180) geopotential model with the terrestrial, airborne and altimetry-derived gravity anomalies using least squares adjustment estimation technique. Reference [22] explains the computation method of EGM2008 in detail and reports the absolute accuracy of the model as ±5 cm to ±10 cm in well-covered areas with high-quality gravity data. Use of EGM2008 in the tests as the baseline model allowed an objective assessment of the level of improvement achieved by the GOCE mission, specifically at the low to middle frequencies of the gravity field spectrum, since this high-resolution model does not include the GOCE mission data. It should be noted that there is no official report that declares the terrestrial gravity or GPS/leveling data from Turkey that has been included in EGM2008 computation. The validation results in this study showed that the accuracy of EGM2008 ($\ell = 2190$) is at the decimeter level in Turkey.

In addition, EGM2008 has also been employed in this study for spectral enhancement of the validated GGMs for approximating the omission error in comparisons with the GPS/leveling data. The spectral content of the tested GGMs is naturally not comparable with the high-frequency GPS/leveling data. This is because of the truncated degrees in the spherical harmonic expansion equation. Recovering the omitted part of the signal, the spectral enhancement method (SEM) was applied. The SEM is one of the options that is given in the literature to overcome this handicap in validations [19,23]. Low-pass filtering of high-frequency data is another recommended option by [24–26]. In numerical tests, the short-wavelength components of the gravity field were taken into account by employing a high-frequency content of EGM2008 in addition to the residual terrain effect (RTE) from satellite-based high-resolution topographic data.

In the following section, the applied methods for validating internal and external accuracies of the tested GGMs, as well as for computing the detailed regional geoid models using the optimum and maximum d/o of the tested GGMs, are explained. A brief explanation of the main characteristics of the models is also provided in this section. The third section introduces the terrestrial data, which were employed in validations and regional geoid model computations. The numerical test results obtained from the validations of the GGMs, as well as the detailed geoid models, are provided using graphics and tables. Results are interpreted and discussed in this section. The fourth section summarizes the major outcomes of the study.

2. Materials and Methods

2.1. Practical Need of Global Geopotential Model Validation Results

Testing and evaluating geopotential model performances is important since determining an optimal model, with the highest accuracy and best resolution, has certain practical consequences in specific applications of the Earth science and engineering disciplines. We exemplify some of the applications in which GGM performance is crucial in the following by addressing the applied numerical tests under this study.

The determination of a high-resolution regional geoid using the remove–compute–restore (RCR) method combines low to high-frequency content of the gravity field signal employing the reference GGM, the terrestrial gravity observations and the topographic height and density data [27]. In this strategy, the best-fitting GGM to the gravity field in the region contributes to the accuracy of the high-resolution geoid by improving the long-wavelength component in the error budget [27–29]. The GGM contribution to the regional geoid model accuracy can be even more crucial if there is a lack of high-quality and sufficiently dense terrestrial gravity data under certain modifications of the kernel function and when a few-centimeter accuracy geoid model is desired (see e.g., [30]). Regarding this issue, we evaluated the released GGMs in high-frequency geoid computations using gravity data in Turkey. The computations were carried out with fast Fourier transform (FFT) evaluation of an unmodified Stokes integral. The conclusions depicted a centimeter-level contribution of the GGMs to the accuracy improvement of the regional geoid models.

The independent data provides an objective measure to examine the gravity field mapping performances of the GGMs in a region. Since the gravity field parameters and their gradients are directly estimated with spherical harmonic expansion equations in some applications, their accuracy checks are important. Regarding this, [31] uses the deflection of vertical (DoV) components estimated from GGMs to compare with the astrogeodetic DoVs obtained through the observations with the digital zenith camera system (DZCS); [32] exploits the GRACE-based GGMs derived gravity gradients in interpretation of geophysical or geodynamical patterns in Iran; [32] compares the upward continued airborne gravity data with the GOCE-based gravity gradients in Antarctica; and [33] issues the GGM-derived geoid heights and gravity anomalies to compare with the mean sea surface and free air anomalies derived from altimetry data in open ocean and coastal regions. In some GNSS applications, GGM-derived geoid undulations/height anomalies are considered for vertical control purposes. Before using the estimated undulations in the transformation of GNSS heights, the accuracies and the fit of the GGMs to the regional vertical datum should also be validated. For such a purpose, [34] evaluates five combined GGMs in a very dense GPS/leveling network on the Greece mainland and tested capability of parametric models for fitting the GGMs to the regional vertical datum.

The independent validations of the GGMs and test results are also important for the height system unification (HSU) and world height system (WHS) realization studies, because the precise determination of the offsets among the regional vertical datums, as well as with the WHS surface, is critical to establishing a globally unified height datum [23,35]. In this way, using a high-precision GGM as a reference in the calculations allows for higher accuracy calculation of regional height system offsets [35,36]. Although numerical assessments on the determination of the geopotential value (W_o) of regional vertical datum and height system unification are not explicitly undertaken in this article, [37] is recommended for further reading on the detailed numerical evaluations of a number of GGMs on the height system unification between Turkey and Greece.

2.2. Overview the Tested Global Geopotential Models

The tested GGMs in this study include the recent models that were calculated with different processing strategies from satellite orbits and GOCE gradiometer observations. These models were released (tagged with the release version from RL1 to RL5) as the products of the GOCE High-Level Processing Facility (HPF) of the ESA project. The first releases of the models (RL1) consist of the initial two months' GOCE observations (between November 2009 and January 2010); the RL2 models are calculated employing the eight months' observation cycle (November 2009 to July 2010); and the RL3, RL4 and RL5 releases exploit the 18 months, 26 months and the full mission observations, respectively [38]. The sixth releases of the direct and time-wise models, which were also based on the GOCE full mission dataset, were recently published and made available during the preparation of this article [21]. Therefore, the RL6 models were also employed and tested in the content of the study in order to compare their performances with the RL5 models, which have been already exploited the full mission observations.

Essentially, the models were developed according to three HPF gravity field processing approaches, namely, the direct (DIR) [39], time-wise (TIM) [40] and space-wise (SPW) [41] approaches. Each approach follows a different processing strategy in calculating the releases. These strategies include mathematical models, the adopted a priori models (or no a priori model use), reference information in constraints, spherical cap regularization, filter algorithms of the observations, etc. Table 1 provides a detailed comparison of the validated GGMs including a short report on the differences in their processing strategies. References [18,42] provide a detailed description of the basics of these data processing methods. Reference [18] reviews and discusses the DIR, TIM and SPW methods, their processing philosophies and architectures in computations of the GOCE gravity field models using RL1 models. Reference [16] reviews the second, third and fourth generation GOCE-based Earth gravity field models with their computation strategies, and evaluates their performances with spectral sensitivities over different functionals.

The products used in computations include gravity gradients in gradiometer reference frame (with identifier EGG_NOM_2), common-mode accelerations (EGG_CCD_2C), attitude quaternions (EGG_IAQ_2C), and precise science orbits (including the sub-products: kinematic orbits (SST_PKI_2I), variance–covariance information of orbit positions (SST_PCV_2I), reduced-dynamic orbits (SST_PRD_2I) and quaternions of transformation from Earth-fixed to inertial reference frame (SST_PRM_2I)); see [43] for the product descriptions and standards that were applied during the processes. In addition to the observations and products, depending on the applied processing strategy, the reference gravity field models were also adopted in some of the solutions. These reference gravity field models were either exploited as a priori information or for regularization, internal assessment and validation purposes. EIGEN5C, EIGEN51C, ITG-GRACE2010S [21] were used in the DIR method for internal validation purposes, whereas the GOCE quick-look gravity field model was used in the first release of the SPW approach as a priori information. EGM2008, EIGEN5C, EIGEN6C3stat were also employed for signal and error covariance computation in SPW solutions [21]. However, as the basic difference among the three methods, in the processing strategy of TIM method no gravity field a priori information entered the solution and, hence, this method yields pure GOCE-only models. All the information given here were retrieved from the headers of the spherical harmonic coefficients' files and respective data sheets, published by [21]. The main characteristics of the models are summarized as follows:

The DIR method, as the classical approach, is based on the orbit perturbation theory and combines a priori orbits and an a priori gravity field model, hl-SST observations and common-mode accelerations, while constituting the normal equations in determining the gravity field model coefficients in an iterative way:

- DIR models contain GRACE observations in the lower to medium degrees of the expansions (in addition to GOCE gravity gradient data).

- As a priori gravity field information EIGEN5C ($\leq$d/o 360) was introduced in DIR RL1, whereas the ITG-GRACE2010S ($\leq$d/o 150) solution was the background model in RL2.

- In RL3, RL4 and RL5 models, GRACE was combined with GOCE and SLR data on the basis of the normal equations (including the SLR data (LAGEOS-1 and -2)) in order to improve the gravity field solution, since the very low-degree harmonics (in particular degrees 2 and 3) cannot be estimated with GRACE and GOCE data).

- In order to overcome the polar gap of GOCE's observations caused by the orbit inclination of the satellite (inclined at 96.7° [18]), a spherical cap regularization (SCR) in accordance to [44] was applied using EIGEN-51C ($\leq$d/o 240). In the later releases after RL1, the SCR was applied using GRACE and LAGEOS data. In the RL3, RL4, RL5 of DIR models, additionally, the predecessor release was used as a priori information ($\leq$d/o 240, 260, 300, respectively, in the releases) and a Kaula regularization [44] was used beyond degrees 200 (for RL3, RL4) and 180 (for RL5).

- In the DIR RL1 solution, the gravity tensor elements, which are measured with GOCE on-board SGG, were accumulated with the relative weights of Txx 1.0, Tyy 0.5 and Tzz 1.0. They were equally weighted in combination for the releases after RL1. In RL4, RL5 and RL6 models, the off-diagonal tensor elements Txz were also included.

- In RL4 and RL5 models, the spectral bandwidth of the bandpass filter used to filter the SGG observations was extended by 1.7 MHz toward the lower frequency domain.

- Reference [16] notes that contrary to the TIM approach, the DIR approach employs the GOCE gravity gradient information as restricted in a certain spectral band, which is close to the gradiometer measurement bandwidth (5–100 MHz [45]). However, the filtering procedure in the TIM approach allows the use of information on the gradient observations over the entire spectrum.

The TIM approach exclusively relies on an epoch-wise processing of GOCE SGG and hl-SST data according to the least squares principle. In the method, the SST data analysis is based on the kinematic orbit solutions with the energy integral approach (short-arc integral approach for TIM-RL4, RL5 and RL6). Hence, since TIM solutions are independent of any other gravity field information, they can be used for independent comparisons and combination with other satellite-only models (such as GRACE), terrestrial gravity data and altimetry data on the normal equations' level. However, the TIM solutions are not competitive with the GRACE models in low degrees, since they are based solely on kinematic GOCE orbits but no external (such as GRACE) information:

- The TIM models were constrained using Kaula's rule regularization applied to (near-) zonal coefficients, improving the signal-to-noise ratio (constraints for high degree coefficients).

- The off-diagonal tensor element Vxz was included in the models TIM RL3, RL4, RL5 and RL6. The stochastic models for the gravity gradient observations rely on optimum weighting based on variance component estimation.

- TIM RL6e includes additional terrestrial gravity field observations over GRACE's polar gap areas (>83° S, N) [46].

The revision in 2012 of the processing procedure of the Level-1b products led to the expectation of performance improvement of the GOCE's SGG observations and consequently a quality improvement of the fourth, fifth and sixth releases of the models. Reference [47] reports that the gradiometry-only gravity field estimates show the largest improvement in recovery of low and medium degree coefficients, due to the new Level-1b products (see [16]).

In the SPW method, spatialized observations are produced from SST and SGG data using Wiener orbital filter [48] and the gravity field coefficients are derived by spherical harmonic analysis after transforming these observations into a spatial grid using least squares collocation [49]:

- The SPW RL1 model includes the first GOCE quick-look model as the a priori information and the EGM2008 was used for error calibration of the estimated gravitational potential along the orbit that affects the low degrees of the solution. However, the later releases, RL2 and RL4, were not corrected using any a priori model, thus they are the GOCE-only models. EIGEN5C and EIGEN6C3stat models were respectively used in SPW RL2 and RL4 models' computations for signal covariance modeling, in addition to FES2004 for ocean tide modeling.

- In all SPW models, the off-diagonal tensor elements Vxz were included.

The maximum d/o of spherical harmonic expansion also changes (see the 2nd column of Table 1). The maximum d/o of the models was determined by the processing teams of HPF, depending on the signal-to-noise ratio, processing issues and the strategy of using additional a priori information [18].

Table 1. Description of direct (GO_CONS_GCF_2_DIR), time-wise (_TIM), space-wise (_SPW) in addition to GOCO05S, GOGRA04S and EGM2008 models (data contents in m: month, y: year) [21,50].

Model	(Max. d/o)	Data	Difference w.r.t. Previous Version	Literature
DIR_RL1 (240)	GOCE (2 m), CHAMP (6 y) The model is more accurate than GRACE models for d/o 130-150 and up, less accurate for the lower degrees.	[18,39]
DIR_RL2 (240)	GOCE (8 m) SST ≤ d/o 130	[18,39]
DIR_RL3 (240)	GOCE (18m), GRACE (6.5y), SLR (6.5y) Combined GRACE as normal equations. and LAGEOS as normal equations., SGG components: equal relative weights 1.0.	[18,39]
DIR_RL4 (260)	GOCE (33 m), GRACE (9 y), SLR (>10 y) SGG: inclusion of Txz-component.8.3–125 MHz filter. GRACE: a blended combination of GRGS-RL02 ≤ d/o 54 and GFZ-RL05 from d/o 55 to 180.	[51]
DIR_RL5 (300)	GOCE (48 m), GRACE (>10 y), SLR (>10 y) The GRACE contribution up to d/o 130. Very low-degree harmonics of (esp. d/o 2 and 3) cannot be estimated accurately with GRACE and GOCE data, therefore LAGEOS-1 and -2 normal equations are used in the combination	[51]
DIR_RL6 (300)	GOCE (48 m), GRACE (>10 y), SLR (>10 y) Based on improved filtering of the reprocessed gradients of the full mission observations	[52]
TIM_RL1 (224)	GOCE (2 m) The model is independent of any other gravity field information. In the low degrees, it is not competitive with GRACE models. Kaula reg. is to improve SNR.	[40]
TIM_RL2 (250)	GOCE (8 m) No change in data processing strategy w.r.t. its predecessor.	[18]
TIM_RL3 (250)	GOCE (18 m) In addition to Vxx, Vyy, Vzz, the inclusion of Vxz component in the SGG observations.	[18]
TIM_RL4 (250)	GOCE (32 m) Applying the short-arc integral method to the kinematic orbits.	[18]
TIM_RL5 (280)	GOCE (48 m) SST ≤ d/o 150, SGG ≤ d/o 280, Kaula ≥ 201	[53]
TIM_RL6 (300)	GOCE (48 m) SGG ≤ d/o 300, digital decorrelation filter to observation equations	[54]
TIM_RL6e (300)	GOCE (48 m) Included terrestrial gravity field observations over GOCES's polar gap areas	[46]

Table 1. *Cont.*

Model | (Max. d/o)	Data | Difference w.r.t. Previous Version	Literature
SPW_RL1 (210)	GOCE (2 m)	[18,41]
	Applying an iterative procedure based on the Wiener orbital filter to reduce time correlated noise of the gradiometer. Including Txz in the gravity gradients.	
SPW_RL2 (240)	GOCE (8 m)	[55]
	No corrections to any a priori model. Using EIGEN5C for signal covariance modeling and FES2004 for ocean tide modeling.	
SPW_RL4 (280)	GOCE (32 m)	[56]
	Using EIGEN6C3stat for signal covariance modeling.	
SPW_RL5 (330)	GOCE (48 m)	[57]
	Using EIGEN6C4 and GOCO05S for signal covariance modeling.	
GOCO05S (280)	GOCE (48 m), GRACE (~10 y), CHAMP (6y), SLR (>10y)	[58]
	Combined satellite gravity field model, estimated with over 800,000,000 observations.	
GOGRA04S (230)	GOCE (48 m), GRACE (>10y)	[59]
	In the solution: the normal equations combine SST ≤ d/o 120, SGG ≤ d/o 230, GRACE ≤ d/o 180	
EGM2008 (2190)	GRACE, Gravity (terrestrial, airborne), Altimetry Data	[22]

2.3. Synthesis of the Gravity Field Parameters Using GGMs Spherical Harmonic Coefficients

The International Center for Global Earth Models [60] database has published 175 geopotential models so far and almost 30% of these models include GOCE mission data. In order to quantify the quality of the GGMs independently, validating the models using external information is necessary. So-called validations require the level-3 (L3) products from the satellite gravity missions to be compared with terrestrial data [24]. The GPS/leveling data is typically used for independent geoid undulations ($N_{GPS/lev.}$) at the benchmarks to compare with the GGM derived functionals (N_{GGM}) using the following equation:

$$N_{GGM} = \frac{GM}{r\gamma} \sum_{\ell=2}^{\ell_{max}} \left(\frac{a}{r}\right)^{\ell} \sum_{m=0}^{\ell} \left(\Delta \bar{C}_{\ell m} \cos m\lambda + \bar{S}_{\ell m} \sin m\lambda\right) \bar{P}_{\ell m}(\cos\theta) \tag{1}$$

where (θ, λ, r) are co-latitude, longitude and geocentric radius of the computation point, respectively; a is the major axis radius of the reference ellipsoid; GM is the product of the gravitational constant and the mass of the Earth; and γ is the mean gravity of the reference ellipsoid. $\Delta \bar{C}_{\ell m} = \bar{C}'_{\ell m} = \bar{C}_{\ell m} - \bar{C}_{\ell m}^{\_ell.}$, and $\bar{S}_{\ell m}$ are the fully normalized spherical harmonic coefficients, related to the normal potential of the reference ellipsoid from a specific degree of ℓ and order m ($m = 0, 1, \dots, \ell$). The coefficients of the ellipsoidal normal potential ($\bar{C}_{\ell m}^{\_ell.}$) contain only terms for order $m = 0$ (rotational symmetry) and degree $\ell =$ even (equatorial symmetry), and $\bar{S}'_{\ell m} = \bar{S}_{\ell m}$. To calculate the normal potential in practice in most cases, it is sufficient to consider only the coefficients $\bar{C}_{00}^{\_ell.}, \bar{C}_{20}^{\_ell.}, \bar{C}_{40}^{\_ell.}, \bar{C}_{60}^{\_ell.}$ and $\bar{C}_{80}^{\_ell.}$ [20]. In the equation, $\bar{P}_{\ell m}(\cos\theta)$ are the fully-normalized associated Legendre functions.

As can be seen, Equation (1) does not include the zero-degree geoid height term (N_0), which is represented by the zero-degree harmonic coefficient in the spherical harmonic expansion model and means the position of the GGM geoid with respect to the geo-center of the Earth. The zero-degree term is expressed as depending on the difference between the constant (GM) value of the GGM and that of the reference ellipsoid (GM_{GRS80}), which is GRS-80 for Turkey (see Equation (2)):

$$N_0 = \frac{GM - GM_{GRS80}}{r\,\gamma} - \frac{W_0 - U_0}{\gamma} \qquad (2)$$

where U_0 corresponds to the normal gravity field of the GRS-80 reference ellipsoid with parameters of $GM_{GRS80} = 398,600.50 \times 10^9 \text{ m}^3/\text{s}^2$; thus, it is $U_0 = 62,636,860.85 \text{ m}^2/\text{s}^2$ [61]. Using $r = 6378.137$ km and $\gamma = 981$ Gal, the zero-degree geoid height (considering solely the first component at the right-hand side of Equation (2)) corresponds to the $N_0 = -0.936$ m that is to be added to the N_{GGM}, derived from Equation (1). The residual zonal-coefficients (shown as $\Delta \bar{C}_{\ell m}$ in Equation (1)) account for the difference between the reference radius of the GGM and the semi-major axis of the GRS-80 ellipsoid [62,63]. In the synthesis of GGMs' spherical harmonic coefficients, the zero-degree term is considered either by taking the zero-degree zonal-harmonic coefficients (as $\bar{C}_{00} = 1.0$, $\bar{S}_{00}$ does not exist, $\bar{C}_{10} = 0.0$, $\bar{S}_{1m} = 0.0$) into account via initiating the computation from the $\ell = 0$ or otherwise separately adding the zero-degree geoid offset, calculated with Equation (2). The second component at the right-hand side of Equation (2), $(-\frac{W_0 - U_0}{\gamma})$, introduces the difference between the gravity potential of the geoid (W_0) and the normal gravity potential on the surface of the normal ellipsoid (U_0). In the ideal case, without irregularities of topographic densities, the two potentials are equal ($W_0 = U_0$). As a primary parameter for the definition of the reference (zero-height) surface with respect to Earth's body (which provides an absolute definition to the vertical datum of a height system), a global estimation of W_0 is required [64].

In addition to the zero-degree term definition, another issue in the synthesis of the gravity field parameters from the GGMs' spherical harmonic coefficients is adopting the proper permanent tide-system to which the GGMs' functionals refer. The permanent tide-systems, commonly adopted in studies investigating Earth dynamics include mean-tide, zero-tide and tide-free [65]. In studies related to the Earth's potential field, investigations of the Earth's crust deformations or determination of the best-fitting ellipsoid to the geoid, a proper permanent tide system is adopted. According to Bruns' definition of geoid undulation [61], the geoid undulation changes from one tide system to another depending on the geometric change in the shape of the geoid (but with a constant $W = W_0$ potential for each tide-system) relative to a fixed reference ellipsoid [65]. In other words, the W_0 value does not change among the systems, but the shape of the geoid changes. The effect of changing the permanent tide-system is defined basically with the second-degree zonal harmonic coefficient $\bar{C}_{20}$ [65].

In this study, the gravity field parameters were synthesized using the GGM coefficients in a tide-free system. In order to have comparable geoid undulations at the GNSS/leveling benchmarks, the orthometric heights in regional vertical datum were converted from the mean-tide to tide-free system according to following the equation [66]:

$$H^{tide\ free} = H^{mean\ tide} - 0.68\left(0.099 - 0.296\ sin^2\varphi\right) \qquad (3)$$

2.4. Assessments of the Global Geopotential Models Using Spectral Enhancement Method (SEM)

The spectral constituents of Earth's gravity field are denoted with the spectral bands of a GGM [15]. The gravity field quantities as observed on the Earth's surface contain the full spectral signal power, while any global geopotential model is limited by its spectral resolution, which leads to omission error in the output functional. When comparing GGMs with independent observational data, one should take into account spectral incompatibility of both data sets. To overcome the spectral band limitation of the mode, the so-called spectral enhancement method (SEM) is suggested and applied by [18]. The fundamental principle of the SEM is based on bridging the spectral gap between the GOCE-based GGM functional and ground-truth data, using the relevant auxiliary data set (such as a higher resolution geopotential model like EGM2008, and the residual terrain data, to account for the ultra-high frequencies) as much as possible. Hence, the ground-truth data for validation purposes are approximated much better, and thus the influences of omission error on the comparison results are consequently minimized. In spectral enhancement, the GGM, expanded to a spherical harmonic degree ℓ_1 (e.g., $\ell_1 = 240$ for DIR_RL1), is enhanced using the spectral bands over the degrees of $(\ell_1 + 1)$ until ℓ_2 from a higher resolution Earth geopotential model (e.g., $\ell_2 = 2190$ when using EGM2008). The very short-wavelength components of the gravity field, represented with spectral bands beyond degree ℓ_2 (e.g., $\ell_2 = 2190$ in our case) until the maximum possible degree (e.g., 216,000), are completed as the residual terrain effect (RTE) of topographic data on the estimated geoid heights (see Equation (4)).

$$N_{enh.} = N_o + \left[N\left|\begin{matrix}\ell_1\\2\end{matrix}\right. + N\left|\begin{matrix}\ell_2\\\ell_1+1\end{matrix}\right. \right] + N_{RTE}\left|\begin{matrix}216000\\\ell_2+1\end{matrix}\right. \tag{4}$$

Equation (4) is a symbolic representation of the enhancement procedure in terms of geoid undulation derived using the GGM, where N_o is zero-degree geoid undulation; $\left[N\left|\begin{matrix}\ell_1\\2\end{matrix}\right. + N\left|\begin{matrix}\ell_2\\\ell_1+1\end{matrix}\right. \right]$ is the geoid undulation calculated with enhanced coefficients according to Equation (1). Here, N_{RTE} is the residual terrain effect on geoid undulation, calculated using the digital topographic data [27]. As the summation of the components that represent each different spectral constituent of the Earth's gravity field, the full spectral content geoid undulation ($N_{enh.}$), which is comparable with ground truth data, is obtained.

In Equation (4), N_{RTE} was calculated using the residual terrain model (RTM) mass reduction method evaluated in the Gravsoft program package, following the processing methodology outlined by [27], *Section 8.4.4 Terrain Effects and High Resolution Global Geopotential Models* in pp. 366–371. In computations, 3 arc-second Shuttle Radar Topography Mission (SRTM) topographic data was used for approximating the real terrain and the spherical harmonic expansion of the DTM2006.0 (up to d/o 2190) was used to model the mean elevation surface [27].

2.5. Spectral Analysis of GGMs

In the assessment of the quality of the GGM solutions, two types of errors are defined. Firstly, the noise in the observations that are used in the calculation of the GGM coefficients propagates from the measurements to the solutions, which is called the commission error. Secondly, an error is revealed because of estimating the projection of the gravity field functional (taking the anomalous potential as an example) on the finite dimensional subspace (generated by linear combinations of solid spherical harmonics up to a limited degree) instead of deriving its full content. This error is known as the omission error. The omission error has an easy formulation that is set by considering the coefficients left out from the truncated convergent series. In fact, this is the sum of the squares of all coefficients of degree higher than ℓ_{max}. Clearly, however, this is an unknown quantity, which cannot be known a priori. However, the so-called degree variances (i.e., the sum over all orders of the squares of the coefficients up to a certain degree) can provide an idea of the GGMs' signal decay, which allows one to judge the commission error and compute the omission error according to an adopted law [27]. Accordingly, the signal and error degree variances of the GGMs provide an insight into the omission/commission error content of a GGM under consideration (see [24]). Hence, they are regarded as internal error estimates for the investigated models.

Using the fully normalized spherical harmonic coefficients, $\bar{C}'_{\ell m}$, $\bar{S}_{\ell m}$ from a specific degree of ℓ_{max} and order m ($m = 0, 1, \ldots \ell_{max}$), the signal degree amplitudes (or square root of power per degree ℓ) of functions of the disturbing potential $T(\varphi, \lambda, R)$ at the Earth's surface are computed as follows [20,21,27]:

$$\sigma_\ell = \sqrt{\sum_{m=0}^{\ell}\left(\bar{C}'^2_{\ell m} + \bar{S}^2_{\ell m}\right)} \tag{5}$$

Equation (5) provides the signal degree amplitude σ_ℓ in terms of unitless coefficients, and the $\sigma_\ell(N)$ in terms of geoid heights (meter) is shown in Equation (6):

$$\sigma_\ell(N) = R\,\sigma_\ell \tag{6}$$

The error spectra of GGMs are investigated depending on the error degree amplitudes, calculated using the estimation errors of the Stokes' coefficients ($\sigma_{\bar{C}_{\ell m}}$, $\sigma_{\bar{S}_{\ell m}}$) of a spherical expansion model:

$$\sigma_{c_\ell}(N) = R\,\sqrt{\sum_{m=0}^{\ell}\left(\sigma_{\bar{C}_{\ell m}}^2 + \sigma_{\bar{S}_{\ell m}}^2\right)} \tag{7}$$

As a function of minimum and maximum degrees ℓ, the accumulated (signal, error) degree amplitudes provide the power spectrum accumulated over a spectral band (between ℓ_1—usually $\ell_1 = 0$ or 2 as minimum degree of the expansion—and ℓ_2—usually $\ell_2 = \ell_{max}$ as the maximum degree of the harmonic expansion equation) and shows the increase in overall power with increasing degree of ℓ:

$$\sigma_{\ell_1\ell_2}(accumulated) = \sqrt{\sum_{\ell=\ell_1}^{\ell_2}\sigma_\ell^2} \tag{8}$$

2.6. GGMs' Contribution in High-resolution Regional Geoid Modeling

For the determination of high-resolution precise regional gravity field models, remove–compute–restore (RCR) is a very well-known approach and has been used in calculation of many countries' geoid models to date [27,61]. In the RCR scheme the terrestrial gravity and topography/bathymetry data are used along with a global geopotential model to smooth the observations for data gridding, transformations, predictions and to eliminate the aliasing effects [27]. The computation steps of the RCR procedure are as follows:

$$\Delta g_{res} = \Delta g_{FA} - \Delta g_{GGM} - \Delta g_{tc} \tag{9}$$

Equation (9) represents the 'remove' step where the long-wavelength component (Δg_{GGM}, from the GGM according to Equation (10)) of the gravity field, as well as the short-wavelength components (Δg_{tc}, the direct topographical effect on gravity, which consists of the subtraction of the attraction of condensed topographic masses from the attraction of all topographic masses above the geoid and is calculated as according to a selected reduction scheme—i.e., Helmert's Second Method of Condensation was used in this study (see *Equations (3–97) of Section 3.9* in [61])—using topographic data), are removed from the free-air gravity anomalies (Δg_{FA}). Hence the calculated and reduced residual gravity anomalies (Δg_{res}) on the geoid surface are evaluated in Stokes' formula.

$$\Delta g_{GGM} = \frac{GM}{r^2}\sum_{\ell=2}^{\ell_{max}}(\ell-1)\sum_{m=0}^{\ell}\left(\Delta\bar{C}_{\ell m}cosm\lambda + \bar{S}_{\ell m}sinm\lambda\right)\bar{P}_{\ell m}(\,cos\theta) \tag{10}$$

where the terms in Equation (10) are as introduced previously in Equation (1).

The 'compute' step includes the computation of the residual co-geoid using Stokes' integral ($N_{\Delta g}$), the long-wavelength geoid undulation (N_{GGM}) using the GGM according to Equation (1) and finally the indirect effect of topographic mass attraction on geoid undulation (N_{ind} using topographic data according to Equation (13)). After computing all constituents separately, their compilation step is called 'restore' and formulated as follows:

$$N = N_{GGM} + N_{\Delta g} + N_{ind} \tag{11}$$

as the high-resolution geoid undulation (N).

The Stokes' formula for calculating $N_{\Delta g}$ is:

$$N_{\Delta g} = \frac{R}{4\pi\gamma} \iint_\sigma S(\psi)\Delta g_{res}d\sigma \tag{12}$$

where σ is the surface element, R is the mean Earth radius and $S(\psi)$ is Stokes' function having the angular distance ψ from the computation point to the differential surface element $d\sigma$. N_{ind} is:

$$N_{ind} = \frac{\pi \, G\rho H_Q^2}{\gamma} - \frac{G\rho R^2}{6\gamma} \iint_\sigma \frac{\left(H_Q^3 - H_P^3\right)}{s^3}d\sigma \tag{13}$$

where ρ is the density of topographic masses; H_Q and H_P are the heights of the running and computation points, respectively; and s is the planar distance between P and Q. Hence, the output of the equation is N_{ind} according to Helmert's Second Method of Condensation reduction scheme [27]. The details of the computation steps followed in this study were explained in [28] pp. 433–435 and Figure 13 given in the cited reference illustrated the adopted computational schema in here.

In the computation of a regional geoid model with hybrid (mixed) data in RCR, the GGM is responsible for the long-wavelength (signal) error content (see Equations (10) and (11)). When calculating a geoid model over a spatially limited region, while having low-accuracy and sparse terrestrial data, the quality of the GGM becomes even more important to compensate for the weakness of the terrestrial data accordingly in the RCR algorithm with specific modifications [30,67].

3. Results and Discussion

3.1. Internal Error Estimates of Tested GGMs

The error degree variances of the tested models were considered for internal uncertainty estimates of the GGMs. The theory and formulations for calculating error degree variances were given in Section 2.5. (see Equations (7) and (8)). Figure 1 shows the degree-wise accumulated error amplitudes of the models. The figures reveal the overall geoid undulation accuracies up to the maximum degree of each model. For d/o 200, which corresponds to the spatial resolution of 100 km, the cumulative errors are 0.8 cm for the DIR RL5 model (see Figure 1a), 2.2 cm for the TIM RL5 model (Figure 1b), 2.7 cm for the SPW RL4 model (Figure 1c), 2.0 cm for the GOCO05S model and 2.9 cm for the GOGRA04S model (Figure 1d), whereas it is 7.1 cm for the EGM2008 model, in terms of geoid undulation.

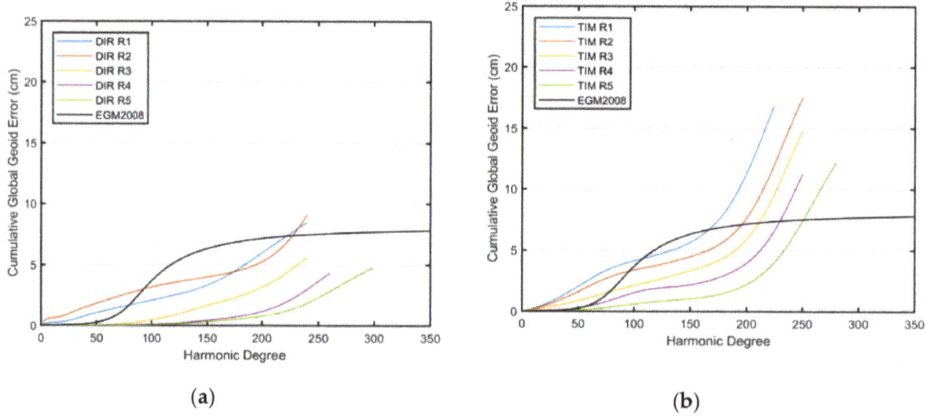

(a)

(b)

Figure 1. *Cont.*

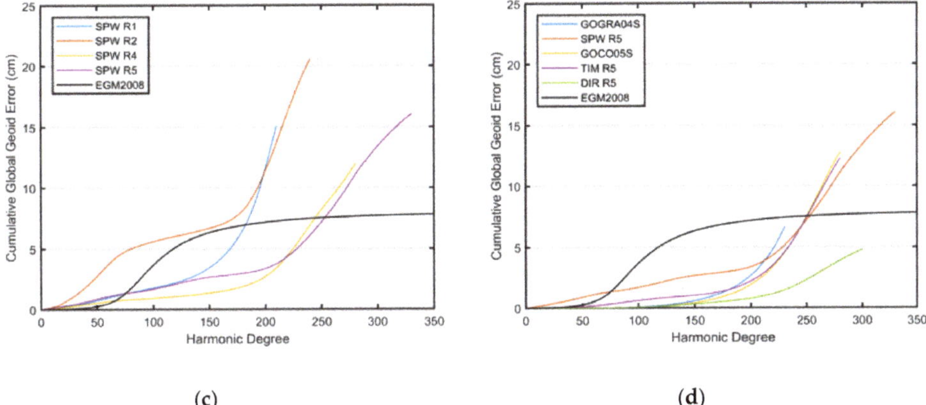

Figure 1. Error amplitudes (accumulated) as a function of the spherical harmonic degrees "ℓ" in terms of geoid heights: (**a**) direct (DIR RLx) models; (**b**) time-wise (TIM RLx) models; (**c**) space-wise (SPW RLx) models; (**d**) GOGRA04S, GOCO05S and EGM2008 models.

When the cumulative error amplitudes are investigated, it is seen that the improvement with the final releases of the GGMs is up to 80% compared to their predecessors. It is also noteworthy that the cumulative error estimates of TIM RL5 and DIR RL5 differ by a factor of three. However, it is arguable if these cumulative error estimates depending on degree variances are realistic and, on the other hand, whether calibration is required or not. In this manner, in a similar study on GGM validations, [51] suggests applying a calibration factor of two for the DIR RL5 model.

Related to concerns about the uncertainty estimates of the GGMs depending on the error degree variances approach, [68] emphasizes these issues: (1) firstly, the correlations among the coefficients are ignored while evaluating the GGMs with degree variances; (2) the effect of the polar-gap problem on the zonal and near-zonal coefficients of the GOCE-only models obviously influences the provided standard deviations of the spherical harmonic coefficients, specifically for the GOCE-only TIM and SPW models; and (3), on the other hand, dependence of the GOCE models' uncertainties and performances on the geographic location cannot be ignored in evaluations. Regarding the mentioned handicaps, beside the degree variances-based assessments, evaluating the GGMs using geographically dependent standard deviations (if possible) with the propagation of full variance–covariance matrices of the spherical harmonic coefficients is suggested by [68].

In summary, although the accumulated error degree variances provide a global mean of the GGM's performance over the entire sphere, which is significantly influenced by the large uncertainties in the polar gap, deriving concluding measures for its performance in a local region is not straightforward. Therefore, Section 3.2 provides results of local validations of tested GGMs using terrestrial data.

3.2. Overview of the Regional Accuracies of the GGMs

The results of evaluating the gravity field models using various data sets in different regions are being published regularly as the new models are released. However, the methodologies applied in these assessments vary to differing degrees. Regarding a number of significant studies from the literature, the following can be noted.

Reference [19] provides a detailed assessment of the first generation GOCE-based models using LEO satellite orbit perturbations and GPS/leveling geoid height differences. The conclusions of their study confirm that the orbital perturbations are mostly sensitive to the long-wavelengths of the gravity signal and hence provide clear results in quality assessments of the GGMs in a global manner. In contrast, the geoid comparisons using GPS/leveling data give the opportunity to assess the model performances at the medium to short-wavelengths. In the study, the regional assessments using the

GPS/leveling data were carried out in Australia, Europe, North America (US and Canada), Japan and Germany. The preliminary conclusions from this study give a global geoid accuracy of about 5–6 cm with a 111 km spatial resolution (at d/o of 180). Here, it is emphasized that they obtain this level of accuracy as a result of just two-month GOCE data, and the goal of the mission (1–2 cm geoid accuracy with 100 km resolution) could be reached with the availability of more data. In terms of geoid residual RMS values, [19] reports the EGM2008 model accuracies as the most improved in Germany (3.5 cm), Japan (10 cm) and Canada (10.5 cm), whereas weaker fitting performances were found in Australia (24 cm), Europe (21.5 cm) and the US (26.5 cm) with comparatively larger residual RMS values. The non-homogeneous control datasets and vertical datum distortions were cited as possible reasons for the larger error levels in the continent-wide assessments.

Reference [21] provides the validation statistics of the released model for the maximum truncation degrees (without spectral enhancement) using similar datasets. Reference [15] evaluated the first generation GOCE-based models globally using EGM2008 and regionally using terrestrial gravity data and astrogeodetic vertical deflections. The analyses by [15] clearly show the improvement in gravity field mapping with GOCE data contribution and identify the reasons for the reported improvement specifically at certain spectral bands. Reference [69] evaluated the first and second releases of the models by means of gravity anomalies and radial gradients from GOCE gradiometer data using least squares collocation at several regions. The evaluations of the models from the first to third releases were published by [70] using the terrestrial gravity data in Norway and [71] over the Earth by means of gravity gradients from the GOCE gradiometer data. There are also other studies that validate the GOCE-based GGMs over different regions using various terrestrial data sets including [72] (in Sudan), [73] (in Brazil), [74] (in Hungary) and [68] (in Germany). Reference [75] provides the recent results of regional validations of all releases over different countries. However, the literature lacks information by means of GGM validation statistics over many areas of Earth. This may stem from the countries' data restriction policies and/or lack of quality terrestrial data for the region. Accordingly, the numerical section of this study focused on regional evaluation of the GGMs in Turkey.

In the following sections, in addition to reporting the models' formal error estimates, the performances of the GOCE-based GGMs were evaluated in terms of absolute accuracies. The GOCE-based models were compared with EGM2008 in order to clarify the GOCE observations' contribution to the GPS/leveling benchmarks in Turkey. In addition, the spectral bands of the spherical harmonic expansion models, where the improvement of tested GGM is the most significant, were determined and employed in the regional geoid modeling.

3.2.1. Assessment of GGMs' Accuracies in Turkey

In accuracy assessments of the HPF-released GGMs in Turkey, a number of numerical tests were carried out in the country at 36°N–42°N latitude and 26°E–45°E longitudes. The test statistics of the geoid height residuals (ΔN) of the observations and the models were calculated for two different high order GPS/leveling network benchmarks, separately. The GPS/leveling networks include 30 and 81 points, respectively. The first dataset covers the whole of Turkey (Dataset I), and the latter is denser but localized in the northwest of the country (Dataset II) [76]. Carrying out the tests individually using both datasets (instead of using a unified dataset) aims to make validation results independent from the effects of possible datum differences between the datasets. A detailed description of the employed data is provided in Table 2. The distributions of the benchmarks on the topographic maps are given in Figures 2 and 3.

Table 2. GPS/leveling datasets description.

Dataset	I	II
Size of area	$6° \times 19°$	$3° \times 4°$
Number of BMs	30	81
BM density	1 BM/200 km	1 BM/45 km
The distribution of BMs	Homogenous and quite sparse	Homogenous and sparse
[1] Coordinate datum	ITRF96	ITRF96
2D coordinate accuracy	±1.0 cm	±1.0 cm
H-accuracy	±2.0 cm	±2.0 cm
[1] Vertical datum	TUDKA99	TUDKA99
H-accuracy	±2.5 cm	±2.0 cm
Topography	0–5000 m	0–2500 m
Reference	[76]	[76]

[1] The datasets include first order GPS network points of Turkey National Fundamental GPS Network with Helmert orthometric heights (third-order leveling network) in regional vertical datum. BM: benchmark.

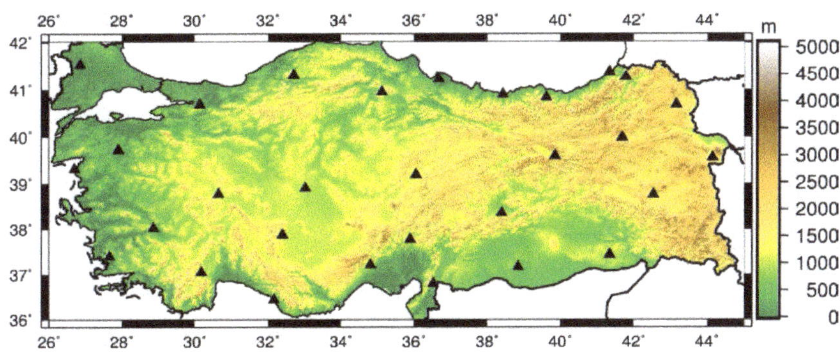

Figure 2. Distribution of 30 benchmarks (Dataset I) on Shuttle Radar Topography Mission (SRTM) topography.

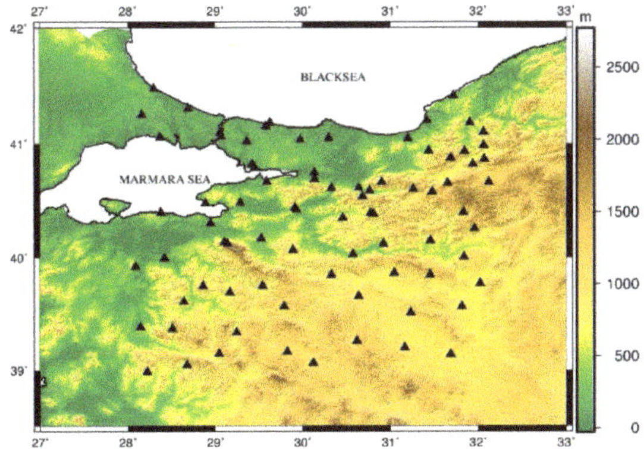

Figure 3. Distribution of 81 benchmarks, northwest of Turkey (Dataset II) on SRTM topography.

The numerical results of the study were interpreted from different perspectives, including: (1) verifying the role of the data amount and the calculation strategy of the model on its accuracy and performance; (2) determining an optimum d/o of the models available in the study area (whereas the term of the "optimum" possibly sounds vague and its definition is dependent on the application, the "optimum d/o" was used here to mean the degree of the spherical harmonic expansion at which the highest accuracy is obtained in the tests performed with the terrestrial data in the study area); and, (3) clarifying the contribution of a best-fitting geopotential model with its optimum d/o to the development of a precise regional geoid model using RCR approach.

In the following, Figures 4a and 5a show the evaluation results of the DIR RL1, RL2, RL3, RL4 and RL5 models, which are enhanced using the high-resolution EGM2008 model and high-resolution DTM data (SRTM3), at GPS/leveling networks with 30 benchmarks throughout Turkey and 81 benchmarks in the northwest of the country, separately. The graphics show the standard deviations of the geoid height residuals, calculated at the co-located GPS/leveling benchmarks with the GGM-derived geoid undulations, which were truncated and enhanced degree by degree from two to the maximum harmonic expansion d/o of the tested models (see Equations (1) and (4)). Looking at the statistics of the two datasets in the graphics, the improvement of the model performances from the first release to the last can be verified.

Similarly, Figures 4b and 5b show the performances of the TIM models, depending on the standard deviations of the geoid undulation residuals at co-located GPS/leveling benchmarks. Considering the results in graphics, the improvement of the TIM models as the release number increases is also clear.

The SPW models were also compared for the given two geodetic network datasets in Figures 4c and 5c. The statistics of SPW model validations also show the progress with increasing release number.

The numerical tests, which were carried out fulfilling the first objective of the current section, show improvement in the models with the increasing amount of data (by means of the observation cycle; please see Table 1). This conclusion has been drawn independently from the GGM computation strategy. Carefully looking at the graphics given in Figures 4 and 5, the direct models reveal almost similar performances to each other and with the EGM2008 model, of up to approximately 70 d/o of the spherical harmonic expansions; however, this is not the case for the TIM and SPW models. This situation can be explained as a consequence of the differences among their computation strategies (compare the strategies and the a priori information in computations in Table 1). Since EIGEN-51C combined GGM [39] was employed for the regularization of the GOCE-based models in the direct approach, they were reinforced by this background model. In the long-wavelengths, the used background model utilizes the information from GRACE and CHAMP, and the medium to short-wavelengths are provided by DNSC08GRA [77] and thus (indirectly) the terrestrial gravity (as contained by EGM2008) over the land areas. Regarding the information on the determination of the direct models, it is not possible to strictly quantify which spectral bands provide improvement, and the extent to which improvement is provided by the GOCE observations when considering solely these models.

The processing strategy for computation of the time-wise models relies exclusively on GOCE gradiometry and hl-SST observations, and only Kaula's regularization is employed to constrain the model at the short-wavelengths. Hence, the assessments of time-wise models can provide a realistic measure of the GOCE-based contribution to the GGMs without the intervention of other data sets. When the evaluation results of these models are considered, the contribution of the GOCE signal to the improvement of gravity field mapping in Turkey is clarified (see Figures 4b and 5b). The maximum improvement is seen at the spectral bands between d/o 100 and 250 as is promised by the GOCE Earth gravity field satellite mission.

The improvement in gravity field mapping with a GOCE-based contribution is shown to be significant in Figure 5. However, it is not highly significant in the test results obtained with 30 benchmarks, distributed over the country (see Figure 4). This may be because of the insufficient density and distribution of the test benchmarks (as seen in Figure 1) and relatively lower accuracy of the GPS/leveling heights at these benchmarks. Hence, it may be concluded that the first dataset, with 30 benchmarks, is not sufficient by means of data quality and distribution for carrying out a detailed assessment of the geopotential models regarding different spectral bands. However, the evaluation results using this data set were nonetheless considered as a country-wide look at the accuracy level of the tested GGMs. In this manner, the test results that were obtained through the locally distributed 81 benchmarks (Dataset II) revealed the significant improvement of the GOCE-based models in the medium-wavelength spectral bands.

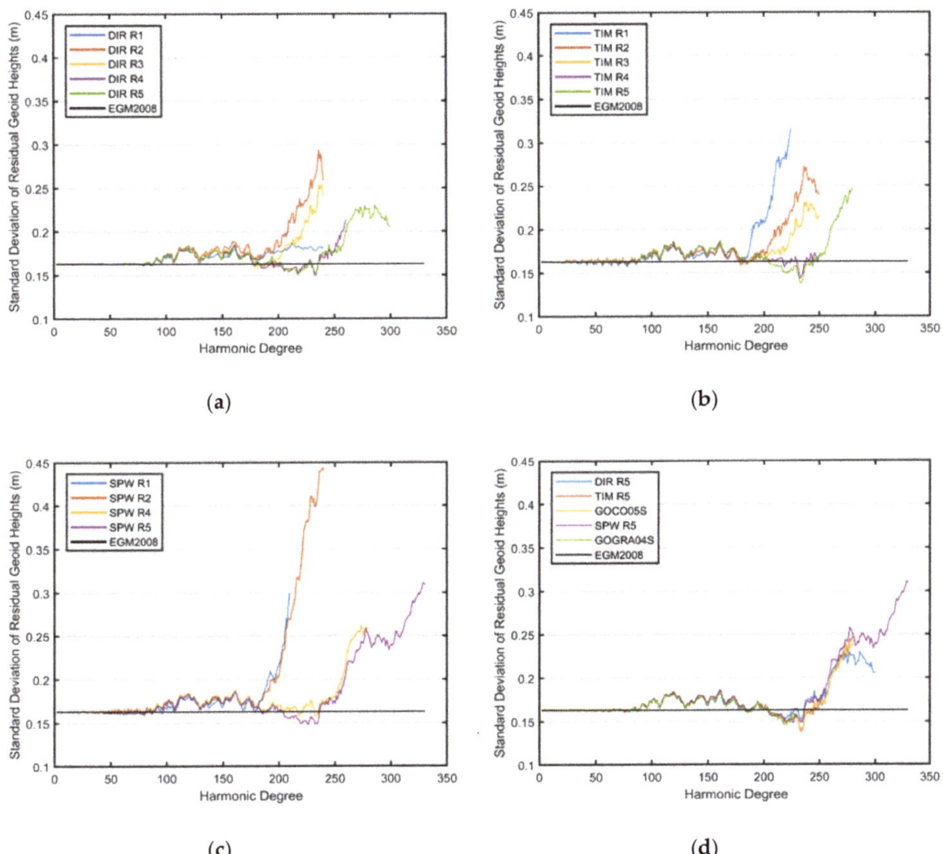

Figure 4. Validation of global geopotential models (GGMs) regarding the standard deviations of the geoid undulation residuals ($N_{GPS/lev.} - N_{GGM}^{enh.}$) in meters (for Dataset I): (**a**) DIR RLx models; (**b**) TIM RLx models; (**c**) SPW RLx models; (**d**) comparison among the last versions of the satellite-based and combined models.

Considering the entire statistics from the evaluations of direct, time-wise and space-wise models (namely, GO_CONS_GCF_2_ DIR, TIM, SPW) for the two datasets, it is concluded that the computation strategy of the GGMs does not have a significant role in the improvement of the model accuracies. In this part of the study, in addition to the GOCE-based DIR, TIM and SPW models (actually computed with the contribution of the GOCE and GRACE missions' data), the final releases of the two other combined GGMs, namely, the GOGRA04S and GOCO05S models, were also evaluated. The GOGRA04S model combined approximately four years of GOCE gradiometry data with GRACE observations from a cycle of satellite tracking in excess of 10 years. Despite the complementary advantage of the two missions in the solution, the GOGRA04S model did not reveal significant improvement in mapping the regional gravity field compared to the last releases of the GOCE models (see graphics in Figure 4d). The GOCO05S model combined four years of GOCE gradiometry data with GRACE, CHAMP and SLR observations (with the GRACE normal equations from the ITSG-GRACE2014S model). In the assessment results, the performance of the model is better than that of GOGRA04S and almost equal to the final releases of the GOCE-only models up to d/o of 155 and, in relative assessment, it performs better than the other models for degrees between 230 and 270 (Figures 4d and 5d).

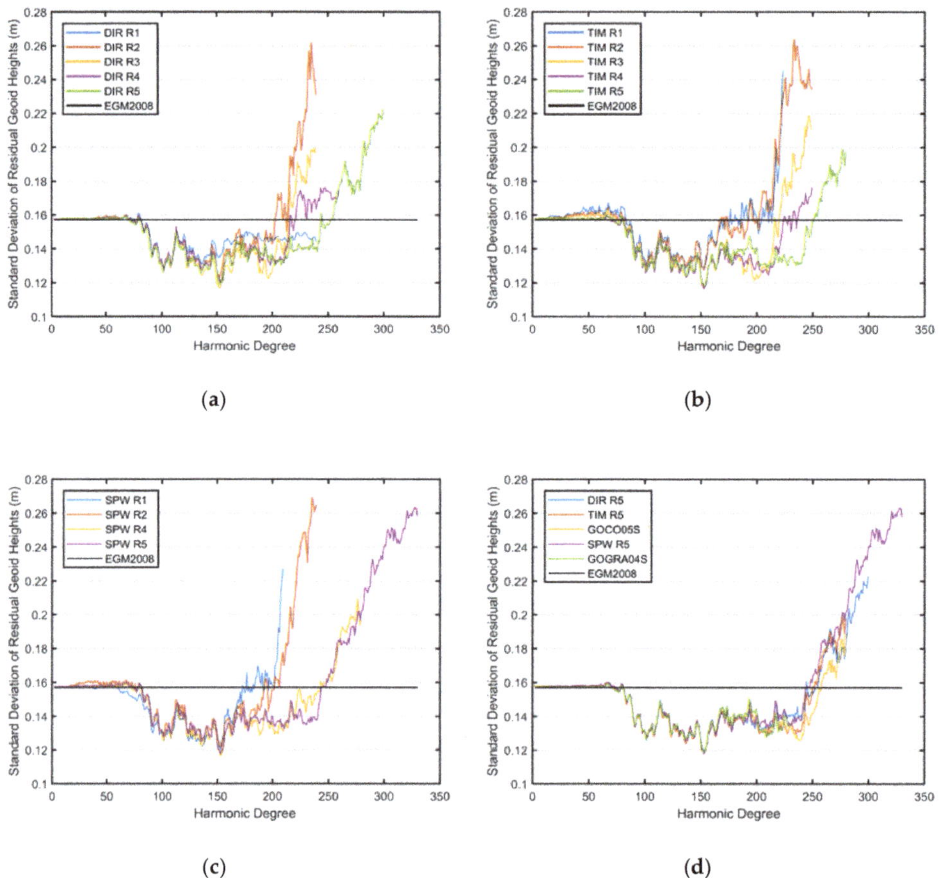

Figure 5. Validation of GGMs regarding the standard deviations of the geoid undulation residuals ($N_{GPS/lev.} - N_{GGM}^{enh.}$) in meters (Dataset II): (**a**) DIR RLx models; (**b**) TIM RLx models; (**c**) SPW RLx models; (**d**) comparison among the last versions of the satellite-based and combined models.

Accordingly, the accuracies of the GGMs in terms of standard deviations of the geoid height residuals, obtained through their maximum and optimum degrees (optimum degree corresponds to spherical harmonic expansion degree of the model with the highest accuracy with a minimum standard deviation of the geoid undulation differences; see Figure 5), are shown in Figure 6a,b, separately computed for each test area, respectively. Given statistics in figures are derived using the enhanced coefficients of the GGMs over optimum (blue bars) and maximum (orange bars) expansion degrees of the models using the EGM2008 model and high-resolution DTM. The accuracies that are obtained through employing the full-expansion of the EGM2008 model are shown with a black color bar at the end of each graphic.

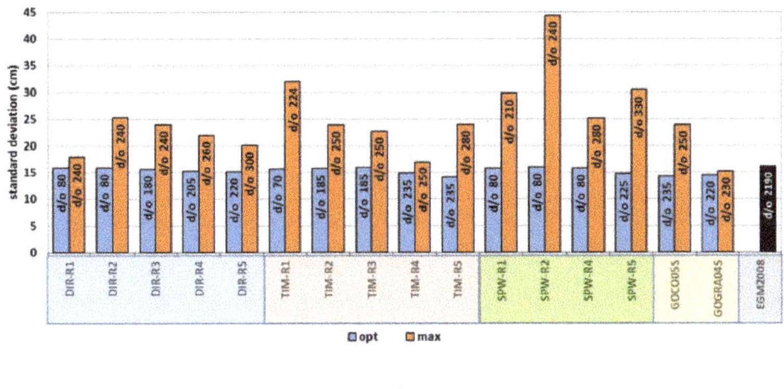

(**a**)

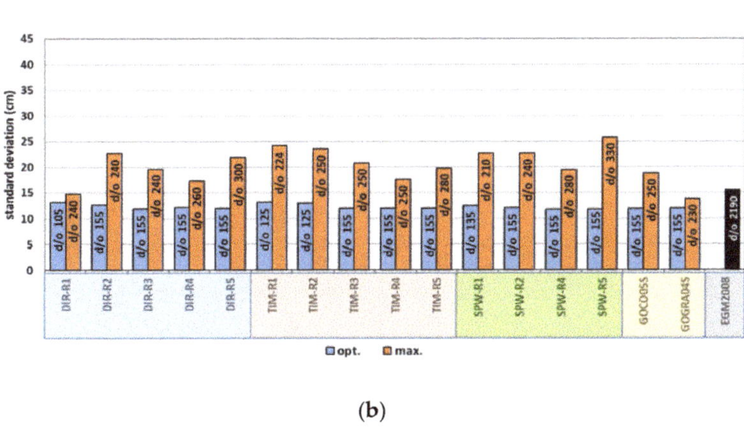

(**b**)

Figure 6. Standard deviations of the geoid height residuals in cm (note that the geoid heights ($N_{GGM}^{enh.}$), derived from the enhanced GGMs over optimum (blue bar) and maximum (orange bar) degrees for (**a**) the first dataset; (**b**) the second dataset.

The given statistics show that the GGMs, enhanced on their optimum d/o, revealed an almost 50% better fit comparing the results obtained from the enhanced models on their maximum degrees. Comparing with the EGM2008 model, the maximum improvement was recorded by the SPW RL5 model (a 23.8% improvement with respect to the EGM2008 model) in the test results using the second data set. Regarding the overall test results, obtained using both datasets, separately, it was seen that the GOCE-based GGMs had accuracies between 10 cm and 15 cm in Turkey. While interpreting the reasons for obtaining different fitting statistics of the GGMs in the tests with two datasets, it is worth noting the possible role of the tectonic structure in the region. Turkey is in the collision zone between three plates (African, Arabian and Eurasian) (see [78] for the active faults in Turkey) and hence continuously exposes horizontal and vertical crustal movements (see [79,80] for vertical and horizontal velocities in Turkey). Accordingly, the approximate rates in horizontal and vertical positions are reported as 2–3 cm/year in [80]. Moreover, an earthquake of $M_w \geq 6$ occurs every 1–2 years, particularly around the North Anatolian Fault Zone. All these activities continuously distort the geodetic networks by time and even have a minor effect on the change of the gravity field; however, the latter is not significant for the scope of this study. Hence, the first geodetic network (Dataset I) was exposed to the Izmit earthquake of 1999 (7.6 Mw) and the northwest part of this geodetic network, including the 81 benchmarks of Dataset II, have been remeasured and adjusted since then. Although the networks were positioned at the same geodetic datum, the physical establishments of the network stations were distorted by time and these distortions negatively influence their quality as control data in model evaluations. As a conclusion, the ground-truth data employed to validate the qualities and fitting performances of the GGMs should be of high quality, sufficiently dense, homogeneously distributed over the test area and as up-to-date as possible, especially in areas under the influence of dynamic Earth activities, such as Turkey.

In addition, the recently published sixth releases of the direct and time-wise models were considered in evaluations and they were compared with the RL5 models by means of standard deviations of the geoid undulations at the second dataset benchmarks. Figure 7 shows the comparisons of the models. According to obtained statistics, the RL6 models shown similar performances with the RL5 models between the 100 and 200 d/o of the spectrum. However, in degrees between 200 and 280, TIM RL6 models shown slightly better performance comparing with the RL5 and DIR RL6 models and had an improvement in standard deviation less than 1 cm.

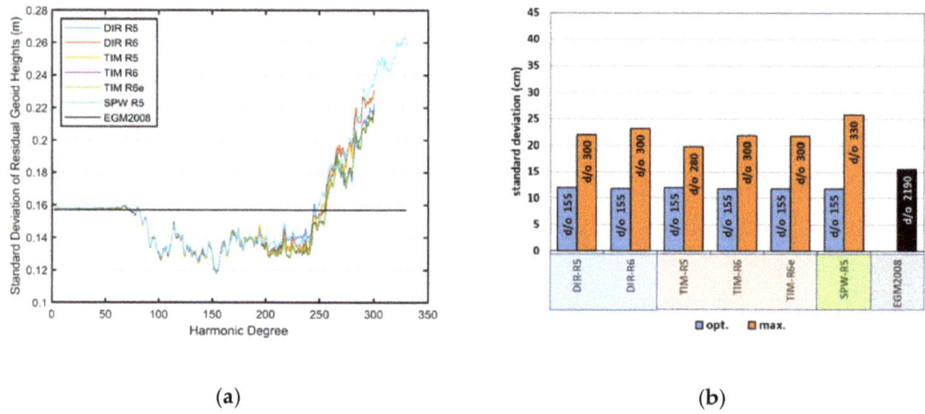

<table>
<tr><td align="center">(a)</td><td align="center">(b)</td></tr>
</table>

Figure 7. Comparison of RL5 and RL6 models: (**a**) the standard deviations of the geoid undulation residuals ($N_{GPS/lev.} - N_{GGM}^{enh.}$) in meters (Dataset II), (**b**) the standard deviations of geoid undulations in cm, and please note that the geoid heights ($N_{GGM}^{enh.}$), derived from the enhanced GGMs over the optimum (blue bar) and the maximum (orange bar) degrees of the models.

3.2.2. Testing the GGMs Contribution in Detailed Geoid Modeling

A high-resolution geoid model has been an essential component of the modern geodetic infrastructure in many countries, particularly in recent decades (see the geoid models' repository at [81] for some of the released regional geoid models). As a consequence of the progress in global positioning systems, many countries, such as Canada, New Zealand and the US, redefined their vertical datum to refer to a precise geoid model utilizing GNSS techniques. Hence, they also aimed to obtain further benefit from the satellite-based positioning technology for obtaining precise physical point heights in real time in many geodetic applications. Equations (11)–(13) provide the fundamental theory in the determination of a high-resolution geoid model using RCR steps. As was seen, the method combines GOCE-based GGM data with terrestrial data. Thus, the error content of the geoid undulation (Equation (11)) includes the errors from the global geopotential model, as well as the terrestrial gravity data and digital height data. Therefore, in order to clarify how the GGM accuracy affects the determination of the high-resolution geoid model, the numerical results from regional geoid modeling studies in Turkey are shared and discussed in this article. The terrestrial gravity data used in the RCR computations of this study are in five arc-minute resolution gravity anomalies in the IGSN-71 datum. Reference [82] reports that the estimated accuracy of the used gravity anomaly grid is at the level of 7–8 mGal based on cross-validations, as well as validations using point-wise gravity observations.

Using the terrestrial gravity data and the identical computation parameters, twelve experimental detailed geoid models (expTG-) were calculated for Turkey. Each of these models relies on a different GGM with either the optimum ($\ell_{opt} = 155$) (called expTG-Opt models) or the maximum (called expTG-M models, having $\ell_{max} = 330/300/280$) degrees of the spherical harmonic expansions. In the tests, in addition to the best performing DIR RL5, TIM RL5, SPW RL5 and GOCO05S models, the EGM2008 coefficients up to the identical expansion degrees ($\ell_{max} = 155, 280, 300, 330$) of the other tested models were also considered (see the test statistics of the experimental models in Table 3).

The twelve experimental models were validated at 81 GPS/leveling benchmarks of the geodetic network in northwest Turkey. Regarding the obtained results (see Table 3), all the validated experimental geoid models (relying on the reference GGM with optimum d/o of spherical harmonic expansion) provided an accuracy of 16.5 cm in terms of standard deviations of the residual geoid undulations without removing the tilt/trend at GPS/leveling benchmarks. However, employing the GGMs with the maximum degrees of their expansions, the calculated experimental geoid models revealed an accuracy of approximately 20 cm. This means that the use of optimum d/o of the geopotential model led to an improvement in the accuracy of the regional geoid model using Stokes' integral without any modification and any cap size limitation. The table also includes the statistics of the experimental geoid models after de-trending the models' surface simply using a third-order polynomial equation. The cross-validation of the de-trended models gave a so-called improved accuracy of around 9.7 cm for the models. The expTG-Opt-1 geoid model surface is shown in Figure 8.

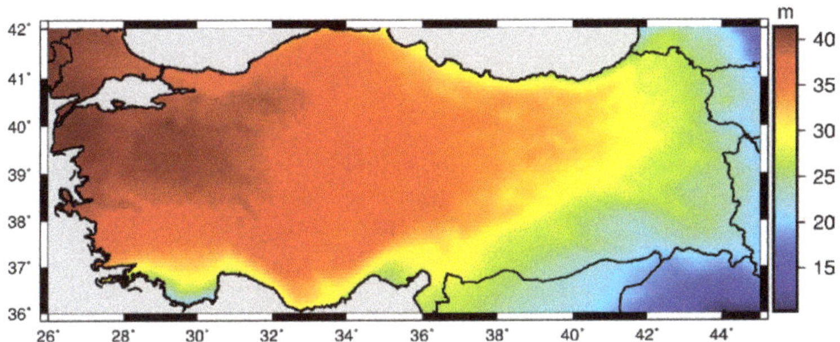

Figure 8. The experimental Turkey geoid, calculated using the remove–compute–restore (RCR) approach based on the reference geopotential model DIR-RL5.

As also seen in Table 3 (statistics before fit), the detailed geoid models, which were calculated using the classical RCR algorithm with GOCE-based reference models, are not more accurate than the enhanced EGM2008 model. This is a consequence of insufficient accuracy and density of the mean terrestrial gravity anomalies that were used in the computations. The role of the terrestrial gravity data accuracy for the quality of the calculated geoid model is explained by [83,84]. In accordance with a 1–2 mGal accuracy of the terrestrial gravity data, having 1–2 km spatial resolution with homogenous distribution is required if 1–2 cm accuracy of the geoid undulations is aimed. Compared to this criterion, the accuracy and resolution of the used mean gravity anomalies (with 6 mGal accuracy and 5′ grid spacing) in computations do not satisfy the requirements for determining a centimeter accuracy regional geoid. The accuracy of the used GPS/leveling data (~1.5–2.0 cm three-dimensional position accuracy and ~2.0 cm leveling height accuracy) in validations was sufficient for the purpose of this study. Although it is not officially declared in technical reports explaining the calculation strategy and data content of EGM2008, if the terrestrial gravity data of Turkey or the surrounding area was used in EGM2008 computations, this could explain the superior performance of this model in the validation results.

It is possible to obtain slightly improved accuracy in regional geoid models by applying modifications to the Stokes' integral in order to achieve a more rigorous combination of the geopotential models with terrestrial data by also considering their variances [38]. Hence, it will be possible to compensate for the random errors that are contained in the terrestrial data with the improved quality of the geopotential models up to a certain amount.

Table 3. Validation statistics of experimental geoid models at 81 GPS/leveling benchmarks (in cm).

GGMs and Detailed Geoid Model	Ref. GGM (d/o)		Statistics			
			Min	Max	Mean	Std.
DIR RL5 (155)	–	–	−192.8	61.8	−72.8	61.0
expTG-Opt-1	DIR RL5 (155)	Before fit	−174.1	−94.4	−127.9	16.5
		After fit	−37.2	31.6	0.0	9.8
TIM RL5 (155)	–	–	−192.8	61.2	−72.7	61.1
expTG-Opt-2	TIM RL5 (155)	Before fit	−174.5	−94.6	−128.2	16.5
		After fit	−37.2	31.5	0.0	9.8
SPW RL5 (155)	–	–	−204.8	62.7	−72.6	63.7
expTG-Opt-3	SPW RL5 (155)	Before fit	−164.9	−83.8	−119.6	16.3
		After fit	−34.3	29.7	0.0	11.4
GOCO05S (155)	–	–	−192.8	61.1	−72.9	60.8
expTG-Opt-4	GOCO05S (155)	Before fit	−174.3	−94.4	−128.0	16.5
		After fit	−37.2	31.6	0.0	9.8
EGM2008 (155)	–	–	−89.9	171.0	23.1	63.1
expTG-Opt-5	EGM2008 (155)	Before fit	−170.3	−89.7	−124.2	16.5
		After fit	−37.2	31.6	0.0	9.8
DIR RL5 (300)	–	–	−42.2	67.4	8.7	26.4
expTG-M-1	DIR RL5 (300)	Before fit	−162.2	−66.6	−112.9	19.4
		After fit	−35.0	30.2	0.0	9.7
TIM RL5 (280)	–	–	−50.2	67.3	8.6	27.2
expTG-M-2	TIM RL5 (280)	Before fit	−164.8	−69.8	−115.2	19.6
		After fit	−34.3	30.0	0.0	9.6
SPW RL5 (330)	–	–	−57.5	88.8	9.1	31.3
expTG-M-3	SPW RL5 (330)	Before fit	−161.8	−75.4	−115.9	18.4
		After fit	−31.1	27.7	0.0	10.7
GOCO05S (280)	–	–	−48.2	62.7	8.0	26.2
expTG-M-4	GOCO05S (280)	Before fit	−163.6	−68.2	−113.9	19.7
		After fit	−34.4	30.9	0.0	9.6
EGM2008 (280)	–	–	−44.7	69.3	8.3	26.4
expTG-M-5	EGM2008 (280)	Before fit	−163.6	−68.2	−113.9	19.7
		After fit	−34.6	30.8	0.0	9.6
EGM2008 (300)	–	–	−43.1	66.6	8.7	25.7
expTG-M-6	EGM2008 (300)	Before fit	−163.6	−68.2	−113.9	19.7
		After fit	−33.0	33.6	0.0	11.7
EGM2008 (330)	–	–	−38.6	59.8	8.3.	23.8
expTG-M-7	EGM2008 (330)	Before fit	−151.0	−63.7	−104.5	18.6
		After fit	−31.1	29.0	0.0	10.8

expTG: experimental Turkey Geoid, Opt: optimal d/o of reference GGM, -M: max. d/o of ref. GGM.

4. Conclusions

The purpose of this study is to investigate the progress of GOCE High-Level Processing Facility (HPF) released geopotential models that include data from gravity field satellite missions, and provide assessment results of these models for Turkey. The analyses were carried out by means of both spectral and spatial approaches. Regarding the numerical results and obtained statistics through these analyses, the following conclusions were drawn:

GOCE HPF-released GOCE- and GRACE-based global geopotential models were validated using the cumulative degree variances. Regarding the spherical harmonic expansion degrees (between 100 and 200 degrees of SH expansions, corresponding to the medium-wavelengths of the gravity field, where a significant contribution of the GOCE mission is expected), the cumulative formal errors (in terms of the geoid undulations) were 0.8 cm, 2.2 cm and 2.7 cm for DIR RL5, TIM RL5 and SPW RL4 models, respectively, whereas they were 2.0 cm, 2.9 cm and 7.1 cm for GOCO05S, GOGRA04S and EGM2008 models, respectively.

Considering the cumulative error amplitudes at the corresponding degrees for 100 km spatial resolution, the significant improvement toward the final releases of the geopotential models was verified (the improvement of final releases is almost up to 80% with respect to the first releases). On the other hand, the cumulative formal error estimate of the TIM RL5 model is three times worse than that of the DIR5 model. Though the difference seems to be quite significant, it is arguable if such a comparison without any calibration is objective.

The regional validations of the tested geopotential models against the GPS/leveling data were carried out with the spectral enhancement method. Hence, the higher frequencies in the content of the tested GGMs were synthesized from ultra-high degree coefficients of the EGM2008 model and the regional DTM data. As a conclusion, the spectral enhancement method revealed meaningful results and is recommended to be used for GGM validations using terrestrial data. In the results of the comparisons among the model releases (in every group of models as DIR, TIM, SPW), it is concluded that the model accuracy improves as its release increases, and that this improvement is independent of the computation strategy.

The spatial analyses of GGMs were carried out using GPS/leveling data of high-order geodetic networks in Turkey. The accuracy of the used control data, their spatial distribution and density, as well as the size of the validation area, are critical for the model validations and affect the reported performances. In this manner, the second test dataset can be said to be more suitable and hence the obtained validation statistics are more realistic in terms of the detailed analyses of the GGMs in the study area. The spatial analyses results clarified the GOCE mission's contribution to the improvement of the GGMs (with respect to the EGM2008 model), especially in the corresponding spectral bands of 100 and 250 degrees. Moreover, comparing standard deviations of the geoid undulation differences calculated with an increasing degree of the spherical harmonic expansion model, the optimum degree was determined with a minimum standard deviation value. Accordingly, the optimal value for d/o of spherical harmonic expansion is around 155 in the study area. In the test results explained in Section 3.2.1, an approximate accuracy for the tested GGMs is about 12 cm. EGM2008 model accuracy is calculated to be around 15 cm in Turkey. This is three times greater than the reported absolute accuracy of this model for well-covered areas with quality terrestrial data on Earth. This result indicates that either EGM2008 does not include the terrestrial data of Turkey or the quality of the included data is low. In the analyses, the GOGRA04S and GOCO05S models were also validated. However, since the obtained statistics did not reveal any improvement, it can be concluded that the complementary advantage of the GOCE gradiometry data with the observations from other gravity field satellite missions and SLR is not significant.

In the final part of the study, the role of the reference geopotential model accuracy, as well as its optimal expansion degree in determination of the detailed hybrid geoid model, were validated based on the RCR computation algorithm.

The obtained results revealed a 15% improvement in the accuracy of the gravimetric geoid model (considering the results without corrector surface fitting) when the GGM is employed with its optimum degree instead of the maximum degree (~16.5 cm vs. 19.5 cm for DIR RL05 as seen in Table 3). This conclusion was drawn using an unmodified Stokes kernel. On the other hand, a combination of the GGMs with terrestrial gravity data is possible using the RCR algorithm with stochastic modification of the Stokes kernel. Various approaches for the modification can be found in the literature. Hence, a more rigorous combination of the data may contribute to further improvement of the detailed geoid accuracy via compensating for the error budget, and realizing the GGMs' improvement to a certain extent would be possible. A research project, which was recently carried out in Turkey and the present study is a part of it (with The Scientific and Technological Research Council of Turkey, TUBITAK, support for contract number 114Y581), performed different methodologies related to detailed geoid modeling using hybrid data, and clarified the role of the modifications on a rigorous combination of GGMs with terrestrial data in order to improve the accuracy of a regional geoid model.

As a summary, global geopotential models have significantly progressed due to the data contribution from recent Earth gravity field satellite missions. The improved GGMs, in terms of accuracy and resolution, hence benefit many engineering and Earth science disciplines that use gravity field parameters in their analyses. The most significant impact of the improvement in GGMs' quality is on the vertical datum definition and realization, and global unification of the height systems. This study provided the methodological steps for the assessment of GGMs and quantified the improvement of the models in various regions, with cited statistics from the literature and numerical test results in Turkey.

Author Contributions: Conceptualization, B.E. and S.E.; methodology, B.E.; software, M.S.I.; validation, B.E., M.S.I. and S.E.; formal analysis, M.S.I.; investigation, M.S.I.; resources, B.E.; data curation, B.E., M.S.I. and S.E.; writing—original draft preparation, B.E.; writing—review and editing, S.E. and M.S.I.; visualization, M.S.I.; supervision, S.E.; project administration, B.E.; funding acquisition, B.E. All authors have read and agreed to the published version of the manuscript.

Funding: This research was funded by The Scientific and Technological Research Council of Turkey (TÜBİTAK), grant number 114Y581 and Istanbul Technical University Scientific Research Projects (ITU BAP) funds, grant number 38828.

Acknowledgments: We acknowledge International Center for Global Earth Models (ICGEM) at GFZ German Research Center for Geosciences that the used geopotential models were obtained from the service's data center.

Conflicts of Interest: The authors declare no conflicts of interest. The funders had no role in the design of the study; in the collection, analyses, or interpretation of data; in the writing of the manuscript, or in the decision to publish the results.

References

1. Tapley, B.D.; Schutz, B.E.; Eanes, R.J.; Ries, J.C.; Watkins, M.M. Lageos laser ranging contributions to geodynamics, geodesy, and orbital dynamics. *Contrib. Space Geod. Geodyn. Earth Dyn. Geodyn. Ser.* **1993**, *24*, 147–173.
2. Huang, J.; Kotsakis, C. External Quality Evaluation Reports of EGM08. *Newton's Bulletin* **2009**, *4*, 1–331.
3. Jin, S.; Van Dam, T.; Wdowinski, S. Observing and understanding the Earth system variations from space geodesy. *J. Geodyn.* **2013**, *72*, 1–10. [CrossRef]
4. Wolff, M. Direct measurements of the Earth's gravitational potential using a satellite pair. *J. Geophys. Res. Space Phys.* **1969**, *74*, 5295–5300. [CrossRef]
5. Rummel, R.; Balmino, G.; Johannessen, J.; Visser, P.; Woodworth, P. Dedicated gravity field missions—Principles and aims. *J. Geodyn.* **2002**, *33*, 3–20. [CrossRef]
6. Drinkwater, M.R.; Floberghagen, R.; Haagmans, R.; Muzi, D.; Popescu, A. VII: CLOSING SESSION: GOCE: ESA's First Earth Explorer Core Mission. *Space Sci. Rev.* **2003**, *108*, 419–432. [CrossRef]
7. Tapley, B.D.; Bettadpur, S.; Watkins, M.; Reigber, C. The gravity recovery and climate experiment: Mission overview and early results. *Geophys. Res. Lett.* **2004**, *31*. [CrossRef]

8. Rummel, R.; Gruber, T.; Flury, J.; Schlicht, A. ESA's gravity field and steady-state ocean circulation explorer GOCE. ZfV-Zeitschrift für Geodäsie. *Geoinf. Landmanag.* **2009**, *134*, 125–130.

9. CHAMP Satellite Mission, GFZ German Research Centre for Geosciences. Available online: http://op.gfz-potsdam.de/champ/ (accessed on 21 September 2019).

10. GRACE (Gravity Recovery and Climate Experiment) Satellite Mission, University of Texas at Austin Center 795 for Space Research. Available online: http://www.csr.utexas.edu/grace/ (accessed on 7 February 2020).

11. Tapley, B.D.; Bettadpur, S.; Ries, J.C.; Thompson, P.F.; Watkins, M.M. GRACE Measurements of Mass Variability in the Earth System. *Science* **2004**, *305*, 503–505. [CrossRef]

12. Flechtner, F.; Morton, P.; Watkins, M.; Webb, F. *Status of the GRACE Follow-on Mission, in Gravity, Geoid and Height Systems*; Springer: Berlin/Heidelberg, Germany, 2014; pp. 117–121.

13. Rummel, R. *Satellite Gradiometry*; Springer: Berlin/Heidelberg, Germany, 1986.

14. Rummel, R.; Colombo, O.L. Gravity field determination from satellite gradiometry. *J. Geod.* **1985**, *59*, 233–246. [CrossRef]

15. Hirt, C.; Gruber, T.; Featherstone, W.E. Evaluation of the first GOCE static gravity field models using terrestrial gravity, vertical deflections and EGM2008 quasigeoid heights. *J. Geod.* **2011**, *85*, 723–740. [CrossRef]

16. Rexer, M.; Hirt, C.; Pail, R.; Claessens, S. Evaluation of the third- and fourth-generation GOCE Earth gravity field models with Australian terrestrial gravity data in spherical harmonics. *J. Geod.* **2014**, *88*, 319–333. [CrossRef]

17. ESA. *The Four Candidate Earth Explorer Core Missions–Gravity Field and Steady-State Ocean Circulation*; Battrick, B., Ed.; ESA: Noordwijk, The Netherlands, 1999.

18. Pail, R.; Bruinsma, S.; Migliaccio, F.; Förste, C.; Goiginger, H.; Schuh, W.-D.; Höck, E.; Reguzzoni, M.; Brockmann, J.M.; Abrikosov, O.; et al. First GOCE gravity field models derived by three different approaches. *J. Geod.* **2011**, *85*, 819–843. [CrossRef]

19. Gruber, T.; Visser, P.N.A.M.; Ackermann, C.; Hosse, M. Validation of GOCE gravity field models by means of orbit residuals and geoid comparisons. *J. Geod.* **2011**, *85*, 845–860. [CrossRef]

20. Barthelmes, F. Definition of Functionals of the Geopotential and Their Calculation from Spherical Harmonic Models. Available online: http://icgem.gfz-potsdam.de/str-0902-revised.pdf (accessed on 7 February 2020).

21. ICGEM—International Centre for Global Earth Models (ICGEM), GFZ Potsdam, Germany. Available online: http://icgem.gfz-potsdam.de/ (accessed on 7 February 2020).

22. Pavlis, N.K.; Holmes, S.A.; Kenyon, S.C.; Factor, J.K. The development and evaluation of the Earth Gravitational Model 2008 (EGM2008). *J. Geophys. Res. Space Phys.* **2012**, *117*. [CrossRef]

23. Gruber, T.; Gerlach, C.; Haagmans, R. Intercontinental height datum connection with GOCE and GPS-levelling data. *J. Géod. Sci.* **2012**, *2*, 270–280. [CrossRef]

24. Gruber, T. Validation concepts for gravity field models from new satellite missions. In Proceedings of the 2nd International GOCE User Workshop: GOCE The Geoid and Oceanography, Bratislava, Slovakia, 8–10 March 2004.

25. Ihde, J.; Wilmes, H.; Müller, J.; Denker, H.; Voigt, C.; Hosse, M. Validation of Satellite Gravity Field Models by Regional Terrestrial Data Sets. In *Advanced Technologies in Earth Sciences*; Springer Science and Business Media: Berlin/Heidelberg, Germany, 2010; pp. 277–296.

26. Erol, S.; Isik, M.S.; Erol, B. Assessments on GOCE-based Gravity Field Model Comparisons with Terrestrial Data Using Wavelet Decomposition and Spectral Enhancement Approaches. In Proceedings of the EGU General Assembly 2016, Vienna, Austria, 17–22 April 2016.

27. Sansò, F.; Sideris, M.G. *Geoid Determination: Theory and Methods*; Springer: San Francisco, CA, USA, 2013; ISBN 978-3-540-74699-7.

28. Erol, B.; Sideris, M.G.; Çelik, R.N. Comparison of global geopotential models from the champ and grace missions for regional geoid modelling in Turkey. *Stud. Geophys. Geod.* **2009**, *53*, 419–441. [CrossRef]

29. Isık, M.S.; Erol, B. Geoid determination using GOCE-based models in Turkey. In Proceedings of the EGU General Assembly 2016, Vienna, Austria, 17–22 April 2016.

30. Sjöberg, L.E.; Bagherbandi, M. *Gravity Inversion and Integration*; Springer Science and Business Media: Berlin/Heidelberg, Germany, 2017.

31. Halicioglu, K.; Deniz, R.; Ozener, H. Digital astro-geodetic camera system for the measurement of the deflections of the vertical: Tests and results. *Int. J. Digit. Earth* **2016**, *9*, 914–923. [CrossRef]

32. Kiamehr, R.; Eshagh, M.; Sjoberg, L.E. Interpretation of general geophysical patterns in Iran based on GRACE gradient component analysis. *Acta Geophys.* **2008**, *56*, 440–454. [CrossRef]

33. Mintourakis, I. Adjusting altimetric sea surface height observations in coastal regions. Case study in the Greek Seas. *J. Géod. Sci.* **2014**, *4*. [CrossRef]

34. Kotsakis, C.; Katsambalos, K. Quality Analysis of Global Geopotential Models at 1542 GPS/levelling Benchmarks Over the Hellenic Mainland. *Surv. Rev.* **2010**, *42*, 327–344. [CrossRef]

35. Ihde, J.; Sánchez, L.; Barzaghi, R.; Drewes, H.; Foerste, C.; Gruber, T.; Liebsch, G.; Marti, U.; Pail, R.; Sideris, M. Definition and Proposed Realization of the International Height Reference System (IHRS). *Surv. Geophys.* **2017**, *38*, 549–570. [CrossRef]

36. Gerlach, C.; Rummel, R. Global height system unification with GOCE: A simulation study on the indirect bias term in the GBVP approach. *J. Geod.* **2013**, *87*, 57–67. [CrossRef]

37. Vergos, G.S.; Erol, B.; Natsiopoulos, D.A.; Grigoriadis, V.N.; Işık, M.S.; Tziavos, I.N. Preliminary results of GOCE-based height system unification between Greece and Turkey over marine and land areas. *Acta Geod. Geophys.* **2018**, *53*, 61–79. [CrossRef]

38. Van Der Meijde, M.; Pail, R.; Bingham, R.; Floberghagen, R. GOCE data, models, and applications: A review. *Int. J. Appl. Earth Obs. Geoinf.* **2015**, *35*, 4–15. [CrossRef]

39. Bruinsma, S.; Marty, J.; Balmino, G.; Biancale, R.; Förste, C.; Abrikosov, O.; Neumayer, H. GOCE gravity field recovery by means of the direct numerical method. In Proceedings of the ESA Living Planet Symposium 2010, Bergen, Norway, June 27–2 July 2010.

40. Pail, R.; Goiginger, H.; Mayrhofer, R.; Höck, E.; Schuh, W.D.; Brockmann, J.M. GOCE gravity field model derived from orbit and gradiometry data applying the time-wise method. In Proceedings of the ESA Living Planet Symposium 2010, Bergen, Norway, 27 June–2 July 2010.

41. Migliaccio, F.; Reguzzoni, M.; Sansò, F.; Tscherning, C.C.; Veicherts, M. GOCE data analysis: The space-wise approach and the first space-wise gravity field model. In Proceedings of the ESA Living Planet Symposium 2010, Bergen, Norway, 27 June–2 July 2010.

42. Rummel, R.; Gruber, T.; Koop, R. High level processing facility for GOCE: Products and processing strategy. In Proceedings of the 2nd International GOCE User Workshop GOCE, The Geoid and Oceanography, ESA SP-569, Bratislava, Slovakia, 8–10 March 2004.

43. Gruber, T.; Rummel, R.; Abrikosov, O.; van Hees, R. *GOCE Level 2 Product Data Handbook*; The European GOCE Gravity Consortium EGG-C: Paris, France, 2010.

44. Metzler, B.; Pail, R. GOCE Data Processing: The Spherical Cap Regularization Approach. *Stud. Geophys. Geod.* **2005**, *49*, 441–462. [CrossRef]

45. ESA. Gravity Field and Steady-State Ocean Circulation Explorer (GOCE) Mission. Available online: http://www.esa.int/Our_Activities/Observing_the_Earth/GOCE (accessed on 21 September 2019).

46. Zingerle, P.; Brockmann, J.M.; Pail, R.; Gruber, T.; Willberg, M. The polar extended gravity field model TIM_R6e. In *GFZ Data Services*; GFZ: Potsdam, Germany, 2019.

47. Pail, R.; Fecher, T.; Murböck, M.; Rexer, M.; Stetter, M.; Gruber, T.; Stummer, C. Impact of GOCE Level 1b data reprocessing on GOCE-only and combined gravity field models. *Studia Geophys. Geod.* **2013**, *57*, 155–173. [CrossRef]

48. Reguzzoni, M. From the time-wise to space-wise GOCE observables. *Adv. Geosci.* **2003**, *1*, 137–142. [CrossRef]

49. Migliaccio, F.; Reguzzoni, M.; Sansò, F.; Tselfes, N. On the use of gridded data to estimate potential coefficients. In Proceedings of the 3rd International GOCE User Workshop, Frascati, Italy, 6–8 November 2006.

50. Gruber, T.; Rummel, R. GOCE gravity field models-Overview and performance analysis. In Proceedings of the 3rd International Gravity Field Service (IGFS) General Assembly, Shanghai, China, 30 June–6 July 2014.

51. Bruinsma, S.L.; Förste, C.; Abrikosov, O.; Lemoine, J.-M.; Marty, J.-C.; Mulet, S.; Rio, M.-H.; Bonvalot, S. ESA's satellite-only gravity field model via the direct approach based on all GOCE data. *Geophys. Res. Lett.* **2014**, *41*, 7508–7514. [CrossRef]

52. Förste, C.; Abrykosov, O.; Bruinsma, S.; Dahle, C.; König, R.; Lemoine, J.M. ESA's Release 6 GOCE gravity field model by means of the direct approach based on improved filtering of the reprocessed gradients of the entire mission. In *GFZ Data Services*; GFZ: Potsdam, Germany, 2019.

53. Brockmann, J.M.; Zehentner, N.; Höck, E.; Pail, R.; Loth, I.; Mayer-Gürr, T.; Schuh, W.-D. EGM_TIM_RL05: An independent geoid with centimeter accuracy purely based on the GOCE mission. *Geophys. Res. Lett.* **2014**, *41*, 8089–8099. [CrossRef]

54. Brockmann, J.M.; Schubert, T.; Mayer-Gürr, T.; Schuh, W.D. The Earth's gravity field as seen by the GOCE satellite—An improved sixth release derived with the time-wise approach. In *GFZ Data Services*; GFZ: Potsdam, Germany, 2019.

55. Migliaccio, F.; Reguzzoni, M.; Gatti, A.; Sansò, F.; Herceg, M. A GOCE-only global gravity field model by the space-wise approach. In Proceedings of the 4th International GOCE User Workshop, Munich, Germany, 31 March–1 April 2011.

56. Gatti, A.; Reguzzoni, M.; Migliaccio, F.; Sansò, F. Space-wise grids of gravity gradients from GOCE data at nominal satellite altitude. In Proceedings of the 5th International GOCE User Workshop, Paris, France, 25–28 November 2014.

57. Gatti, A.; Reguzzoni, M.; Migliaccio, F.; Sansò, F. Computation and assessment of the fifth release of the GOCE-only space-wise solution. In Proceedings of the the the 1st Joint Commission 2 and IGFS Meeting, Thessaloniki, Greece, 19–23 September 2016.

58. Mayer-Guerr, T. The combined satellite gravity field model GOCO05s. In Proceedings of the EGU General Assembly Conference, Vienna, Austria, 12–17 April 2015.

59. Yi, W.; Rummel, R.; Gruber, T. Gravity field contribution analysis of GOCE gravitational gradient components. *Stud. Geophys. Geod.* **2013**, *57*, 174–202. [CrossRef]

60. Ince, E.S.; Barthelmes, F.; Reißland, S.; Elger, K.; Förste, C.; Flechtner, F.; Schuh, H. ICGEM—15 years of successful collection and distribution of global gravitational models, associated services, and future plans. *Earth Syst. Sci. Data* **2019**, *11*, 647–674. [CrossRef]

61. Hofmann-Wellenhof, B.; Moritz, H. *Physical Geodesy*; Springer: New York, NY, USA, 2006.

62. Ellmann, A.; Kaminskis, J.; Parseliunas, E.; Jürgenson, H.; Oja, T. Evaluation Results of the Earth Gravitational Model EGM08 over the Baltic Countries. *Newton's Bulletin* **2009**, *4*, 110–121.

63. Kirby, J.; Featherstone, W. A study of zero-and first-degree terms in geopotential models over Australia. *Geomat. Res. Australas.* **1997**, *66*, 93–108.

64. Sánchez, L.; Čunderlík, R.; Dayoub, N.; Mikula, K.; Minarechová, Z.; Šíma, Z.; Vatrt, V.; Vojtíšková, M. A conventional value for the geoid reference potential $$ W_{0} $$. *J. Geod.* **2016**, *90*, 815–835. [CrossRef]

65. Smith, D.A. There is no such thing as "The" EGM96 geoid: Subtle points on the use of a global geopotential model. *IGeS Bull. Int. Geoid Serv.* **1998**, *8*, 17–27.

66. Ekman, M. Impacts of geodynamic phenomena on systems for height and gravity. *J. Geod.* **1989**, *63*, 281–296. [CrossRef]

67. Kiamehr, R. A strategy for determining the regional geoid by combining limited ground data with satellite-based global geopotential and topographical models: A case study of Iran. *J. Geod.* **2006**, *79*, 602–612. [CrossRef]

68. Voigt, C.; Denker, H. Validation of GOCE Gravity Field Models in Germany. *Newton's Bulletin* **2015**, *5*, 37–49.

69. Tscherning, C.C.; Arabelos, D. Gravity anomaly and gradient recovery from GOCE gradient data using LSC and comparisons with known ground data. In Proceedings of the 4th International GOCE user workshop, Munich, Germany, 31 March–1 April 2011.

70. Šprlák, M.; Gerlach, C.; Pettersen, B. Validation of GOCE global gravity field models using terrestrial gravity data in Norway. *J. Géod. Sci.* **2012**, *2*, 134–143. [CrossRef]

71. Bouman, J.; Fuchs, M.J. GOCE gravity gradients versus global gravity field models. *Geophys. J. Int.* **2012**, *189*, 846–850. [CrossRef]

72. Abdalla, A.; Fashir, H.; Ali, A.; Fairhead, D. Validation of recent GOCE/GRACE geopotential models over Khartoum state-Sudan. *J. Géod. Sci.* **2012**, *2*, 88–97. [CrossRef]

73. Guimarães, G.; Matos, A.; Blitzkow, D. An evaluation of recent GOCE geopotential models in Brazil. *J. Géod. Sci.* **2012**, *2*, 144–155. [CrossRef]

74. Szűcs, E. Validation of GOCE time-wise gravity field models using GPS-levelling, gravity, vertical deflections and gravity gradient measurements in Hungary. *Period. Polytech. Civ. Eng.* **2012**, *56*, 3–11. [CrossRef]

75. Huang, J.; Reguzzoni, M.; Gruber, T. Assessment of GOCE Geopotential Models. *Newton's Bulletin* **2015**, *5*, 1–192.

76. Ayhan, M.E.; Demir, C.; Lenk, O.; Kılıçoğlu, A.; Aktug, B.; Acikgoz, M.; Firat, O.; Sengun, Y.S.; Cingoz, A.; Gurdal, M.A.; et al. Türkiye Ulusal Temel GPS Ağı. *Harit. Derg.* **2002**, *16*, 1–80.

77. Andersen, O.B.; Knudsen, P.; Berry, P.A. The DNSC08GRA global marine gravity field from double retracked satellite altimetry. *J. Geod.* **2010**, *84*, 191–199. [CrossRef]

78. Şaroğlu, F.; Emre, Ö.; Kuşçu, İ. *Turkish Active Faults Map*; Directorate of Mineral Research and Exploration: Ankara, Turkey, 1992.
79. Kılıçoğlu, A.; Direnç, A.; Simav, M.; Lenk, O.; Aktuğ, B.; Yıldız, H. Evaluation of the Earth Gravitational Model 2008 in Turkey. *Newton's Bulletin* **2009**, *4*, 164–171.
80. Aktuğ, B.; Kilicoglu, A.; Lenk, O.; Gürdal, M.A.; Aktuğ, B. Establishment of regional reference frames for quantifying active deformation areas in Anatolia. *Stud. Geophys. Geod.* **2009**, *53*, 169–183. [CrossRef]
81. International Service for the Geoid (ISG) at DICA Politecnico di Milano. Available online: http://www. isgeoid.polimi.it/ (accessed on 7 February 2020).
82. Erol, B.; Işık, M.S.; Erol, S. Assessment of Gridded Gravity Anomalies for Precise Geoid Modeling in Turkey. *J. Surv. Eng.* **2020**. [CrossRef]
83. Kearsley, A.H.W. Data requirements for determining precise relative geoid heights from gravimetry. *J. Geophys. Res. Space Phys.* **1986**, *91*, 9193–9201. [CrossRef]
84. Farahani, H.H.; Klees, R.; Slobbe, C. Data requirements for a 5-mm quasi-geoid in the Netherlands. *Stud. Geophys. Geod.* **2017**, *25*, 17–702. [CrossRef]

Article

Next-Generation Gravity Missions: Sino-European Numerical Simulation Comparison Exercise

Roland Pail [1,*], Hsien-Chi Yeh [2], Wei Feng [3], Markus Hauk [1], Anna Purkhauser [1], Changqing Wang [3], Min Zhong [3], Yunzhong Shen [4], Qiujie Chen [4], Zhicai Luo [5], Hao Zhou [5], Bingshi Liu [6], Yongqi Zhao [6], Xiancai Zou [6], Xinyu Xu [6], Bo Zhong [6], Roger Haagmans [7] and Houze Xu [3]

[1] Institute of Astronomical and Physical Geodesy, Technical University of Munich, 80333 Munich, Germany; markus.hauk@tum.de (M.H.); anna.purkhauser@tum.de (A.P.)
[2] School of Physics and Astronomy, Sun Yat-sen University (Zhuhai Campus), Zhuhai 519082, China; yexianji@mail.sysu.edu.cn
[3] State Key Laboratory of Geodesy and Earth's Dynamics, Institute of Geodesy and Geophysics, APM, Chinese Academy of Sciences, Wuhan 430077, China; fengwei@whigg.ac.cn (W.F.); whiggsdkd@whigg.ac.cn (C.W.); zmzm@whigg.ac.cn (M.Z.); hsuh@whigg.ac.cn (H.X.)
[4] College of Surveying and Geo-informatics, Tongji University, Shanghai 200092, China; yzshen@tongji.edu.cn (Y.S.); qiujie.chen@uni-bonn.de (Q.C.)
[5] Institute of Geophysics and PGMF, Huazhong University of Science and Technology, Wuhan 430074, China; zcluo@hust.edu.cn (Z.L.); zhouh@hust.edu.cn (H.Z.)
[6] School of Geodesy and Geomatics, Wuhan University, Wuhan 430079, China, and Key Laboratory of Geospace Environment and Geodesy, Ministry of Education, Wuhan 430079, China; bs.liu@whu.edu.cn (B.L.); yqzhao@whu.edu.cn (Y.Z.); xczou@sgg.whu.edu.cn (X.Z.); xyxu@sgg.whu.edu.cn (X.X.); bzhong@sgg.whu.edu.cn (B.Z.)
[7] ESA–European Space Agency, 2201 AZ Noordwijk, The Netherlands; roger.haagmans@esa.int
[*] Correspondence: roland.pail@tum.de; Tel.: +49-89-289-23199

Received: 10 October 2019; Accepted: 11 November 2019; Published: 13 November 2019

Abstract: Temporal gravity retrieval simulation results of a future Bender-type double pair mission concept, performed by five processing centers of a Sino-European study team, have been inter-compared and assessed. They were computed in a synthetic closed-loop simulation world by five independent software systems applying different gravity retrieval methods, but were based on jointly defined mission scenarios. The inter-comparison showed that the results achieved a quite similar performance. Exemplarily, the root mean square (RMS) deviations of global equivalent water height fields from their true reference, resolved up to degree and order 30 of a 9-day solution, vary in the order of 10% of the target signal. Also, co-estimated independent daily gravity fields up to degree and order 15, which have been co-estimated by all processing centers, do not show large differences among each other. This positive result is an important pre-requisite and basis for future joint activities towards the realization of next-generation gravity missions.

Keywords: next-generation gravity mission; temporal gravity field; numerical closed-loop simulation; satellite mission constellations; mass transport

---

1. Introduction

Next Generation Gravity Missions (NGGMs), expected to be launched in the midterm future, have set high anticipations for an enhanced monitoring of mass transport in the Earth system, establishing that their products are applicable to new scientific fields and serving societal needs. Past and current gravity missions such as Gravity Recovery and Climate Experiment (GRACE, [1]) and GRACE Follow-On [2] have improved our understanding of many mass change processes, such as the global water cycle, ice mass melting of ice sheets and glaciers, changes in ocean mass closely related to eustatic sea level rise, which are subtle indicators of climate change, but also gravity changes related to solid Earth processes such as big earthquakes or glacial isostatic adjustment (GIA). The Gravity field and steady-state Ocean Circulation Explorer (GOCE) mission [3] has improved our knowledge of long-term mass distribution and has provided the physical reference surface of the geoid, with a resolution down to 70–80 km.

The main objective of NGGMs is not only the continuation and extension of existing mass transport time series, but also a significant gain in spatial and temporal resolution. Correspondingly, the science and user needs of NGGMs were defined in [4]. In several previous studies, different mission concepts have been studied in detail, with emphasis on orbit design and resulting spatial-temporal ground track pattern, enhanced processing and parameterization strategies, and improved post-processing/filtering strategies in order to reduce temporal aliasing effects, which are one of the main error sources of current temporal gravity field solutions [5]. The typical GRACE-type concept of two satellites (satellite pair) following each other on the same orbit with an inter-satellite distance of about 200 km observes only the along-track component of the Earth's gravity field, and therefore leads to a very anisotropic error structure and the typical striping patterns in the resulting temporal gravity solutions. An alternative mission concept is a satellite pair in pendulum configuration, where the trailing satellite performs a pendulum motion with respect to the leading satellite, thus observing not only the along-track, but also parts of the cross-track component. This can be realized by a shift of the right-ascension of the satellites of the second pair, resulting in a shifted orbit plane of the second satellite pair compared to the first one. Based on this concept, in 2010 the mission proposal "e.motion —Earth System Mass Transport Mission" [6] was submitted in response to ESA's Earth Explorer 8 call. Another promising mission scenario is a double-pair constellation being composed of two in-line pairs, the so-called Bender configuration [7]. It is composed of a polar pair similar to the GRACE-type concept, and an inclined pair with an inclination of 65–70 degrees. An example of the resulting ground-track pattern of such a Bender constellation is shown in Figure 1.

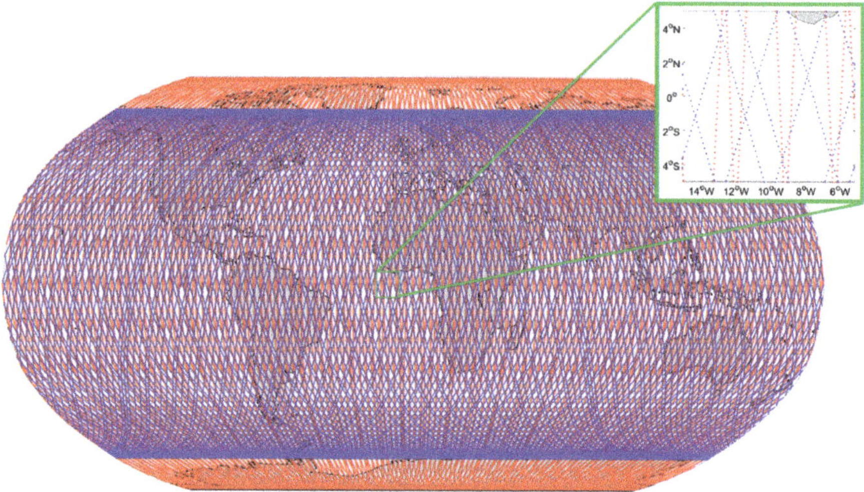

Figure 1. Ground track pattern of Bender configuration of Scenario A (cf. Table 1). The red curve shows the 9-day near-repeat ground track pattern of the polar satellite pair, and the blue curve the ground track of the inclined pair with an inclination of 70°. The zoom-in illustrates the resulting regular spatial sampling pattern as well as the fact that the direction of the inter-satellite ranges is different.

Table 1. Main orbit parameters of numerical simulation scenarios.

Simulation Scenario	Semi-major Axis Minus 6378 km	Inclination (°)	Cycle/Retrieval Period (days)	Start Epoch	Drift Rate (°/cycle)
A	~355/~390	70/89	9	1 Jan 2002	1.5
B	~355/~340	70/89	7	1 Jan 2002	1.3

Extensive numerical simulations have shown that this set-up also results in a significantly improved isotropy and reduction of stripes, e.g., [8,9], which is mainly due to the different directions of inter-satellite ranging observations, and the different orbital frequencies of the two pairs. Based on this concept, in 2016 e.motion^2 was proposed as ESA Earth Explorer 9 mission [10]. Another innovative observation concept is high-precision high-low inter-satellite ranging, which was proposed as the MOBILE mission in response to ESA's Earth Explorer 10 call, e.g., [11] and [12]. On US side, the United States National Academies of Sciences, Engineering, and Medicine published the decadal strategy for Earth observation from space [13], where mass change was identified as one of the top five designated observables to be implemented by future US Earth observation missions, in order to ensure continuity and enable long-term mass budget analyses of the Earth system.

A NGGM concept was also proposed in China, and the key technologies have been developed to meet the requirements of NGGM [14–16]. The objectives of Chinese NGGM are not only to perform global gravity measurements, but also to verify the maturity of key technologies required for space-based gravitational-wave detection [17] to a certain level of precision. The scientific objectives and the system requirements of Chinese NGGM are under investigation in these years. This mission is expected to be approved soon, and is planned to launch in the middle 2020s.

The existence of this study roots back to a Round Table China–Europe meeting on Satellite Gravity Exploration, which took place in September 2013 in Beijing. There it was agreed to evaluate options for further joint activities on NGGMs, especially science studies on numerical mission simulations. This was eventually realized in the frame of the European Space Agency's "Additional Constellation and Scientific Analysis Studies of the Next Generation Gravity Mission (ADDCON)" study. ESA's ADDCON study team and a Chinese study team joined forces in an initiative to perform numerical studies on NGGMs based on double-pair and multi-pair constellations of GRACE/GRACE-FO-type satellites.

A main prerequisite for the performance assessment and interpretation of joint numerical simulations is to ensure comparability of the results of the contributing groups. Therefore, it was decided early in the project to jointly set up common test scenarios and corresponding data sets, which shall be processed by all contributing processing centers. This paper reports on results of this inter-comparison exercise. The contributing processing centers are Technical University of Munich (TUM), Institute of Geodesy and Geophysics, Chinese Academy of Sciences (IGG-CAS), Tongji University (Tongji), Wuhan University (WHU), and Huazhong University of Science and Technology (HUST).

The paper is structured as follows. Chapter 2 describes the basic idea of a numerical closed-loop NGGM simulation, and describes the main input data, which have been jointly used by all groups. Chapter 3 provides an overview of different methods applied by the study teams. Chapter 4 reports the numerical simulation results, their comparison, assessment, and inter-comparison. The final Chapter 5 contains main conclusions as well as an outlook to future work.

2. Closed-Loop Simulations and Input Data

The basis of the inter-comparison exercise are so-called numerical closed-loop simulations. Compared to real-data processing, the advantage of these closed-loop simulations, which are performed in a completely synthetic simulation world, is that the truth is perfectly known. Therefore, the effect of any data set, instrument noise assumption, or processing strategy is directly reflected by deviation of the final result from the "true world", and can be studied separately from all other error contributors to the total error budget.

Figure 2 shows an overview of the numerical simulation set-up. The simulation starts with the definition of the "true world" (upper left), using force models for the static gravity field as well as tidal and non-tidal temporal variations. All gravity-related models are parameterized as coefficients of a spherical harmonic (SH) series expansion. In parallel, specific orbit constellations and the corresponding orbit parameters have to be defined, in our case Bender-type double pairs. The corresponding orbits are generated by numerical integration. Based on the a-priori force models, error-free observations are computed along the orbit tracks, with a sampling period of 5 s. They are then superimposed by instrument errors with pre-specified spectral noise characteristics, resulting in "true observations", which mimic as realistically as possible the real observations. On the other hand, a reference world is built up (upper right). This is necessary because the functional model of gravity field estimation from low–low inter-satellite ranging observations is highly non-linear, and this requires a Taylor point as approximate prior knowledge. In the same way as for the "true observations", so-called "reference observations" are computed, which are based on force models that are usually different from those used for the "true world" scenario. The main goal of the gravity field determination is then to estimate residuals with respect to these a-priori models. Adding the residuals to the a-priori models should result in the ideal case (no errors in the system) in the true world models. In the case that errors do exist, such as instrument errors, but also aliasing errors due to the specific space-time sampling properties of the chosen orbit constellation, the estimated gravity field will deviate from the true world, and thus provides a clear hint on the achievable performance of the specific mission design.

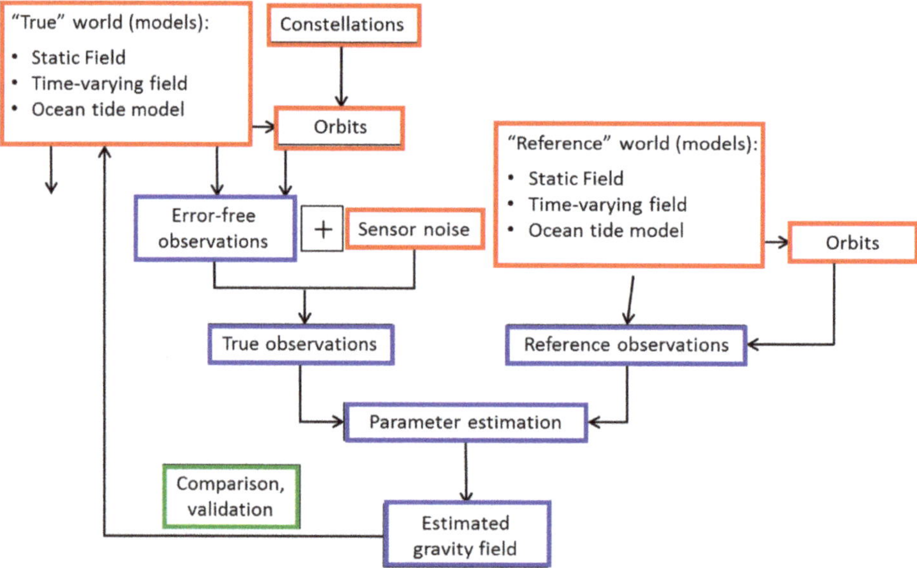

Figure 2. Scheme of a closed-loop simulation set-up. The red blocks are related to orbit design and input data generation, the blue blocks are tightly linked to the gravity parameter estimation process, while the green block refers to the comparison and validation of the resulting gravity models.

For the inter-comparison exercise, two Bender double-pair scenarios have been defined (Table 1). They are based on extensive prior studies on optimum orbit design of double-pair configuration with the goal to optimize the space-time sampling. The two scenarios A and B mainly differ in the orbit altitude of the polar satellite (390 vs. 340 km) and their repeat cycle (9 vs. 7 days). The orbits were designed such that ground tracks of the polar and inclined satellite pair drift at the same rate within the sub-cycle period, and (b) the retrieval period is always equal to the length of the sub-cycle, so that the gravity field retrieval is always supported by the densest possible ground track pattern. As a side aspect, these drifting orbits have the additional advantage that they produce a dense ground-track distribution over longer periods of time, which might improve the spatial resolution for the gravity retrieval of long-term static gravity solutions.

Table 2 provides an overview of the force and noise models of the true and reference world. Static, tidal, and non-tidal gravity models are used up to SH degree and order (d/o) 120.

Table 2. Force and noise models of the "true" and "reference" world used in the Next Generation Gravity Mission (NGGM) simulations.

Model	True World	Reference World
Static gravity field	GOCO05s [18]	GOCO05s
Non-tidal time-variable gravity field	ESA AOHIS [19]	—
Ocean tides	EOT11a [20]	GOT4.7 [21]
Inter-Satellite Link (ISL) noise	Equation (1)	—
Accelerometer (ACC) noise	Equations (2) and (3)	—
Star tracker (STR) noise	—	—
Orbit noise	1 cm white noise	—

The ISL is assumed to be realized by a Laser Ranging Interferometer (LRI). The assumed spectral noise behavior of the LRI in terms of range rate is expressed by the following Amplitude Spectral Density (ASD) [22]:

$$d_{range\text{-}rates} = 2 \times 10^{-8} \times 2\pi f \times \sqrt{\left(\frac{10^{-2}\text{Hz}}{f}\right)^2 + 1} \frac{\text{m}}{\text{s} \sqrt{\text{Hz}}} \tag{1}$$

where f is the linear frequency. The resulting error spectrum is shown in Figure 3a (dark blue curve), and the corresponding noise realization in light blue color.

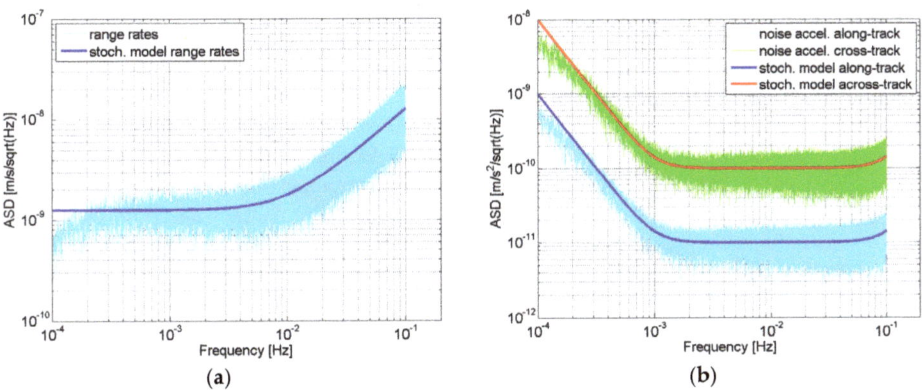

Figure 3. Amplitude spectral density (ASD) of (**a**) relative distant measurement errors in terms of range-rates, and (**b**) accelerometer errors. The dark blue and red curves show the analytical noise models in accordance with Equations (1) to (3), and the light blue and green curves the corresponding noise realizations, which have been generated scaling the spectrum of normally distributed random time-series with their individual spectral model.

The non-gravitational forces are typically sensed by the on-board accelerometers located in the center-of-mass of the satellite. In our simulations, an electrostatic accelerometer is assumed with two high-sensitive axes oriented in the flight direction (largest signal) and radial direction, and one low-sensitive axis in the cross-track direction (see Figure 3b). The accuracy level in terms of accelerations is expressed by [22]:

$$d_{acc.\ along} = d_{acc.\ radial} = 10^{-11} \sqrt{\left(\frac{10^{-3}\mathrm{Hz}}{f}\right)^4 \bigg/ \left(\left(\frac{10^{-5}\mathrm{Hz}}{f}\right)^4 + 1\right) + 1 + \left(\frac{f}{10^{-1}\mathrm{Hz}}\right)^4} \frac{\mathrm{m}}{\mathrm{s}^2 \sqrt{\mathrm{Hz}}} \tag{2}$$

$$d_{acc.\ across} = 10 \times d_{acc.\ along}, \tag{3}$$

It has a noise level of 10^{-11} m/s^2 in the measurement bandwidth, and a $1/f^2$ slope at frequencies from 10^{-3} Hz and lower. Equation (3) expresses the fact that electrostatic accelerometers, as they have been used in all previous and current satellite gravity missions, usually have two high-sensitive axes and one less-sensitive axis. In our set-up, we assume a degraded performance of the less-sensitive axis by one order of magnitude, and that it is oriented in an across-track direction. In this study, we further assume that the attitude, which is usually a minor error contributor compared to ISL and ACC, is known perfectly. Orbit errors are simulated by white noise with a standard deviation of 1 cm. The generation of the ISL and ACC noise time-series was done by scaling the spectrum of normally distributed random time-series with their individual spectral analytical noise model, as given in Equations (1) to (3).

3. Methods

The five contributing processing centers used their specific software systems for NGGM simulation and gravity field recovery. However, as a joint spatial–temporal parametrization, low-degree daily gravity field parameters were co-estimated together with the higher-degree temporal gravity solution, where the latter covers a retrieval period of 9 days (Scenario A) and 7 days (Scenario B). This co-estimation was first proposed by [23] and investigated in detail in [9]. There it could be shown that by this "Wiese parameterization" temporal aliasing can be reduced significantly, because short-term temporal gravity signals (mainly atmosphere and ocean signals), which would alias in standard processing into the solutions, are explicitly estimated, thus avoiding their aliasing effect. On the one hand this "self-dealiasing" approach improves the gravity field solution over the whole spectral range, and on the other hand, as a positive side effect daily gravity field products, which are independent of each other, are produced. This might be an important aspect for operational service applications, where usually a high temporal resolution and short latencies are required [24].

In all five processing centers, the non-tidal gravity field signals along the orbits were derived from the ESA Earth System Model (ESM, [19]) for all five components: atmosphere (A), ocean (O), hydrology (H), ice (I), and solid Earth (S), together abbreviated by AOHIS. Based on the 6-hourly AOHIS input products, the signals at certain measurement epochs were derived by linear interpolation. In the recovery, the full AOHIS signal was estimated, i.e., no a-priori non-tidal de-aliasing is performed, which is possible due to the co-parameterization of daily gravity field parameters [8,9]. Therefore, also no additional errors were assumed for the AOHIS signal. The true world is composed of this AOHIS signal itself, and the reference world does not contain any non-tidal temporal gravity signal (cf. Table 2).

In the following, we describe briefly the different methods applied by the five processing centers. Detailed information can be found in the cited references.

<u>TUM</u>: All simulations were executed with a full-scale numerical mission simulator [25,26], which has already been successfully applied in real data applications to recover satellite-only gravitational field models for CHAMP (Challenging Mini-Satellite Payload), GRACE and GOCE. The simulation environment is based on numerical orbit integration following a multistep method [27], which applies a modified divided difference form of the Adams-Bashforth-Moulton Predict-Evaluate-Correct-Evaluate (PECE) formulas and local extrapolation. The adopted gravity field approach is based on a modification of the integral equation approach from [28] where the orbit is divided into continuous short arcs of 6 h length. The position vectors at the arc node points are set up as unknown parameters, which are estimated together with the gravity field coefficients. As a stochastic

observation model, autocovariances computed from an auto-regressive moving average (ARMA) filter model, approximating the spectral characteristics of the residuals of instrument errors, are used [9].

IGG-CAS: Simulation work was done with a full numerical mission simulator, which is modified from the software tool developed in IGG-CAS for recovering satellite-only gravitational field models from GRACE Level1b observations [29]. The simulation environment is based on numerical orbit integration applying the multistep Gauss-Jackson method [30], which uses start-up value using Runge-Kutta single step integral method. The gravity field recovery is based on the variational equation approach or dynamic method, where the orbit is divided into continuous arcs of 6 h length, and initial state vectors at the beginning of the arc are set up as unknown parameters, which are estimated together with the gravity field coefficients. The 1-cpr effect in the residuals of range-rate observation are modelled [31], which are also estimated together with the gravity field coefficients.

Tongji: The simulations were carried out by using the Satellite Gravimetry Analysis Software (SAGAS) developed by Tongji University [32–35], which has been successfully applied to determine gravity field models from GRACE Level-1b observations. The simulation environment is based on numerical orbit integration, where Adams and Kiogh-Shampine-Gordon (KSG) numerical integration methods are jointly used. The adopted gravity field approach is the modified short-arc approach [32–34], where the observation equation is directly linearized with respect to the real orbit observations and the integral arc is divided into continuous short arcs of 2 h in length. Non-conservation acceleration observation errors are estimated together with the gravity field coefficients according to [33,34]. Variance–covariance matrices for both orbit and range-rate observations are constructed from residuals of the instrument errors on the basis of [35].

WHU: WHU team has developed a software platform which was used in the real data processing of the modern satellite gravity missions such as GRACE and GOCE [36,37]. For the current work on the evaluation of the NGGM schemes described in this paper, the brute-force dynamic method was selected as the basic approach in the numerical simulations [38]. For the non-conservative forces there was only accelerometer noise provided, so the orbit was integrated with the non-gravitational forces being constant zeros. In the parameter estimation step, the arc length is adjusted to 1.5 h for an optimal estimation. The initial satellite states of every arc and low degree geo-potential coefficients up to d/o 15 were processed as the local parameters, and the other geopotential coefficients to degree 70 were estimated as the global ones [23]. Regarding the stochastic model, a filtering strategy of processing band-limited measurements proposed by Schuh [39] was applied to deal with the colored observation noise.

HUST: All simulation works were implemented with a full numerical mission simulator, which has been successfully used to develop the HUST gravity field models [40,41]. The simulation environment is based on numerical orbit integration. The Gauss-Jackson multistep method is used for the numerical integration, which applies start-up values using the Runge-Kutta single step integral method. The adopted gravity field approach is based on the variational equation approach or dynamic method. The orbit is divided into continuous arcs of 6 h length, and initial state vectors at the begin epoch are set as unknown parameters, which are estimated together with the gravity field coefficients. Kinematic empirical parameters for range-rates are estimated to reduce the 1-cpr effect in the residuals of range-rate observations. The filtered predetermined strategy (FPS) was applied to process these kinematic empirical parameters [40]. No stochastic observation model was used in this simulation work.

Table 3 gives an overview of these methods and the specific settings chosen in the gravity retrieval.

Table 3. Gravity retrieval methods and specific settings of the 5 processing centers.

Proc. Center	Numerical Integration Method	Gravity Retrieval Method	Max. Degree of Input Models	Max. Degree of Retrievals (Subcycle/Daily)	Co-estimated Parameters (in addition to Daily Gravity Fields)	Stochastic Model
TUM	Adams-Bashforth-Moulton (5 s)	Integral equations (short-arc; 6-h arcs)	120	70/15	Initial position and velocity (PV) per arc + boundary condition	Yes, autocovariances based on residuals
IGG-CAS	Gauss Jackson 12 Start: Runge-Kutta-Nyström 12/10	Dynamic method (6-h arcs)	70	Sc.12: 70/15 Sc.13: 65/15	Initial PV per arc	No
Tongji	Adams (5 s) and Kiogh-Shampine-Gordon 7	Integral equations (short-arc; 2-h arcs)	120	70/15	Initial positions, daily polynomials of 5th order, bias and scale parameter for acceleration correction per day	Yes, full covariance based on residuals
WHU	Adams-Bashforth-Moulton (5 s) Start: Runge-Kutta 8	Dynamic method (1.5-h arcs)	120	70/15	Initial PV per arc	Yes, ARMA filter model based on residuals
HUST	Gauss-Jackson 14 Start: Runge-Kutta 8	Dynamic method (6-h arcs)	70	70/15	Initial PV per arc	No

All processing centers exploit both orbit data for Global Positioning System (GPS) satellite-to-satellite tracking (SST) in high-low mode, as well as the inter-satellite ranging observations (satellite-to-satellite tracking in low-low mode), by adding the corresponding normal equation (NEQ) systems. No additional constraints such as regularization are applied to the NEQ. As shown in Table 3, the methods mainly differ in the co-estimation of additional parameters and the stochastic modelling of instrument errors. WHU, HUST, and IGG-CAS co-parameterize, in addition to daily gravity parameters, only initial state vectors (orbit positions and velocities) per arc. At TUM, additional boundary conditions between arcs are introduced in order to avoid jumps of consecutive arcs. At Tongji, additional daily parameters (polynomials, bias, scale) are co-estimated.

Table 3 shows that the processing centers have used different maximum degree and order for the computation of the observations (70 vs. 120), but all of them have resolved gravity fields up to d/o 70. However, the spectral leakage effect of the signals beyond d/o 70 is minor compared to the other error sources. If there were a significant spectral leakage effect, it would clearly show up in a severely degraded performance of the very highest resolved degrees [42], which is obviously not the case, as will be shown in Figure 5.

4. Results and Discussions

The simulation scenarios described in Chapter 2 have been processed by all contributing groups applying the methods described in Chapter 3. In this chapter, the results will be compared, assessed, and discussed.

Figure 4 shows the differences of the gravity field solutions of Scenarios A and B to their corresponding references solutions, which are represented by AOHIS mean fields over the 9- and 7-day period, respectively. They are expressed in terms of the degree root mean square (RMS) error of equivalent water heights (EWH), which is computed from fully normalized coefficients of a spherical harmonic series expansion $\left\{\overline{C}_{nm}, \overline{S}_{nm}\right\}$ of degree n and order m, by [43]

$$\sigma_n(EWH) = \frac{a\rho_e}{3\rho_w} \frac{2n+1}{1+k_n} \sqrt{\sum_{m=0}^{n} \overline{C}_{nm}^2 + \overline{S}_{nm}^2} \tag{4}$$

where ρ_w and ρ_e represent the average densities of water and Earth, respectively, a the semi-major axis of the Earth, and k_n the load Love number of degree n. $\left\{\overline{C}_{nm}, \overline{S}_{nm}\right\}$ can either represent the full signal, or the differences between estimated and true coefficients. The EWH expresses the height of a mass-equivalent column of water per unit area. It is used here for comparing the results of five processing centres, because it is one of the standard quantities to express mass change processes in the Earth system, and was also used in the reference document [4] for this purpose. Figure 4 shows all 9- and 7-daily solutions obtained by the TUM processor for a two-month period as thin lines, as well as the mean of the two-month period as solid lines. Even though Scenario B is based on an inclined satellite with a lower orbit altitude, the performance of these two scenarios is very similar. This can be explained by the fact that Scenario A comprises 9-day solutions, while in Scenario B the solutions are based only on 7 days each. This lower number of observations seems to counteract the lower orbit altitude of the second pair. Since the performance of these two scenarios is very similar, in the following, when we will inter-compare the solutions of the five processing centers, we will concentrate on Scenario A.

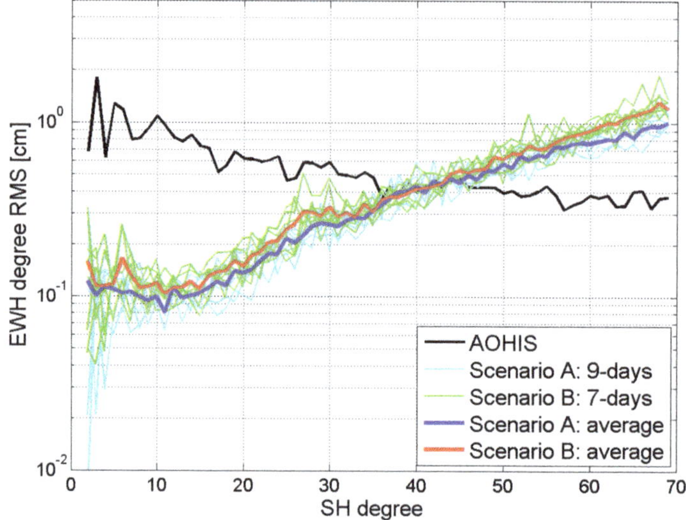

Figure 4. Degree root mean square (RMS) of Scenarios A and B computed with the TUM simulator. The light blue and green lines show the individual 9- and 7-day solutions, respectively, and the blue and red solid lines are the average performance over the two-month period.

Figure 5a shows the degree RMS of the solutions with respect to the AOHIS input signal of the five processing centers for the first 9-day period. Even though there are several differences, they perform reasonably similarly. Apart from a slightly different error level over the whole SH spectrum, evidently the Tongji solutions perform better in the higher degrees, followed by TUM and WHU. A possible reason could be that both processing centers are using full covariance information as a stochastic model (cf. Table 3). In order to analyze the variability of the solutions among themselves, Figure 5b shows again the degree RMS, but now with the TUM solution as the reference. Evidently, the internal variability of the solutions is almost at the same level as their deviation from the input AOHIS signal.

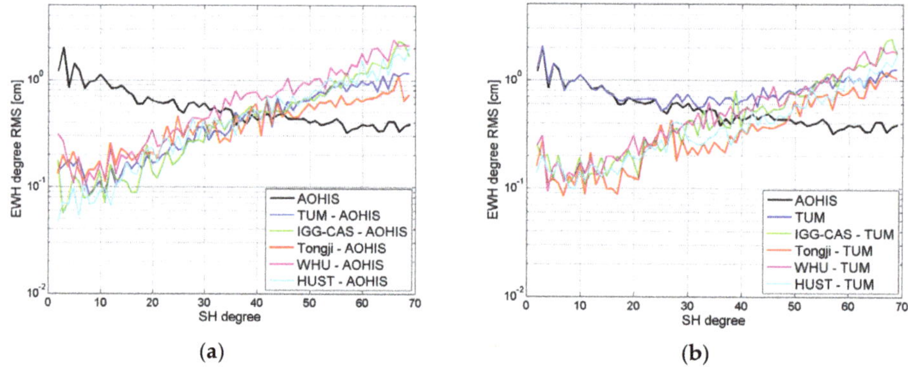

(a)   (b)

Figure 5. Degree RMS of first 9-day period of Scenario A of the five numerical simulators: (**a**) degree RMS with respect to the input AOHIS signal; (**b**) degree RMS with respect to the TUM solution.

In order to analyze these differences in more detail, Figure 6 shows the SH coefficients of the input AOHIS signal averaged over the first 9 days of the analysis period (Figure 6a), as well as the coefficient differences of the five individual solutions from this AOHIS signal (Figure 6b–f). In general, there is a weakness in the estimation of the sectorial and near-sectorial coefficients, which is intrinsic to the in-line inter-satellite ranging concept. It measures mainly in a North–South direction, while the cross-track direction, which is mainly represented by the sectorial coefficients, is determined worse. This general weakness of in-line pairs is also nicely reflected in the estimated formal coefficient standard deviations (Figure 7), representing the main diagonal elements of the variance–covariance matrix.

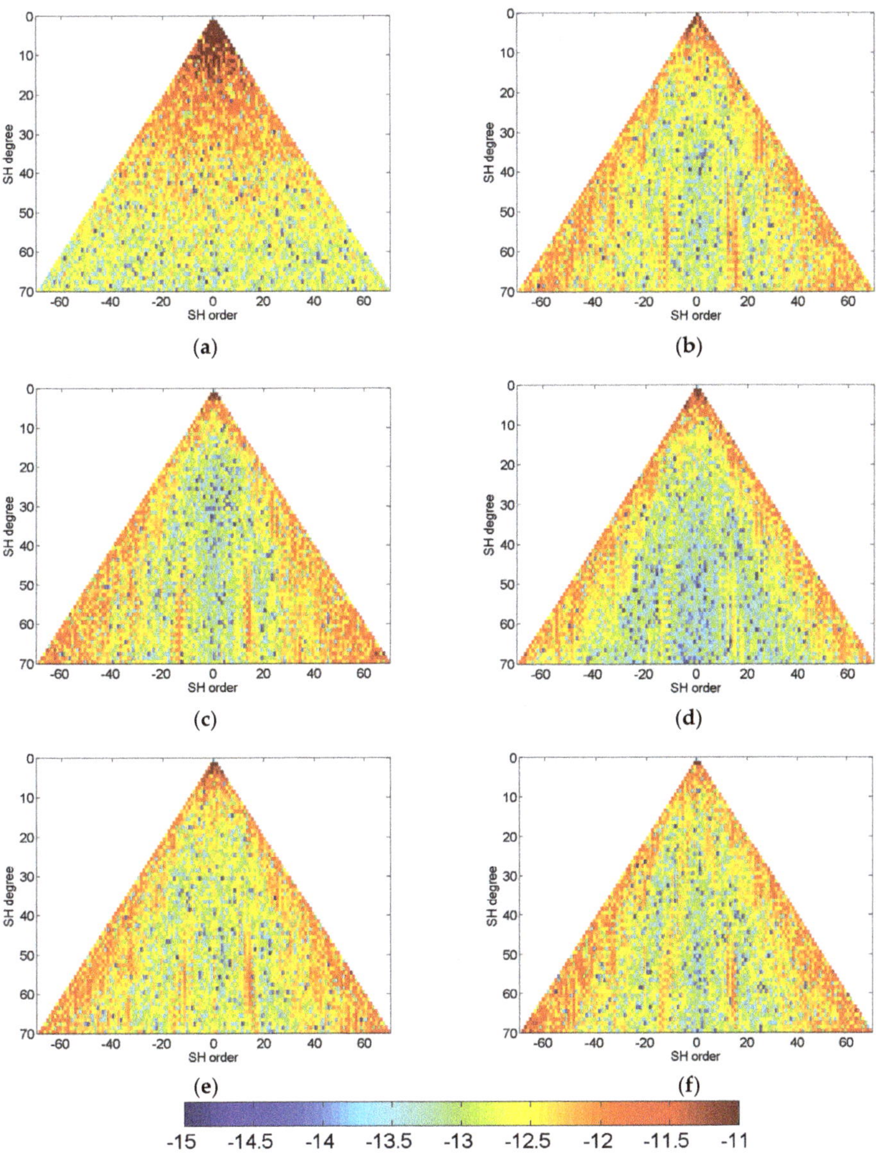

Figure 6. (**a**) Spherical harmonic (SH) coefficients of the AOHIS model, and SH differences to the AOHIS model of the (**b**) TUM; (**c**) IGG-CAS; (**d**) Tongji; (**e**) WHU; (**f**) HUST solutions. Shown is the first 9-day period of Scenario A. Colorbar scale: $\log_{10}(|\ldots|)$.

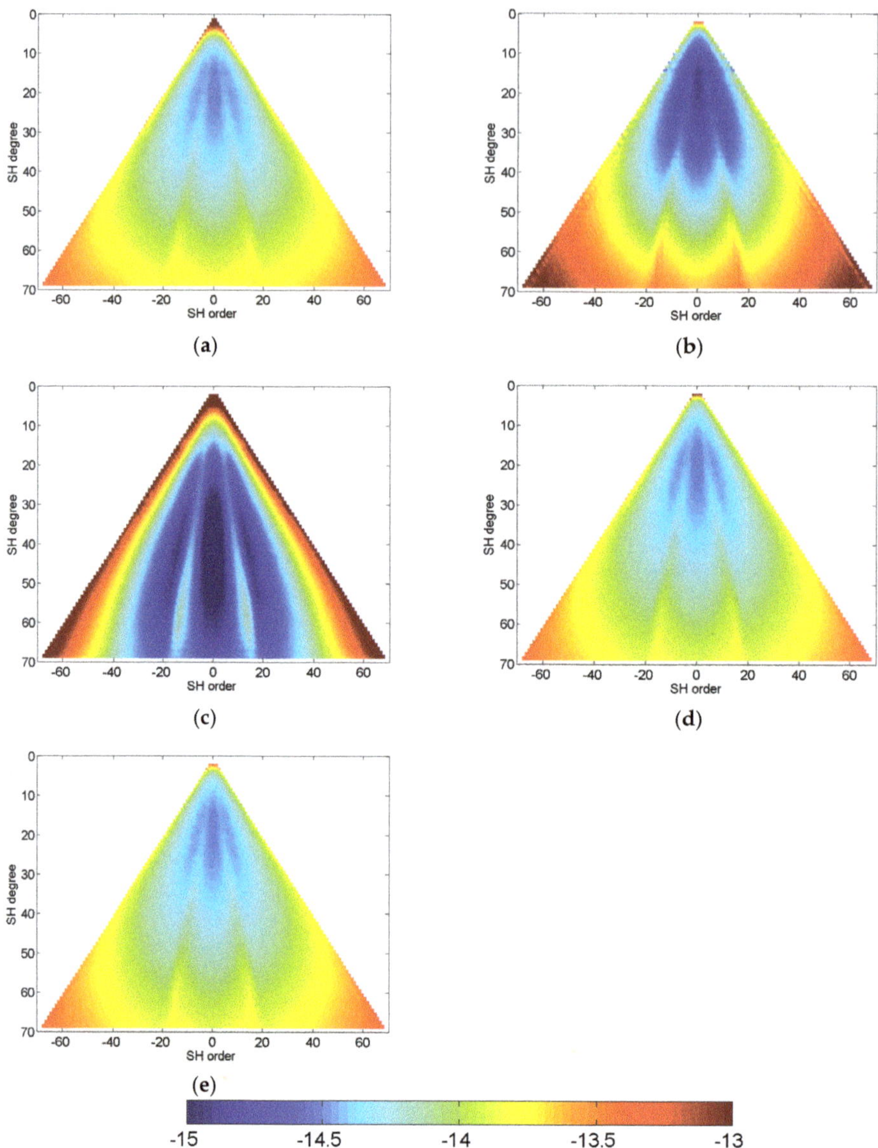

Figure 7. Estimated formal error standard deviations of SH coefficients of the (**a**) TUM; (**b**) IGG-CAS; (**c**) Tongji; (**d**) WHU; (**e**) HUST solutions (empirically scaled). Shown is the first 9-day period of Scenario A. Colorbar scale: $\log_{10}(|\ldots|)$.

A typical feature of Bender-type solutions are the error bands forming an inner triangle, which is directly related to the inclination of the second pair, expressing the fact that the near-polar areas are covered only by observations from the polar satellite pair. Since this feature results from the orbit configuration of the double pair, it is also visible in the estimated error standard deviations (Figure 7). In general, these error standard deviations mainly reflect the observation type, the location of the observations, and the stochastic model of observation errors, but they are blind with respect to signal-related temporal aliasing effects, which affect only the right-hand side of the NEQ system.

An exception would be if these systematic effects are included in the stochastic observation model, as it is done in the Tongji solution. Therefore, the coefficient differences to the true solutions (Figure 6) show a higher error level than the formal error estimates and also additional features, because they are affected by temporal aliasing effects.

In Figure 7, the formal error estimates were scaled by empirically derived factors to make them comparable among each other. This was necessary because the processing centers followed different strategies regarding stochastic modelling (cf. Table 3). Of course, in principle the scaling (calibration) of the formal errors could have been done also by means of a-posteriori variance estimates from the post-fit residuals. However, on the one hand this information was not available for all processing centers, and on the other hand these variance estimates would be hampered by the systematic errors related to temporal aliasing effects anyway.

Analyzing the coefficient differences (Figure 6) and the corresponding scaled statistical error estimates (Figure 7) of the various processing centers in more detail, a striking fact is that the triangular error structure related to the Bender-type constellation of the Tongji solution in Figure 6d is much weaker than for the other processing centers, even though it is clearly visible in the error estimate (Figure 7c). This error triangle of the Tongji solution is hardly visible in the coefficient differences to the true AOHIS model of the Tongji solution and looks generally very different from the other solutions (Figure 7c), with much larger degradation of the sectorial and near-sectorial coefficients. As already addressed above, this could be explained by the fact that the covariance information is derived from the post-fit residuals, which contain also temporal aliasing effects. In the IGG-CAS error estimates (Figure 7b), the contrast between the well-estimated near-zonals of lower SH degrees and worse-estimated near-sectorials of high degree is much larger than in the TUM (Figure 7a), WHU (Figure 7d), and HUST (Figure 7e) solutions. Another issue to mention in the error standard deviations is the transition at harmonic degree 15, which is related to the maximum degree of the daily co-parameterization, which is slightly visible in the IGG-CAS formal errors (Figure 7b), while the other four solutions show a very smooth transition. Analyzing the error estimates of the low-degree SH coefficients also reveals significant differences. The error estimates of the IGG-CAS (Figure 7b) and HUST (Figure 7e) solutions predict much lower errors of the low-degree components relative to the high-degree coefficients than the other solutions, which might indicate that the GPS-SST high-low component has been given a higher relative weight when combining the NEQs. In this sense, the analysis of the error estimates allows deeper insights into the gravity recovery strategies, leading, however, as discussed above to very comparable results in terms of SH coefficient retrieval.

Figure 8a shows the global EHW field derived from the full AOHIS signal based on the ESA AOHIS model [19] (cf. Table 2) of the first 9-day period of Scenario A, and Figure 8b–f show EWH difference grids associated to the coefficient differences shown in Figure 6. These global grids have been evaluated up to the full SH resolution of d/o 70. All EWH difference fields show the typical striping pattern which is mainly resulting from residual temporal aliasing. The Tongji and TUM solutions show the lowest stripiness. In order to quantify these results, Table 4 provides an overview of the main statistical parameters of these EWH (difference) fields. Also here, the Tongji solution shows the smallest RMS deviation from the "true" AOHIS model, followed by the TUM result. Evidently, these statistics are dominated by the remaining high-frequency errors expressed by stripes in the EHW fields. Therefore, we re-do the analysis for fields resolved only up to SH d/o 30. According to Figure 5, at this degree the signal-to-noise ratio is safely above one for all five solutions. The results are shown in Figure 9 and Table 5. Please notice that the scale is reduced from 20 cm to 10 cm for the EWH (difference) fields. In this d/o 30 analysis, now the IGG-CAS solution performs best, followed by the results of TUM, WHU, and HUST. These three solutions are practically on the same error level. This is also consistent with the degree RMS results shown in Figure 5.

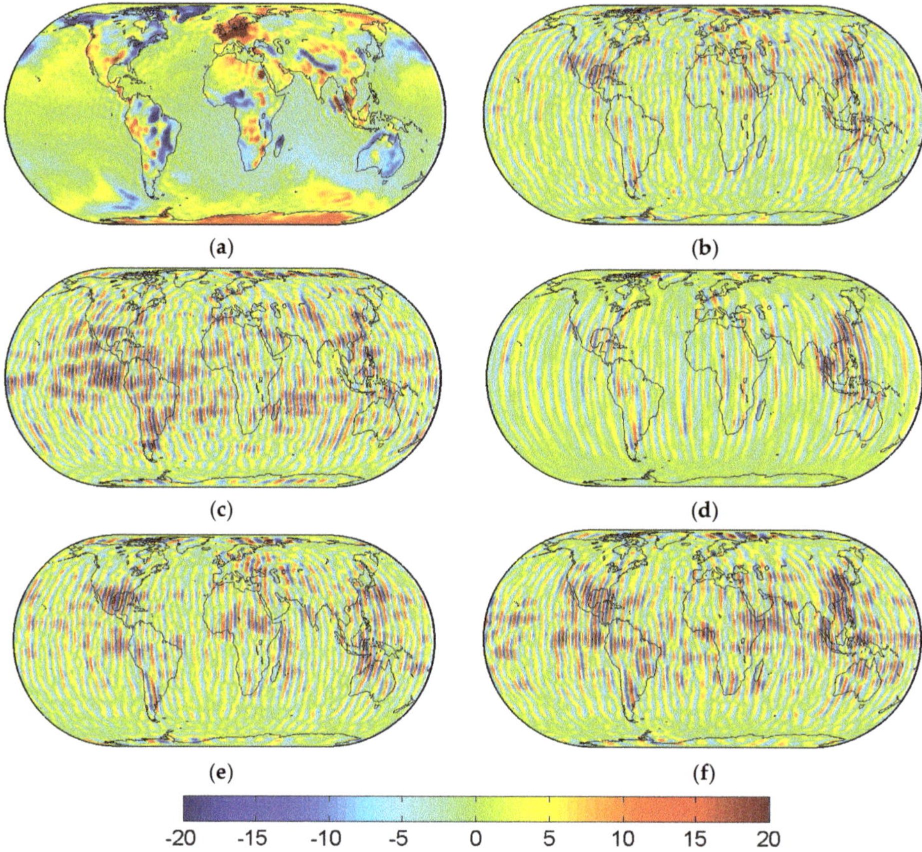

Figure 8. (**a**) Equivalent water heights (EWH) (cm) grid related to the input AOHIS model up to d/o 70, and EWH (cm) differences to this AOHIS model of the (**b**) TUM; (**c**) IGG-CAS; (**d**) Tongji; (**e**) WHU; (**f**) HUST solutions. Shown is the first 9-day period of Scenario A.

Table 4. Main statistical parameters of EWH input and EWH difference fields up to SH d/o 70 of the first 9-day period of Scenario A.

Model (d/o 70)	Min (cm)	Max (cm)	RMS (cm)
AOHIS	−50.38	66.84	5.46
TUM – AOHIS	−37.62	36.69	4.80
IGG-CAS – AOHIS	−29.62	30.43	6.36
Tongji – AOHIS	−31.46	26.71	3.91
WHU – AOHIS	−31.97	32.58	5.41
HUST – AOHIS	−40.68	29.56	6.11

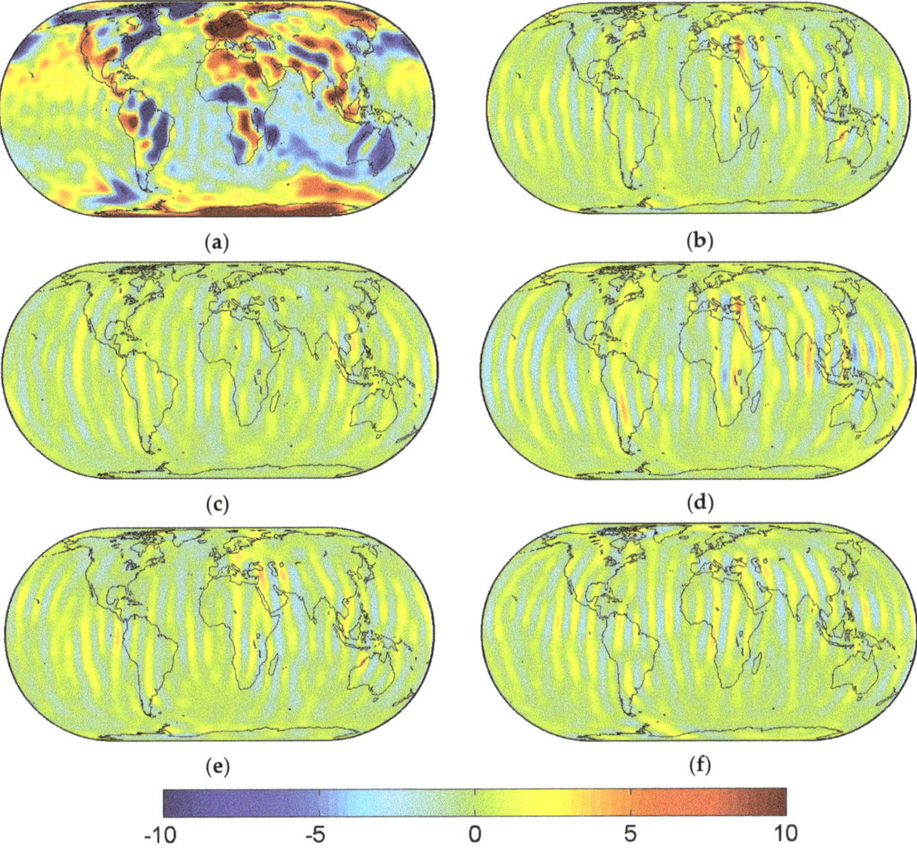

Figure 9. (**a**) EWH (cm) grid related to the input AOHIS model up to d/o 30, and EWH (cm) differences to this AOHIS model of the (**b**) TUM; (**c**) IGG-CAS; (**d**) Tongji; (**e**) WHU; (**f**) HUST solutions. Shown is the first 9-day period of Scenario A.

Table 5. Main statistical parameters of EWH input and EWH difference fields up to SH d/o 30 of the first 9-day period of Scenario A.

Model (d/o 30)	Min (cm)	Max (cm)	RMS (cm)
AOHIS	−25.83	24.90	4.74
TUM − AOHIS	−3.63	5.42	0.98
IGG-CAS − AOHIS	−3.14	3.66	0.83
Tongji − AOHIS	−5.46	5.43	1.27
WHU − AOHIS	−3.76	4.55	1.01
HUST − AOHIS	−3.76	4.85	1.02

As a general conclusion, at degree 30 the RMS variation among the five solutions is in the order of 10% of the total signal RMS. Solutions for other time periods show very similar behavior to the ones presented above, and are therefore not explicitly presented and discussed in this paper.

As already mentioned in Chapter 3, all processing centers co-estimated daily gravity field parameters up to SH d/o 15 in order to reduce temporal aliasing effects. Figure 10 shows the resulting degree error RMS curves, which represent differences of daily estimates to the daily AOHIS "true" signals (black curve), averaged over the whole recovery period. Also here, the performance of these daily solutions are very comparable among the processing centers, where IGG-CAS shows a slightly better performance than TUM, HUST, and WHU, and only the Tongji daily estimates are worse.

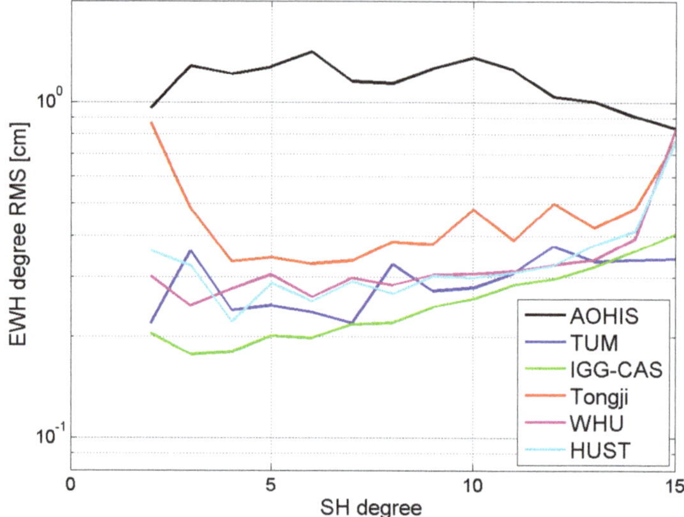

Figure 10. Degree RMS of co-parameterized daily gravity solutions up to d/o 15, averaged over the whole recovery period.

5. Conclusions and Outlook

Temporal gravity retrieval results of a future Bender-type double pair mission concept performed by five processing centers of a Sino-European NGGM study team based on jointly defined mission scenarios have been inter-compared and assessed. In spite of some remaining differences, the solutions of the contributing teams TU Munich, Institute of Geodesy and Geophysics, Tongji University, Wuhan University, and Huazhong University of Science and Technology show quite similar performances. In general, these results demonstrate that one of the main goals of this inter-comparison exercise, the consistency of simulation results among the contributing groups, could be achieved to a large extent. Remaining differences might be explained by different processing strategies and partially different parameterizations, such as the co-parameterization of accelerometer biases and drifts, or empirical parameters to compensate for long-wavelength errors in the observation time series. Some remaining differences also exist in the error level of the co-estimated daily solutions.

As already mentioned above, the main study goal was not to identify a "winner" among the contributing groups, but rather to achieve consistency of numerical simulation results of NGGM concepts to the best possible extent, even though five independent software packages, which are based partly on different gravity retrieval methods, have been applied. This positive result is an important pre-requisite and basis for future joint activities towards the realization of NGGMs.

Based on this successful inter-comparison exercise of numerical simulators and in view of the fact that in the future several satellite pairs might be in orbit in parallel, further joint studies on optimized constellations of triple- and multi-pairs will be performed, and their impact on the main fields of applications, such as hydrology, glaciology, and solid Earth physics, will be quantified. The results of these studies will be presented in a follow-up paper.

Author Contributions: Conceptualization, R.P., H.-C.Y., W.F., Z.L., Y.S., M.Z., H.X.; Methodology, R.P., Q.C., M.H., A.P., X.X., B.Z., H.Z., X.Z.; Software, Q.C., M.H., A.P., X.X., B.Z., H.Z., B.L., Y.Z., X.Z.; Validation of results, A.P., R.P.; Writing—original draft preparation, R.P.; Writing—review and editing, R.P., Q.C., W.F., R.H., M.H., Z.L., A.P., Y.S., C.W., X.X., B.Z., H.Z., B.L., Y.Z., X.Z., H.Y., M.Z.; visualization, R.P.

Funding: This research was performed in the framework of the study "Assessment of satellite constellations for monitoring the variations in Earth gravity field (ADDCON)", ESA-ESTEC, Contract AO/1-7317/12/NL/AF funded by the European Space Agency. This study was also funded by the National Natural Science Foundation of China (projects no. 41704012, 41931074 and 41731069), and the Strategic Priority Research Program of the Chinese Academy of Sciences (grant nos. XDB23030100 and XDA15017700).

Acknowledgments: We further acknowledge the contribution of the ADDCON project partners Pieter Visser, Nicolas Sneeuw, Wei Liu, Johannes Engels, Qiang Chen, Tonie van Dam and Thomas Gruber for a successful completion of the ESA project, as well as Christian Siemes and Luca Massotti for their valuable support during this project.

Conflicts of Interest: The authors declare no conflict of interest.

References

1. Tapley, B.D.; Bettadpur, S.; Watkins, M.; Reigber, C. The gravity recovery and climate experiment experiment, mission overview and early results. *Geophys. Res. Lett.* **2004**, *31*. [CrossRef]
2. Flechtner, F.; Neumayer, K.H.; Dahle, C.; Dobslaw, H.; Fagiolini, E.; Raimondo, J.C.; Günter, A. What can be expected from the GRACE-FO Laser Ranging Interferometer for Earth Science applications? *Surv. Geophys.* **2016**, *37*, 453–470. [CrossRef]
3. Drinkwater, M.R.; Floberghagen, R.; Haagmans, R.; Muzi, D.; Popescu, A. GOCE: ESA's first Earth Explorer Core mission. In *Earth Gravity Field from Space - from Sensors to Earth Science*; Space Sciences Series of ISSI; Beutler, G., Drinkwater, M.R., Rummel, R., Steiger, R., Eds.; Kluwer Academic: Dordrecht, The Netherlands, 2003; Volume 18, pp. 419–432. ISBN 1-4020-1408-2.
4. Pail, R.; Bingham, R.; Braitenberg, C.; Dobslaw, H.; Eicker, A.; Güntner, A.; IUGG Expert Panel. Science and user needs for observing global mass transport to understand global change and benefit society. *Surv. Geophys.* **2014**, *36*, 743–772. [CrossRef]
5. Murböck, M.; Pail, R.; Daras, I.; Gruber, T. Optimal orbits for temporal gravity recovery regarding temporal aliasing. *J. Geod.* **2015**, *88*, 113–126. [CrossRef]
6. Panet, I.; Flury, J.; Biancale, R.; Gruber, T.; Johannessen, J.; van den Broeke, M.R.; van Dam, T.; Gegout, P.; Hughes, C.W.; Ramillien, G.; et al. Earth System Mass Transport Mission (e.motion): A Concept for future Earth Gravity Field Measurements from Space. *Surv. Geophys.* **2012**, *34*, 141–163. [CrossRef]
7. Bender, P.L.; Wiese, D.N.; Nerem, R.S. A possible Dual-GRACE mission with 90 degree and 63 degree inclination orbits. In Proceedings of the 3rd International Symposium on Formation Flying, Missions and Technologies, Noordwijk, The Netherlands, 23–25 April 2008.
8. Wiese, D.N.; Nerem, R.S.; Han, S.C. Expected improvements in determining continental hydrology, ice mass variations, ocean bottom pressure signals, and earthquakes using two pairs of dedicated satellites for temporal gravity recovery. *J. Geophys. Res. Solid Earth* **2011**, *116*. [CrossRef]
9. Daras, I.; Pail, R. Treatment of temporal aliasing effects in the context of next generation satellite gravimetry missions. *J. Geophys. Res.* **2017**, *122*, 7343–7362. [CrossRef]
10. Gruber, T.; Panet, I.; E.motion2 Team. Proposal to ESA's Earth Explorer Call 9: Earth System Mass Transport Mission—E.motion2. Deutsche Geodätische Kommission der Bayerischen Akademie der Wissenschaften, Reihe B, Angewandte Geodäsie, Series B. 2015. Available online: https://dgk.badw.de/fileadmin/user_upload/Files/DGK/docs/b-318.pdf (accessed on 12 November 2019).
11. Pail, R.; Bamber, J.; Biancale, R.; Bingham, R.; Braitenberg, C.; Eicker, A.; Flechtner, F.; Gruber, T.; Güntner, A.; Heinzel, G.; et al. Mass variation observing system by high low inter-satellite links (MOBILE)—a new concept for sustained observation of mass transport from space. *J. Geod. Sci.* **2019**. [CrossRef]
12. Hauk, M.; Pail, R. Gravity field recovery by high-precision high-low inter-satellite links. *Remote Sens.* **2019**, *11*, 537. [CrossRef]
13. NAS—National Academies of Sciences, Engineering, and Medicine. *Thriving on Our Changing Planet: A Decadal Strategy for Earth Observation from Space*; The National Academies Press: Washington, DC, USA, 2018. Available online: https://essp.nasa.gov/essp/files/2018/02/2017-Earth-Science-Decadal-Survey.pdf (accessed on 12 November 2019).
14. Lu, Y.; Li, H.; Yeh, H.C.; Luo, J. A self-analyzing double-loop digital controller in laser frequency stabilization for inter-satellite laser ranging. *Rev. Sci. Instrum.* **2015**, *86*, 044501. [CrossRef]

15. Liang, Y.R.; Duan, H.Z.; Xiao, X.L.; Wei, B.B.; Yeh, H.C. Inter-satellite laser range-rate measurement by using digital phase locked loop. *Rev. Sci. Instrum.* **2015**, *86*, 016106. [CrossRef] [PubMed]

16. Zhang, J.Y.; Ming, M.; Jiang, Y.Z.; Duan, H.Z.; Yeh, H.C. Inter-satellite laser link acquisition with dual-way scanning for Space Advanced Gravity Measurements mission. *Rev. Sci. Instrum.* **2018**, *89*, 064501. [CrossRef] [PubMed]

17. Luo, J.; Chen, L.S.; Duan, H.Z.; Gong, Y.G.; Hu, S.; Ji, J.; Liu, Q.; Mei, J.; Milyukov, V.; Sazhin, M.; et al. TianQin: A space-borne gravitational wave detector. *Class. Quantum Gravity* **2016**, *33*, 035010. [CrossRef]

18. Mayer-Gürr, T.; Pail, R.; Gruber, T.; Fecher, T.; Rexer, M.; Schuh, W.D.; Kusche, J.; Brockmann, J.M.; Rieser, D.; Zehentner, N.; et al. The combined satellite gravity field model GOCO05s. Geophysical Research Abstracts, EGU2015-12364. In Proceedings of the European Geosciences Union General Assembly 2015, Vienna, Austria, 12–17 April 2015; Volume 17.

19. Dobslaw, H.; Flechtner, F.; Bergmann-Wolf, I.; Dahle, C.; Dill, R.; Esselborn, S.; Sasgen, I.; Thomas, M. Simulating high-frequency atmosphere-ocean variability for dealiasing of satellite gravity observations: AOD1b RL05: ATMOSPHERE-OCEAN MASS VARIABILITY: AOD1b. *J. Geophys. Res. Oceans* **2013**, *118*, 3704–3711. [CrossRef]

20. Savcenko, R.; Bosch, W. *EOT11A—Empirical Ocean Tide Model from Multi-Mission Satellite Altimetry*; Deutsches Geodätisches Forschungsinstitut (DGFI): München, Germany, 2012.

21. Ray, R.D. *A Global Ocean Tide Model from Topex/Poseidon Altimetry: Got99.2*; Technical Report; NASA Technical Memorandum 209478; NASA: Washington, DC, USA, 1999.

22. Iran Pour, S.; Reubelt, T.; Sneeuw, N.; Daras, I.; Murböck, M.; Gruber, T.; Pail, R.; Weigelt, M.; van Dam, T.; Visser, P.; et al. *Assessment of Satellite Constellations for Monitoring the Variations in Earth Gravity Field—SC4MGV*; ESA—ESTEC Contract No. AO/1-7317/12/NL/AF, Final Report; University of Stuttgart: Stuttgart, Germany, 2015.

23. Wiese, D.N.; Visser, P.; Nerem, R.S. Estimating low resolution gravity fields at short time intervals to reduce temporal aliasing errors. *Adv. Space Res.* **2011**, *48*, 1094–1107. [CrossRef]

24. Purkhauser, A.; Pail, R. Next generation gravity missions: Near-real time gravity field retrieval strategy. *Geophys. J. Int.* **2019**, *217*. [CrossRef]

25. Daras, I.; Pail, R.; Murböck, M.; Yi, W. Gravity field processing with enhanced numerical precision for LL-SST missions. *J. Geod.* **2015**, *89*, 99–110. [CrossRef]

26. Daras, I. *Gravity Field Processing Towards Future LL-SST Satellite Missions*; DGK Series C, Dissertation No. 770; Verlag der Bayerischen Akademie der Wissenschaften: Munich, Germany, 2016; pp. 23–39. ISSN 0065-5325.

27. Shampine, L.F.; Gordon, M.K. *Computer Solution of Ordinary Differential Equations: The Initial Value Problem*; W.H. Freeman: San Francisco, CA, USA, 1975.

28. Schneider, M. Outline of a general orbit determination method. In *Space Research IX, Proceedings of the Open Meetings of Working Groups (OMWG) on Physical Sciences of the 11th Plenary Meeting of the Committee on Space Research (COSPAR), Tokyo, Japan, 9–21 May 1968*; Mitteilungen aus dem Institut für Astronomische und Physikalische Geodäsie, Nr. 51; Champion, K.S.W., Smith, P.A., Smith-Rose, R.L., Eds.; North Holland Publ. Company: Amsterdam, The Netherlands, 1969; pp. 37–40.

29. Wang, C.; Xu, H.; Zhong, M.; Feng, W. Monthly gravity field recovery from GRACE orbits and K-band measurements using variational equations approach. *Geod. Geodyn.* **2015**, *6*, 253–260. [CrossRef]

30. Berry, M.M.; Healy, L.M. Implementation of Gauss-Jackson Integration for Orbit Propagation. *J. Astronaut.* **2004**, *52*, 331–357.

31. Kim, J. Simulation Study of a Low-Low Satellite-to-Satellite Tracking Mission. Ph.D. Thesis, The University of Texas at Austin, Austin, TX, USA, 2000.

32. Chen, Q.; Shen, Y.; Zhang, X.; Hsu, H.; Chen, W.; Ju, X.; Lou, L. Monthly gravity field models derived from GRACE level 1b data using a modified short arc approach. *J. Geophys. Res. Solid Earth* **2015**, *120*, 1804–1819. [CrossRef]

33. Chen, Q.; Shen, Y.; Chen, W.; Zhang, X.; Hsu, H. An improved GRACE monthly gravity field solution by modeling the non-conservative acceleration and attitude observation errors. *J. Geod.* **2016**, *90*, 503–523. [CrossRef]

34. Chen, Q.; Shen, Y.; Francis, O.; Chen, W.; Zhang, X.; Hsu, H. Tongji-Grace02s and Tongji-Grace02k: High-precision static GRACE-only global Earth's gravity field models derived by refined data processing strategies. *J. Geophys. Res. Solid Earth* **2018**, *123*, 6111–6137. [CrossRef]

35. Chen, Q.; Shen, Y.; Chen, W.; Francis, O.; Zhang, X.; Chen, Q.; Li, W.; Chen, T. An optimized short-arc approach: Methodology and application to develop refined time series of Tongji-Grace2018 GRACE monthly solutions. *J. Geophys. Res.* **2019**, *124*, 6010–6038. [CrossRef]

36. Zou, X.C.; Zhong, L.P.; Li, J.C. Simultaneous solution for precise satellite orbit and earth gravity model. *Chin. J. Geophys.* **2016**, *59*, 2413–2423.

37. Xu, X.; Zhao, Y.; Reubelt, T.; Tenzer, R. A GOCE only gravity model GOSG01S and the validation of GOCE related satellite gravity models. *Geod. Geodyn.* **2017**, *8*, 260–272. [CrossRef]

38. Reigber, C. Gravity field recovery from satellite tracking data. In *Theory of Satellite Geodesy and Gravity Field Determination*; Lecture Notes in Earth Sciences; Sansò, F., Rummel, R., Eds.; Springer: Berlin/Heidelberg, Germany, 2005; Volume 25, pp. 197–234.

39. Schuh, W.D. The processing of band-limited measurements; filtering techniques in the least squares context and in the presence of data gaps. *Space Sci. Rev.* **2003**, *108*, 67–78. [CrossRef]

40. Zhou, H.; Luo, Z.; Zhou, Z.; Li, Q.; Zhong, B.; Lu, B.; Hsu, H. Impact of different kinematic empirical parameters processing strategies on temporal gravity field model determination. *J. Geophys. Res. Solid Earth* **2018**, *123*, 252–276. [CrossRef]

41. Zhou, H.; Zhou, Z.; Luo, Z. A new hybrid processing strategy to improve temporal gravity field solution. *J. Geophys. Res. Solid Earth* **2019**, *124*. [CrossRef]

42. Goossens, S. Applying spectral leakage corrections to gravity field determination from satellite tracking data. *Geophys. J. Int.* **2010**, *181*, 1459–1472. [CrossRef]

43. Wahr, J.; Moleneer, M.; Bryan, F. Time variability of the Earth's gravity field: Hydrological and oceanic effects and their possible detection using GRACE. *J. Geophys. Res. Solid Earth* **1998**, *103*, 30205–30229. [CrossRef]

Article

Combination Analysis of Future Polar-Type Gravity Mission and GRACE Follow-On

Yufeng Nie [1], Yunzhong Shen [1,*] and Qiujie Chen [1,2]

[1]   College of Surveying and Geo-Informatics, Tongji University, Shanghai 200092, China; yufeng.nie@tongji.edu.cn

[2]   Institute of Geodesy and Geo-information, University of Bonn, 53115 Bonn, Germany; chenqiujie2009@163.com

*   Correspondence: yzshen@tongji.edu.cn

Received: 20 December 2018; Accepted: 18 January 2019; Published: 21 January 2019

Abstract: Thanks to the unprecedented success of Gravity Recovery and Climate Experiment (GRACE), its successive mission GRACE Follow-On (GFO) has been in orbit since May 2018 to continue measuring the Earth's mass transport. In order to possibly enhance GFO in terms of mass transport estimates, four orbit configurations of future polar-type gravity mission (FPG) (with the same payload accuracy and orbit parameters as GRACE, but differing in orbit inclination) are investigated by full-scale simulations in both standalone and jointly with GFO. The results demonstrate that the retrograde orbit modes used in FPG are generally superior to prograde in terms of gravity field estimation in the case of a joint GFO configuration. Considering the FPG's independent capability, the orbit configurations with 89- and 91-degree inclinations (namely FPG-89 and FPG-91) are further analyzed by joint GFO monthly gravity field models over the period of one-year. Our analyses show that the FPG-91 basically outperforms the FPG-89 in mass change estimates, especially at the medium- and low-latitude regions. Compared to GFO & FPG-89, about 22% noise reduction over the ocean area and 17% over land areas are achieved by the GFO & FPG-91 combined model. Therefore, the FPG-91 is worthy to be recommended for the further orbit design of FPGs.

Keywords: gravity field recovery; GRACE Follow-On; orbit configuration; synergistic observation

---

1. Introduction

The Earth is undergoing complicated and continuous mass transport [1]. To better quantify the mass changes, the Gravity Recovery and Climate Experiment (GRACE) mission was launched in 2002, which consists of the twin satellites flying in the same near-polar orbit at about 500-km altitude and separated by 220 km [2]. The key technology that enables GRACE to observe large-scale mass variations near the Earth surface is the onboard K-band microwave ranging system (KBR), which delivers inter-satellite range measurements with an accuracy at micron level to precisely capture orbit perturbation differences between the two GRACE satellites caused by heterogeneous mass distribution of the Earth [2,3]. Along with the KBR, other important sensors onboard GRACE satellites include Global Positioning System (GPS) receivers for precise orbit determination (POD), high-precision accelerometers for measuring non-gravitational forces and star cameras for attitude determination [4–6]. Based on the precise data collected by the above sensors, many scientific institutions have produced high-quality static and temporal gravity field models [7–11], which are successfully adopted for quantitative researches concerning the Earth's mass changes, such as global mean sea-level variations [12,13], continental water balance [14,15], and the Tibetan Plateau and Antarctic regional ice losses [16,17].

In view of such a great success achieved by GRACE in terms of the Earth's mass change exploration, its successor GRACE Follow-On (GFO) was launched on 22 May 2018 to continue providing the geodetic observation records [18,19]. As a demonstrator, GFO carries a laser ranging interferometer (LRI) for measuring inter-satellite ranges with an accuracy level of a few tens nanometers in addition to the sensors onboard the GRACE satellites [18–20]. However, even with such a dramatic accuracy improvement for the range observations, some full-scale pre-launch simulations indicated that only a moderate improvement on gravity field estimation can be reached by GFO as compared to GRACE [18,21]. The dominant barrier to further improvement on gravity field determination by GFO is the temporal aliasing issue, which is more intractable than other limiting factors in the GFO error budget [18,21].

From a digital signal processing aspect, aliasing is due to undersampling of the signal to be recovered. According to the Nyquist sampling theorem, when sampling frequency is smaller than twice of the original signal frequency, aliasing definitely occurs [22]. In terms of GRACE monthly gravity field recovery, temporal aliasing is mainly attributed to those high-frequency signals originating from the oceanic tidal variations at daily or sub-daily scale, as well as the short-term energy exchange process (generally within a few hours) between atmosphere and ocean [23,24]. Because of the limited bandwidth for observations and sparse ground track of GRACE, most part of these signals typically need to be reduced by priori background models during gravity field recovery process [25], which are, however, subject to errors [26,27]. Therefore, the remaining high-frequency signals result in the temporal aliasing errors in monthly-average gravity solutions, which manifest themselves as the typical north-south stripes in spatial pattern of unfiltered GRACE models [28,29].

To mitigate the troublesome temporal aliasing effects, post-processing filtering techniques are generally applied as a standard procedure [28–30]. However, the fundamental solution should be based on the enhanced signal sampling and gravity field modelling strategy [31–34]. One of the most straightforward and efficient ways is to increase the number of satellites in space for denser sampling of gravity signals. Thereby, many researchers have investigated novel concepts of satellite formation flight (SFF) or satellite constellations dedicated to gravity field recovery, including the cartwheel, pendulum, and Bender-type configurations [35–39]. Among all, the Bender-type SFF, formed by two pairs of GRACE-type inline satellites (in polar and inclined orbits), is most promising from both technological and economic aspects. Therefore, it is selected as the primary candidate for the Next Generation Gravity Mission (NGGM) concepts [40,41] by the European Space Agency (ESA). During the service period of GFO, other Future Polar-type Gravity mission (FPG) will be probably launched [42]. Therefore, the primary motivation of this article is to investigate the optimal FPG orbit configuration in combination with GFO, under the precondition that the FPG has the same payload accuracy as GRACE and comparable capability of recovering global static and temporal gravity fields to GRACE. With this in mind, the search space for FPG's key orbit elements (i.e., orbit altitude and inclination) is greatly reduced. Considering a five-year nominal mission lifetime for FPG, its orbit altitude is similar to GRACE. As such, the emphasis of FPG orbit adjustment will be put on the orbit inclination selection issue.

The rest of this article is organized as follows. Section 2 presents the details of full-scale simulations for FPG orbit selection and performance evaluation. Gravity field results from the standalone FPG, with different inclinations, and jointly with GFO are demonstrated in Section 3. In Section 4, we explain the underlying mechanism of the presented results and further discuss the practical consideration of FPG orbit design. Conclusions are given in Section 5.

2. Numerical Simulation Scheme

2.1. Mathematical Model for Gravity Field Recovery

The dynamic approach is adopted in our gravity field model recovery [43], which is widely used in real GRACE data processing by different scientific institutions [7,8,44–46]. In this section, we outline basic formulas of the dynamic approach used in the article, as well as relevant references for more detailed descriptions.

The Earth gravity field model can be expressed via a set of Spherical Harmonic Coefficients (SHC) $u = \{\overline{C}_{lm}, \overline{S}_{lm}\}$ with following Spherical Harmonic (SH) expansion,

$$V(\rho, \varphi, \lambda; u) = \frac{GM}{\rho}\left[1 + \sum_{l=2}^{L_{MAX}} \sum_{m=0}^{l} \left(\frac{R_E}{\rho}\right)^l (\overline{C}_{lm}\cos m\lambda + \overline{S}_{lm}\sin m\lambda) \cdot \overline{P}_{lm}(\sin\varphi)\right] \tag{1}$$

where V is the gravitational potential of an arbitrary point on or outside the Earth with the spherical coordinate of radius ρ, latitude φ and longitude λ, in the Earth-fixed frame. GM is the product of the gravitational constant and the Earth mass, and R_E is the Earth's mean equatorial radius; l and m stand for the degree and order of SH expansion, respectively, with L_{MAX} being the maximum degree of the expansion; $\overline{P}_{lm}$ is the fully normalized associate Legendre function of degree l and order m.

The estimation of SHC u from satellite observations is based on the Newton's law of motion [47], which is formulated as a three-dimensional second-order nonlinear ordinary differential equation (ODE),

$$\ddot{r}(t) = a(t; r(t), \dot{r}(t), p, u) \tag{2}$$

where $r(t)$, $\dot{r}(t)$, and $\ddot{r}(t)$ denote the vectors of satellite's position, velocity and acceleration at epoch t, respectively. a is the force acting on the unit mass of the satellite, where p represents the parameters vector for the force models except the Earth gravity field. In the case of GRACE, p is referred to accelerometer calibration parameters (e.g., bias and scale parameters) [48]. When the force models are given, the position and velocity of satellite's orbit are determined as follows [49],

$$\begin{cases} r(t) = r(t_0) + (t - t_0) \cdot \dot{r}(t_0) + \int_{t_0}^{t}\int_{t_0}^{\tau} a(\tau'; r(\tau'), r(\tau'), p, u)d\tau'd\tau \\ \dot{r}(t) = \dot{r}(t_0) + \int_{t_0}^{t} a(\tau; \dot{r}(\tau), r(\tau), p, u)d\tau \end{cases} \tag{3}$$

where $(r^T(t_0), \dot{r}^T(t_0))^T$ denote the position and velocity vectors at initial epoch t_0 of an orbit arc. According to Equation (3), the position and velocity vectors at any epoch t within the orbit arc are the function of parameter vector $x = (w^T, u^T)^T$, where $w = (r^T(t_0), \dot{r}^T(t_0), p^T)^T$ is the arc-specific parameters vector. Linearizing the Equation (3) with respect to the approximate vector x_0 of parameters, we have,

$$\begin{cases} r(t) = r^{REF}(t) + \frac{\partial r^{REF}(t)}{\partial x_0}\delta x \\ \dot{r}(t) = \dot{r}^{REF}(t) + \frac{\partial \dot{r}^{REF}(t)}{\partial x_0}\delta x \end{cases} \tag{4}$$

where the vectors of position $r^{REF}(t)$ and velocity $\dot{r}^{REF}(t)$, named as the reference orbit, are numerically integrated with Equation (3) by replacing x with x_0. δx is the correction vector to x_0. Substituting $r(t) = r^{OBS}(t) + v_r(t)$ into the first equation of Equation (4), we get the observational equation for position at epoch t,

$$v_r(t) = \frac{\partial r^{REF}(t)}{\partial x_0}\delta x - \left[r^{OBS}(t) - r^{REF}(t)\right] = \frac{\partial r^{REF}(t)}{\partial x_0}\delta x - \Delta r(t) \tag{5}$$

where $r^{OBS}(t)$ and $v_r(t)$ represent the position observation and its correction at epoch t. The inter-satellite range rate $\dot{\rho}(t)$ between GRACE-A and B can be expressed as,

$$\dot{\rho}(t) = e_{AB}^T(t)\dot{r}_{AB}(t) \tag{6}$$

where, the subscripts A and B denote the values of GRACE-A and B, $\dot{r}_{AB}(t) = \dot{r}_B(t) - \dot{r}_A(t)$ is the velocity difference vector and $e_{AB}(t) = r_{AB}(t)/\sqrt{r_{AB}^T(t)r_{AB}(t)}$ with $r_{AB} = r_B - r_A$. By substituting

$\dot{\rho}(t) = \dot{\rho}^{OBS}(t) + v_{\dot{\rho}}(t)$ and the second equation of Equation (4) into Equation (6) for both GRACE-A and B, we get the following observational equation for range rate measurement at epoch t,

$$v_{\dot{\rho}}(t) = \frac{\partial \dot{\rho}^{REF}(t)}{\partial x_0} \delta x - \left[\dot{\rho}^{OBS}(t) - \dot{\rho}^{REF}(t) \right] = \frac{\partial \dot{\rho}^{REF}(t)}{\partial x_0} \delta x - \Delta \dot{\rho}(t) \tag{7}$$

where, $\dot{\rho}^{REF}(t)$ is computed with the reference orbit, $\dot{\rho}^{OBS}(t)$ and $v_{\dot{\rho}}(t)$ stand for the range rate observation and its correction, respectively. The partial derivatives $\frac{\partial r^{REF}(t)}{\partial x_0}$ and $\frac{\partial \dot{r}^{REF}(t)}{\partial x_0}$ are numerically integrated by solving the so-called 'variational equation' and $\frac{\partial \dot{\rho}^{REF}(t)}{\partial x_0}$ is the combination of them, which can be referred to [50] (pp. 240–243), [51] (pp. 30–32) and [49] for detailed descriptions. The observational equations in the form of Equations (5) and (7) for all epochs within one arc, i.e., arc j, are collected together and briefly expressed in the following form,

$$v_j = A_j \delta x_j - y_j = \left(\begin{array}{cc} A_{w_j} & A_{u_j} \end{array} \right) \left(\begin{array}{c} \delta w_j \\ \delta u \end{array} \right) - y_j \tag{8}$$

where v_j denotes the vector of all corrections, A_j, A_{w_j} and A_{u_j} are the design matrices, and y_j is the vector of all constant terms of the arc j. The Normal Equation (NEQ) of the arc based least squares adjustment is derived as [48],

$$\left(\begin{array}{cc} N_{w_j w_j} & N_{w_j u_j} \\ N_{w_j u_j}^T & N_{u_j u_j} \end{array} \right) \left(\begin{array}{c} \delta w_j \\ \delta u \end{array} \right) = \left(\begin{array}{c} R_{w_j} \\ R_{u_j} \end{array} \right) \tag{9}$$

where $N_{w_j w_j} = A_{w_j}^T P_j A_{w_j}$, $N_{u_j u_j} = A_{u_j}^T P_j A_{u_j}$, $N_{w_j u_j} = A_{w_j}^T P_j A_{u_j}$ and $R_{w_j} = A_{w_j}^T P_j y_j$, $R_{u_j} = A_{u_j}^T P_j y_j$. P_j denotes the weight matrix constructed based on the precision of observations. To recover δu with observations of multiple arcs, such as one month, the arc-specific parameters δw_j should be pre-eliminated arc-wise, forming the reduced normal equation. After that, reduced normal equations are summed up to J arcs, and then the SHC parameter δu is obtained by solving Equation (10),

$$\widetilde{N}_J \cdot \delta u = \widetilde{R}_J \tag{10}$$

where $\widetilde{N}_J = \sum_J \widetilde{N}_j$ and $\widetilde{R}_J = \sum_J \widetilde{R}_j$. $\widetilde{N}_j$ and $\widetilde{R}_j$ are coefficient matrix and the right-hand side of the reduced normal equation of arc j, with $\widetilde{N}_j = N_{u_j u_j} - N_{w_j u_j}^T N_{w_j w_j}^{-1} N_{w_j u_j}$ and $\widetilde{R}_j = R_{u_j} - N_{w_j u_j}^T N_{w_j w_j}^{-1} R_{w_j}$ [48].

As regard to the GFO and FPG combination, it can be carried out on NEQ level by Equation (11),

$$(\widetilde{N}_J^{GFO} + \widetilde{N}_J^{FPG}) \cdot \delta u = (\widetilde{R}_J^{GFO} + \widetilde{R}_J^{FPG}) \tag{11}$$

where $\widetilde{N}_J^{GFO}, \widetilde{R}_J^{GFO}$ and $\widetilde{N}_J^{FPG}, \widetilde{R}_J^{FPG}$ are established by Equation (10) of GFO and FPG, respectively. It should be mentioned here that the same variance of unit weight must be employed in constructing the weight matrices of GFO and FPG, otherwise the two normal equations cannot be simply summed up as in Equation (11).

2.2. Orbit Configuration Selection for FPG

In gravity field recovery, the orbit altitude and inclination play the key role [52] (pp. 41–43). Since the gravity signals attenuate rapidly with the increase of orbit altitude [47], the ideal gravimetric satellite should fly as low as possible in order to be sufficiently sensitive to the signals. However, the lower altitude of the satellite, the stronger atmospheric drag acting on it, which requires more fuels to maintain the mission operation and thus shortens its lifetime [37]. As regard to the orbit inclination, it could not be better to select the 90-degree inclination for a global coverage of the Earth. However, due to the launch limit, it is a challenge to put satellites in an exact polar orbit. For example, GRACE was flying in a near-polar orbit with 89-degree inclination, which left two 1-degree-radius circles in the south and north poles (namely the polar gaps). In general, a large polar gap will significantly impact the estimation of zonal SHC and eventually corrupt the estimated gravity field. The maximum recoverable degree and order (d/o) of SHC in the case of a given g-degree polar gap can be roughly calculated via dividing 180 (degree) by g [51] (p. 205).

Considering the twin satellites of GFO separated by 220 km both in the near-circular orbit of 490-km altitude and 89-degree inclination [18], its initial orbit configuration in the following simulations will be kept fixed as listed in Table 1. In consideration of a nominal five-year mission lifetime, the FPG initial orbit altitude is currently supposed to be the same as GRACE (namely 500 km) [2]. With regard to the inclination choice of FPG, the relationship between the polar gap and the maximum recoverable SHC d/o, as mentioned above, should be taken into consideration. For typical monthly temporal gravity field model up to 60 d/o, it requires the polar gap smaller than 3 degrees, while for static gravity field models up to 180 d/o, the polar gap should be within 1 degree in theory. Once the polar gap size is determined, there exist two kinds of orbits, namely the prograde orbit (inclination smaller than 90 degrees) and the retrograde orbit (inclination larger than 90 degrees) [53] (p. 99). For example, the previously dedicated gravity missions CHAllenging Mini-satellite Payload (CHAMP) (87-degree inclination) [54] and GRACE (89-degree inclination) are in prograde orbits, while Gravity field and steady-state Ocean Circulation Explorer (GOCE) (96.5-degree inclination) [55] is in a retrograde orbit. In general, the retrograde-orbit satellites require more fuel than the prograde ones during launch in order to overcome the Earth's rotation speed, but it is not a technical problem to launch either prograde or retrograde satellites.

In the following simulation study, we preliminarily select four sets of FPG configurations, which differ only in orbit inclinations. Their initial orbit elements are listed in Table 1 and denoted as FPG-87, FPG-89, FPG-91, and FPG-93 (corresponding to the inclinations of 87, 89, 91, and 93 degrees, respectively). Note that the initial Right Ascension of Ascending Nodes (RAAN) are selected as 0 degree for GFO, FPG-87, FPG-91, and FPG-93 while 180 degrees for FPG-89, which is designed to avoid the nearly full overlay of ground tracks of GFO and FPG-89 when combining the two [32]. The initial inter-satellite distances for all GFO and FPG configurations are fixed as 220 km.

Table 1. Initial orbit configurations of GFO and FPG for simulation study

Orbit Configuration	Altitude (km)	Inclination (deg.)	Eccentricity	Inter-Satellite Distance (km)
GFO	490	89	0.0015	220
FPG-87		87		
FPG-89	500	89	0.0015	220
FPG-91		91		
FPG-93		93		

2.3. Numerical Simulation Procedures

The simulation procedures involving the forward and the backward simulation steps [18] are shown in Figure 1. In the forward simulation step, we apply a set of true background models to generate the simulated observations. The true background models are listed in Table 2, where EIGEN6C4 [56] is used as the true static gravity field model and EOT11a (including main wave and interpolated secondary wave) [57] is treated as the true ocean tide model. For non-tidal variation signals, the AOHIS component of ESM is adopted to represent the integrated signal of atmosphere (A), oceans (O), terrestrial hydrological water (H), continental ice-sheets (I), and the solid earth (S) [24]. The maximum degree and order for all models is chosen as 100 [18]. Given the initial orbit elements defined in Table 1, the satellite orbits for both GFO and FPG can be generated by numerical integration [50] (pp. 117–146). The combined eighth-order Runge–Kutta and eighth-order

Gauss–Jackson integrator with an integration step of 5 s is applied to generate the error-free data, which include satellite orbit positions and inter-satellite range-rates. To produce more realistic observations, above error-free data should be added with sensor noises. In this work, the sensor noises are treated as Gaussian white noise with zero expectation [31,32], while more sophisticated noise models can be found in [58]. The standard deviations (SD) of major onboard sensors' noises, listed in Table 3, are based on their performances in real GRACE data. As shown, 1-cm SD is selected for orbit positions [31–33]. With regard to the range-rate data, we add noise with SD of 0.2 μm/s to KBR measurements in GFO and FPG [18,45], while with SD of 10 nm/s to LRI measurements in GFO [18]. Although the non-gravitational forces are not simulated during orbit integration, the accelerometer's noise with SD of 0.3 nm/s^2 is directly simulated during the integration of orbit [52] (pp. 50–51). We simulate full-year observations in 2006 for all orbit configurations, which are subsequently used in the backward simulation step for gravity field model recovery.

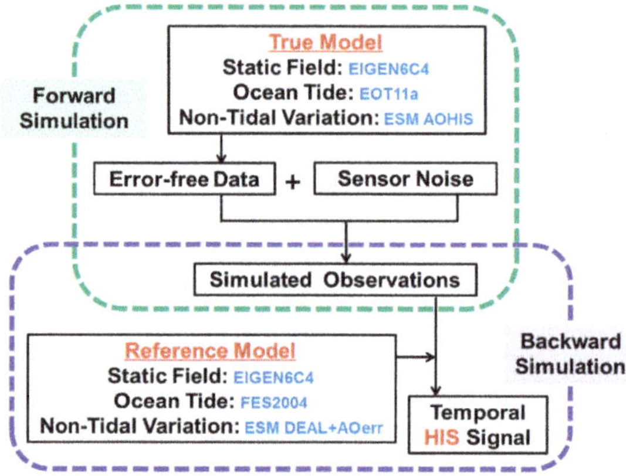

Figure 1. Processing flow of the forward and the backward simulations.

Table 2. Background models for numerical simulation. True models are used for simulated observations generation while the reference models are applied to gravity field recovery.

Force Categories	True Model	Reference Model
Static Field	EIGEN6C4 (100 d/o)	EIGEN6C4 (100 d/o)
Ocean Tide	EOT11a (100 d/o)	FES2004 (100 d/o)
Non-tidal Variation	ESM AOHIS (100 d/o)	ESM DEAL + AOerr (100 d/o)

In the backward simulation step, we recover the monthly gravity field model up to 100 d/o [18] by the dynamic approach using above simulated observations. For parameterizations, we choose the orbit arc length as 6 h for the initial position and velocity estimation [44–46], and estimate the accelerometer parameters (bias and linear drift) every 1.5 h (orbital period) [45]. Observations' weight matrices in Equation (9) are constructed based on corresponding SD values listed in Table 3 with the same variance of unit weight. The reference background models, used for reference orbit and partial derivatives integration in Equations (5) and (7), are listed in Table 2. As shown, FES2004 [59] is applied as reference ocean tide model, and DEAL plus AOerr components of ESM [27] are used to represent the AO components of AOHIS in true background models [24]. Since the emphasis of the article is on temporal gravity model recovery, the true and reference static models are identical [21,33,35]. The differences between true and reference background models are designed to introduce the temporal aliasing errors to the recovered gravity model [18,21,33,37]. As shown in Figure 1, the temporal signal we want to recover in each monthly model is the average HIS of this month, which is the integrated signal of terrestrial hydrological water (H), continental ice-sheets (I), and the solid earth (S) [18,24].

Table 3. The standard deviations (SD) of white Gaussian sensor noises used for the generation of simulated observations

Obs. Type	Orbit Pos.	GFO/FPG-KBR	GFO-LRI	Accelerometer
SD Value	1 cm	0.2 µm/s	10 nm/s	0.3 nm/s^2

2.4. Evaluation Metrics for Recovered Gravity Field Model

In a simulated world, the quality of the recovered gravity field model can be easily evaluated by comparing the true and the recovered fields. In this article, the evaluation metrics for recovered gravity field models are derived based on monthly SHC differences $\{\Delta C_{lm}, \Delta S_{lm}\}$, which are calculated by subtracting the recovered model from the true static model (EIGEN6C4 100 d/o) plus the true monthly HIS signal (100 d/o) as shown in Equation (12) according to [18],

$$\Delta C_{lm} = \overline{C}_{lm}^{EIGEN6C4+HIS} - \overline{C}_{lm}^{REC}, \Delta S_{lm} = \overline{S}_{lm}^{EIGEN6C4+HIS} - \overline{S}_{lm}^{REC} \tag{12}$$

where the superscripts *EIGEN6C4*, *HIS*, and *REC* represent the SHC from EIGEN6C4, monthly HIS and the recovered models in simulations. Given SHC differences, the degree geoid height error (DGHE) and cumulative geoid height error (CGHE) of the recovered model can be calculated by

$$DGHE(n) = R_E \cdot \sqrt{\sum_{m=0}^{l} (\Delta C_{lm}^2 + \Delta S_{lm}^2)}, (l = n)$$
$$CGHE(n) = R_E \cdot \sqrt{\sum_{l=2}^{n} \sum_{m=0}^{l} (\Delta C_{lm}^2 + \Delta S_{lm}^2)} \tag{13}$$

where $DGHE(n)$ represents the DGHE of degree n and $CGHE(n)$ represents the CGHE up to degree n. One can also construct spatial-domain metrics by the spatial differences plot (SDP) [60] (pp. 24–25). For the k-th grid point on the Earth surface, its geoid height difference can be calculated by Equation (14) [60] (p. 22), with (1) and (12),

$$\Delta N_k(\varphi_k, \lambda_k) = R_E \cdot \sum_{l=2}^{L_{MAX}} \sum_{m=0}^{l} (\Delta C_{lm} \cos m\lambda_k + \Delta S_{lm} \sin m\lambda_k) \cdot \overline{P}_{lm}(\sin \varphi_k) \tag{14}$$

where ΔN_k stands for the geoid height difference, and other symbols are the same as those defined in Equation (1); L_{MAX} is chosen as 100. Based on Equation (14), the SDP can be created by plotting all grids' ΔN_k within a specific spatial range Ω on the world map given corresponding geographic locations. Ω can be global or regional, e.g., oceans, Amazon basin. The grid size (interval) is chosen as 1-by-1 degree in the followings. Additionally, the latitude-weighted root mean square (wRMS) [60] (pp. 24–25) of above SDP can be calculated by Equation (15),

$$wRMS(\Omega) = \sqrt{\frac{\sum_{\forall k \in \Omega} (\Delta N_k^2 \cdot \cos \varphi_k)}{\sum_{\forall k \in \Omega} \cos \varphi_k}} \tag{15}$$

where all grids within Ω are summed based on their area weights reflected by latitudes.

Since metrics in (13) and (15) represent absolute errors (positive values) of the recovered model, comparisons between any two recovered models, in terms of DGHE, CGHE, or wRMS, can be carried out by calculating error reduction rate (ERR) of the model i over the model j as shown in Equation (16),

$$ERR_j^i(q) = \frac{q_j - q_i}{q_j} \times 100\% \tag{16}$$

where q can be any value of DGHE, CGHE and wRMS.

3. Results and Analysis

3.1. Standalone Models of FPG

We first evaluate standalone models recovered by different FPG configurations listed in Table 1. The results in January (2006) are demonstrated in Figure 2, while other months show similar results. In Figure 2, the black curve represents the true average HIS signal in January (an integrated signal of terrestrial hydrological water, continental ice-sheets, and the solid earth [24]), while other curves are the degree geoid height errors (DGHE) of different FPG standalone models in this month. Observing signal (in black) and error (in colored) curves, we find that the recovered models are gradually dominated by noises after degree $n = 26$, which is indicated by the intersection point of signal curve and error curves around degree 26. Among four FPG standalone models, the FPG-91 in general produces smaller errors especially for degrees beyond 60. When looking to low-degree SHC near the intersection point, we find that the retrograde FPGs (FPG-91 and FPG-93) outperform the prograde ones (FPG-89 and FPG-87). The expected degradations of FPG-87 and FPG-93 due to their larger polar gaps compared to FPG-89 and FPG-91, however, is not apparent in Figure 2. This can be explained by denser ground tracks gained by FPG-87 and FPG-93 in the mid and low latitudes, which finally compensate the negative effects caused by large missing polar data, as discussed in [51] (pp. 204–214).

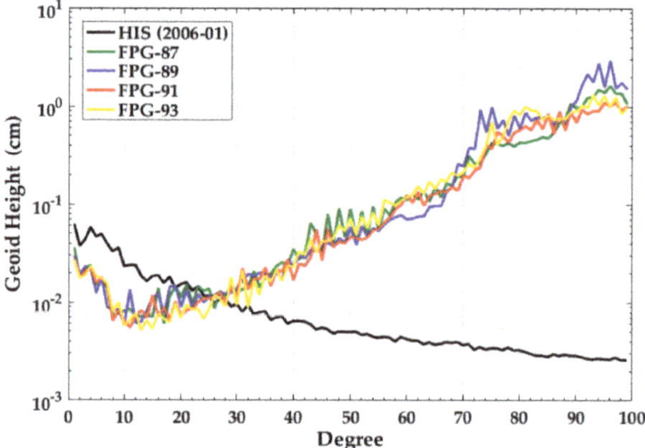

Figure 2. True HIS signal and degree geoid height errors (DGHE) of different FPG standalone models in January 2006.

Further comparisons are carried out in terms of absolute SHC differences ($|\Delta C_{lm}|$, $|\Delta S_{lm}|$) of each degree l and order m, as drawn in Figure 3. Subfigures (a), (b), (c), and (d) stand for results of FPG-87, FPG-89, FPG-93, and FPG-91, respectively. Now we can clearly observe the deficiencies of FPG-87 and FPG-93 in zonal SHC, i.e., C_{l0}, estimation, which are demonstrated by larger $|\Delta C_{l0}|$ values in (a) and (c) compared to that in (b) and (d). Recall the approximate relationship between the polar gap size and the maximum recoverable gravity signal in Section 2.2, the weaknesses of FPG-87 and FPG-93, with 3-degree polar gaps, are expected to become much more pronounced in terms of high-degree, e.g., 180 d/o [11], static gravity model recovery.

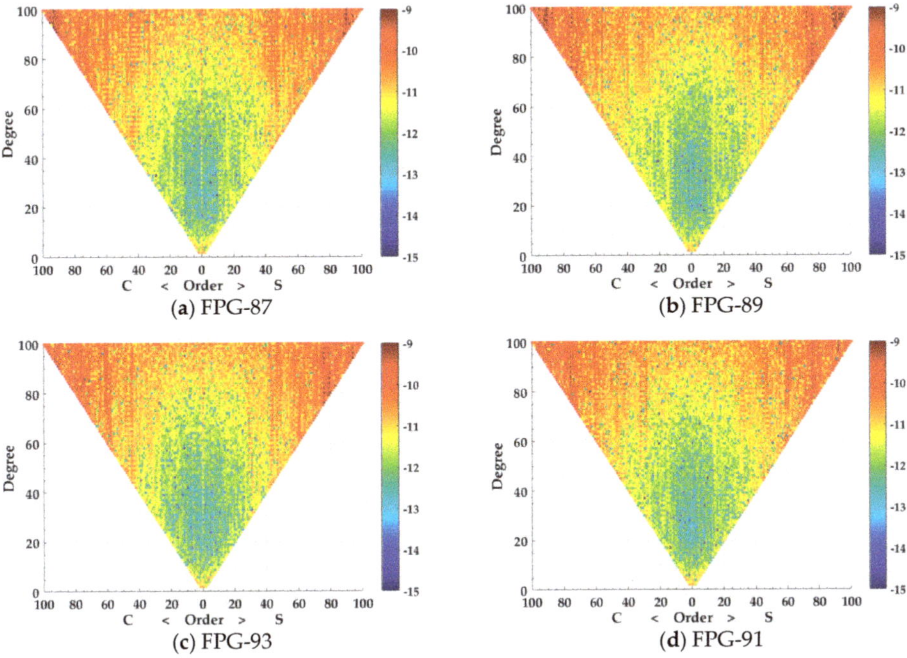

Figure 3. Absolute spherical harmonic coefficients (SHC) differences of different FPG standalone models in January 2006: (**a**) FPG-87; (**b**) FPG-89; (**c**) FPG-93; (**d**) FPG-91.

3.2. Combined Models of GFO-LRI and FPG

Since GFO carries both LRI and KBR, the combinations between GFO and FPG will be carried out in GFO-LRI & FPG and GFO-KBR & FPG modes respectively. In the GFO-LRI & FPG mode, the information from GFO-LRI will dominate the combined model because it has much larger weight in Equations (9) and (11) than the KBR of FPG, based on Table 3. Therefore, we can expect only minor differences among different GFO-LRI & FPG combined models. Even so, comparisons can still be carried out by calculating the error reduction rate (ERR) of different GFO-LRI & FPG models with respect to the GFO-LRI standalone model. Here, we use the ERR in terms of DGHE, and therefore set $q_i = DGHE_{GFO\text{-}LRI \text{ \& } FPG}$ and $q_j = DGHE_{GFO\text{-}LRI}$ in Equation (16). Figure 4 plots the degree-wise $ERR_{GFO\text{-}LRI}^{GFO\text{-}LRI \text{ \& } FPG}(DGHE)$ for all four GFO-LRI & FPG models, which can be regarded as contributions of adding different FPGs' information to the GFO-LRI data.

As expected, the contributions are not significant. Among four combined models, GFO-LRI & FPG-93 has the largest ERR for the majority of SHC, and GFO-LRI & FPG-91 is next after that. Moreover, even with the same polar gap, retrograde FPGs (FPG-91 and FPG-93) outperform the prograde ones (FPG-89 and FPG-87). This can be attributed to the more isotropic integrated system of GFO & retrograde-FPG than GFO & prograde-FPG, as learnt from Bender-type SFF design [37]. More specifically, the orbit inclination differences between the GFO (in prograde orbit) and retrograde FPGs are larger than that in GFO & prograde-FPGs, which enables the GFO & FPG integrated system to observe the gravity signals in a more isotropic sense. However, degradations occur at some very low degrees of the combined models, which is likely due to insufficiently perfect data weighting between GFO-LRI and FPG [61].

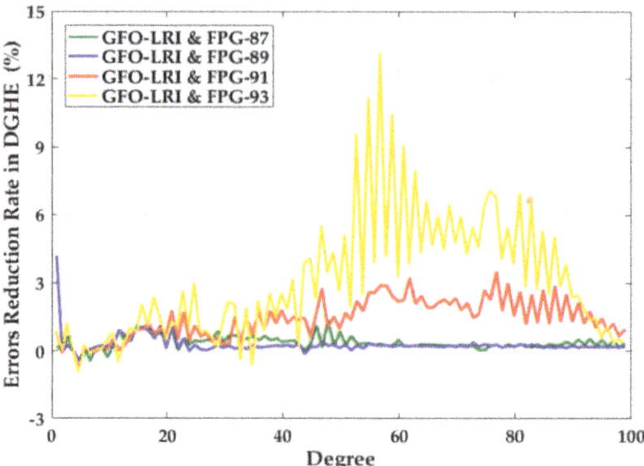

Figure 4. Error reduction rate (ERR) of different GFO-LRI & FPG combined models with respect to the GFO-LRI standalone model in terms of degree geoid height error (DGHE) in January 2006.

3.3. Combined Models of GFO-KBR and FPG

More representative results can be demonstrated by GFO-KBR & FPG combined models. In Figure 5, we plot the true HIS signal (black curve) of January 2006, as well as error curves (in colored) of four GFO-KBR & FPG models in terms of DGHE in this month. Basically, the results are comparable with those in Figure 4, that is, the retrograde FPGs generally outperform prograde ones for their lower DGHE. Table 4 records the ERR of GFO-KBR & FPG combined models with respect to the GFO-KBR standalone model in terms of cumulative geoid height error (CGHE) in Equation (13), which is calculated by setting $q_i = CGHE_{GFO\text{-}KBR\ \&\ FPG}$ and $q_j = CGHE_{GFO\text{-}KBR}$ in Equation (16). It shows that the retrograde FPGs start highlighting their advantages after degree 50.

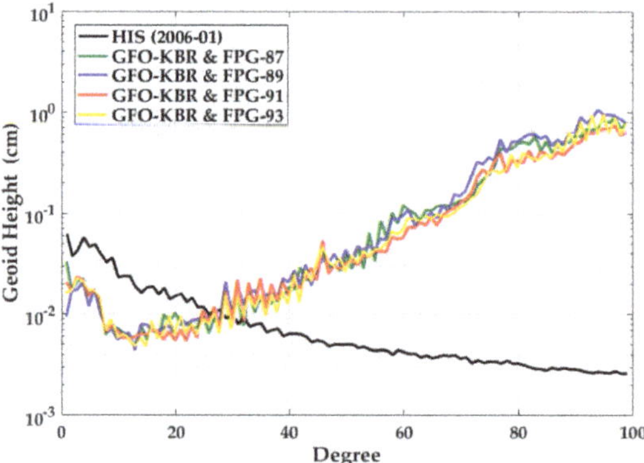

Figure 5. True HIS signal and degree geoid height errors (DGHE) of different GFO-KBR & FPG combined models in January 2006.

Table 4. Error reduction rate (ERR) of different GFO-KBR & FPG combined models with respect to the GFO-KBR standalone model in terms of cumulative geoid height error (CGHE) in January 2006

Degree	GFO-KBR & FPG-87	GFO-KBR & FPG-89	GFO-KBR & FPG-91	GFO-KBR & FPG-93
25	−3.6%	16.7%	7.9%	7.9%
50	17.5%	15.1%	14.6%	16.7%
75	31.1%	16.7%	33.6%	41.6%
100	28.0%	14.7%	39.5%	31.7%

To observe the error patterns in spatial domain, we draw the global spatial differences plots (SDP), introduced in Section 2.4, of above four GFO-KBR & FPG combined models. As shown in Figure 6, obviously fewer stripes can be found in the retrograde-orbit based GFO-KBR & FPG combined models, comparing subfigure (a) GFO-KBR & FPG-87 with (c) GFO-KBR & FPG-93 or (b) GFO-KBR & FPG-89 with (d) GFO-KBR & FPG-91. These typical north-south stripes are, to a large extent, the product of temporal aliasing errors, which are caused by errors in both ocean tide model and Atmosphere and Ocean De-Aliasing (AOD) product [25,26]. It seems that the retrograde FPGs, in combination with GFO, are more effective on reducing the temporal aliasing errors than prograde ones, especially in low-latitude areas within $(30°S, 30°N)$. However, only marginal improvement can be observed over high-latitude regions in Figure 6.

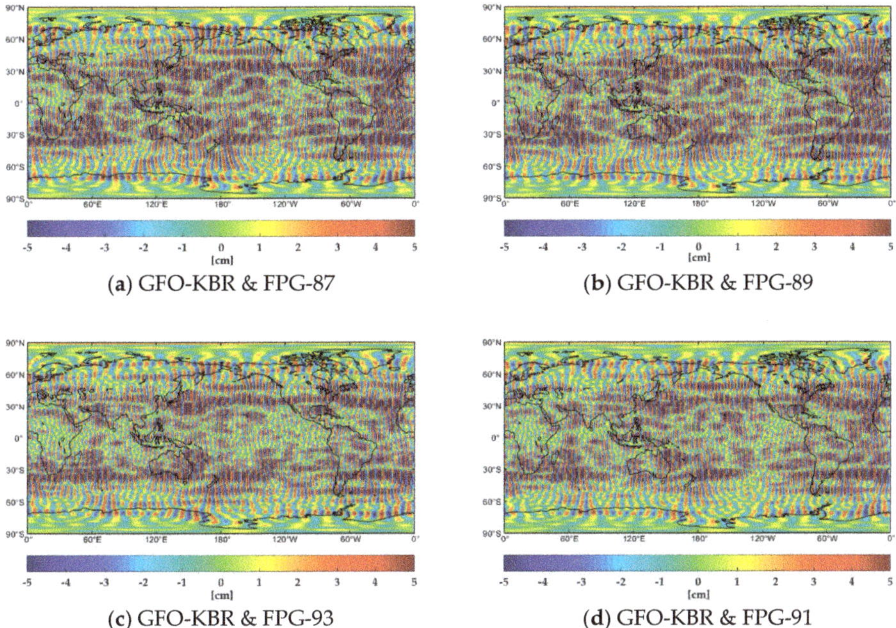

Figure 6. Global spatial differences plots (SDP) of different GFO-KBR & FPG combined models in January 2006: (**a**) GFO-KBR & FPG-87; (**b**) GFO-KBR & FPG-89; (**c**) GFO-KBR & FPG-93; and (**d**) GFO-KBR & FPG-91.

3.4. One-Year GFO-combined Model Series of FPG-89 and FPG-91

In view of FPG-87 and FPG-93's deficiencies in zonal SHC determination, we therefore choose FPG-89 and FPG-91 in further GFO-joint gravity model recovery for the entire year 2006. We produce 12 monthly GFO-KBR & FPG models for FPG-89 and FPG-91, respectively. Monthly wRMS values of their global SDPs are demonstrated in Figure 7. As expected, GFO-KBR & FPG-91 outperforms the GFO-KBR & FPG-89 for its lower wRMS values over the whole year. Furthermore, we can observe slight variations in wRMS values of GFO-KBR & FPG-91 and GFO-KBR & FPG-89 models from month to month. Therefore, April and October are selected for further illustrations with corresponding results plotted in Figure 8.

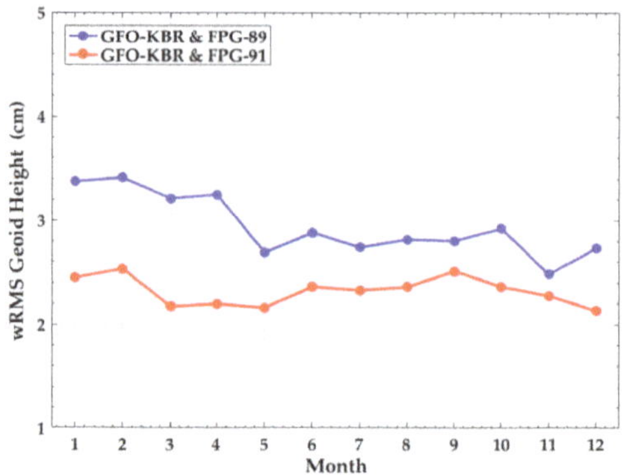

Figure 7. Monthly latitude-weighted root mean square (wRMS) of global spatial differences plots (SDP) for GFO-KBR & FPG-89 and GFO-KBR & FPG-91 in 2006.

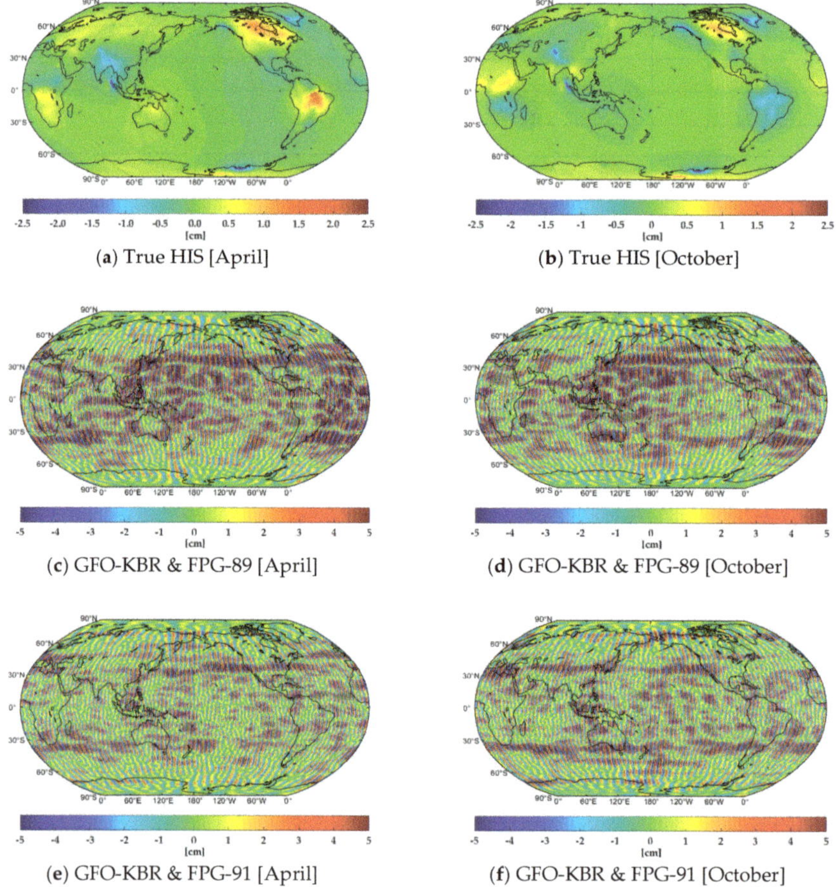

(**a**) True HIS [April]

(**b**) True HIS [October]

(**c**) GFO-KBR & FPG-89 [April]

(**d**) GFO-KBR & FPG-89 [October]

(**e**) GFO-KBR & FPG-91 [April]

(**f**) GFO-KBR & FPG-91 [October]

Figure 8. True monthly HIS signal spatial plots and global spatial differences plots (SDP) of GFO-KBR & FPG-89 and GFO-KBR & FPG-91 in April and October 2006: (**a**) HIS signal in April; (**b**) HIS signal in October; (**c**) SDP of GFO-KBR & FPG-89 in April; (**d**) SDP of GFO-KBR & FPG-89 in October; (**e**) SDP of GFO-KBR & FPG-91 in April; (**f**) SDP of GFO-KBR & FPG-91 in October.

On a global scale, results in April and October are plotted in the left and right columns of Figure 8, respectively. Subfigures in the top row represent the spatial signal plots of monthly true HIS signals, which are created the same way as SDP except that the $(\Delta C_{lm}, \Delta S_{lm}) = (\overline{C}_{lm}^{HIS}, \overline{S}_{lm}^{HIS})$ in Equation (14); The middle and bottom rows represent SDPs of GFO-KBR & FPG-89 and GFO-KBR & FPG-91, respectively. As shown, the GFO-KBR & FPG-91 apparently suffers less from the temporal aliasing errors, especially in low- and medium-latitude areas. The error reductions in these areas are beneficial to the scientific study of global water cycle since the world's major river basins (e.g., the Amazon and Yangtze Rivers), as well as the majority of oceans, are located in the low and medium latitudes.

On a regional scale, yearly-average wRMS of GFO-KBR & FPG-89 and GFO-KBR & FPG-91 are respectively calculated based on their regional SDPs. The regions that we select basically represent typical basins in low, medium, and high latitudes as marked in Figure 9. Besides, the ocean and land areas are also separately analyzed. As listed in Table 5, the superiority of GFO-KBR & FPG-91 over GFO-KBR & FPG-89 is mainly presented in the low- and medium-latitude regions, including the Amazon basin, North Murray–Darling basin and North China Plain (NCP). The last column of Table 5 shows ERR of GFO-KBR & FPG-91 over GFO-KBR & FPG-89 in terms of regional wRMS by setting $q_i = wRMS_{GFO\text{-}KBR\ \&\ FPG\text{-}91}^{\Omega}$ and $q_j = wRMS_{GFO\text{-}KBR\ \&\ FPG\text{-}89}^{\Omega}$ in Equation (16). It demonstrates that the GFO-KBR & FPG-91 reduces about 22% errors in ocean and 17% in land areas with respect to GFO-KBR & FPG-89. However, ERR mainly decreases with higher latitude and is even negative in the Greenland. The series of GFO-LRI & FPG-89 and GFO-LRI & FPG-91 models are not demonstrated here because of their indistinctive differences in spatial domain.

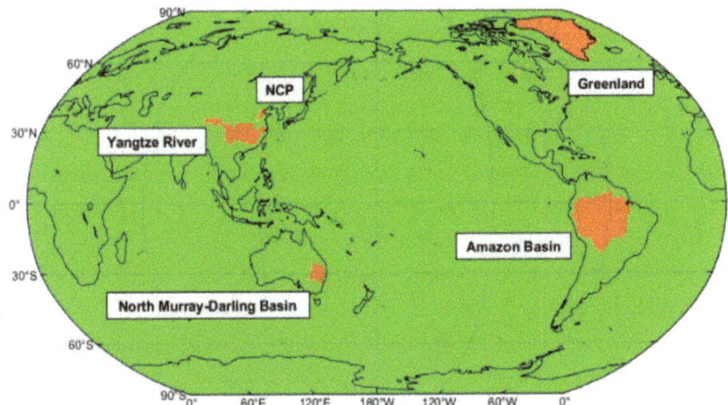

Figure 9. Typical regions selected for GFO & FPG-89 and GFO & FPG-91 models' comparisons.

Table 5. Yearly-average latitude-weighted root mean squares (wRMS) of regional spatial differences plots (SDP) in GFO-KBR & FPG-89 and GFO-KBR & FPG-91 models. The final column shows the error reduction rate (ERR) of GFO-KBR & FPG-91 with respect to GFO-KBR & FPG-89 in terms of regional wRMS.

Area	wRMS of GFO & FPG-89 (cm)	wRMS of GFO & FPG-91 (cm)	GFO & FPG-91 Error Reduction Rate (%)
Amazon	2.76	1.98	28.3
North Murray–Darling	3.56	2.54	28.7
Yangtze	2.63	2.32	11.8
NCP	3.19	2.89	9.4
Greenland	1.01	1.15	-13.9
Ocean	3.09	2.41	22.0
Land	2.53	2.09	17.4

4. Discussion

The advantages of GFO & retrograde-FPGs over GFO & prograde-FPGs are believed to be contributed by two aspects. On one hand, since the prograde and retrograde orbits sample the Earth in different directions, combining the two can obtain an integrated system that observes the Earth in a more multi-perspective view. Furthermore, as GFO is flying in a prograde orbit, adding retrograde FPGs (rather than the prograde ones) can make the GFO & FPG integrated system more isotropic [37]. On the other hand, according to the Kaula's linear perturbation theory [47], the RAANs of prograde and retrograde orbits precess in different directions in the inertial frame. Therefore, when projected into the Earth-fixed frame, the satellite ground tracks of the two evolve with different rates since the Earth rotates eastwards. As such, the signal sampling of the two differs in time domain. Once combined, they contribute to a more homogeneous system for signal sampling in terms of temporal resolution.

Due to the designed polar orbit for polar-type satellites, the low- and medium-latitude areas suffer from much sparser ground tracks than the polar region. The former, however, is just the place where most of temporal aliasing errors appear. Therefore, improved ground tracks in these areas matter much to the final gravity solution, and that is where most of the gains of GFO & FPG-91 come from. However, we also observe the degradation of GFO & FPG-91 at high latitudes such as the Greenland compared with GFO & FPG-89, which is possibly due to the pattern change of GFO & FPG-joint ground tracks at the intersection of prograde FPG-89 and retrograde FPG-91. Similar degradations in the intersection of two Bender-type orbits are also found in [60] (pp. 108–111), which needs further investigations.

5. Conclusions

In this paper, we investigate the orbit configurations for Future Polar-type Gravity missions (FPG) with emphasis on orbit inclination selection via full-scale simulations based on both standalone and joint GFO constellations. Four monthly models in January 2006 under FPG configurations with varying inclinations (87, 89, 91, and 93 degrees) show similar performances except that FPG-87 and FPG-93 suffer from relatively serious deficiencies in the zonal SHC estimation. To improve the GFO-only monthly models, the information from different FPGs are respectively added to them, leading to different GFO & FPG combined models. Among combined models, the retrograde-orbit FPGs (FPG-91 and FPG-93) generally outperform the prograde ones (FPG-89 and FPG-87). The advantages are revealed not only in spectral domain by smaller degree geoid height error (DGHE) and cumulative geoid height error (CGHE), but also in the spatial domain by obvious fewer north–south stripes, especially at low and medium latitudes. More comprehensive investigation is carried out for FPG-89 and FPG-91 by analyzing one-year time series of GFO-KBR joint monthly models, which further highlights the superiority of FPG-91 over FPG-89. Detailed regional analysis shows that improvements mainly reside over those low- and medium-latitude regions. The mechanism behind is explained by more isotropic GFO & FPG observation system that integrates two different orbit types (prograde and retrograde). Based on our study, the FPG configuration with 91-degree inclination is therefore worthy to be recommended for the further orbit design of FPG.

Author Contributions: Y.S. conceived and designed the experiments; Y.N. and Q.C. performed the experiments; Y.N. and Q.C. analyzed the data; Y.N. wrote the paper; Y.S. and Q.C. revised the manuscript.

Funding: This research was funded by the Natural Science Foundation of China (41731069) and Alexander von Humboldt Foundation.

Acknowledgments: We are grateful to Wei Feng for the GRACE Matlab Toolbox (GRAMAT) [62] used in some figures' plotting of the article. The first author would like to thank Weiwei Li from Shandong University of Science and Technology and Tianyi Chen from Tongji University for their valuable suggestions on preparing the article. Two anonymous reviewers and the academic editor are sincerely appreciated for improving the article.

Conflicts of Interest: The authors declare no conflict of interest.

References

1. Kusche, J.; Klemann, V.; Sneeuw, N. Mass distribution and mass transport in the earth system: Recent scientific progress due to interdisciplinary research. *Surv. Geophys.* **2014**, *35*, 1243–1249. [CrossRef]
2. Tapley, B.D.; Bettadpur, S.; Ries, J.C.; Thompson, P.F.; Watkins, M.M. GRACE measurements of mass variability in the Earth system. *Science* **2004**, *305*, 503–505. [CrossRef] [PubMed]
3. Colombo, O.L. *The Global Mapping of Gravity with Two Satellites*; Publications on Geodesy, New Series; Netherlands Geodetic Commission: Delft, Switzerland, 1984.
4. Jäggi, A.; Hugentobler, U.; Bock, H.; Beutler, G. Precise orbit determination for GRACE using undifferenced or doubly differenced GPS data. *Adv. Space Res.* **2007**, *39*, 1612–1619. [CrossRef]
5. Flury, J.; Bettadpur, S.; Tapley, B.D. Precise accelerometry onboard the GRACE gravity field satellite mission. *Adv. Space Res.* **2008**, *42*, 1414–1423. [CrossRef]
6. Inácio, P.; Ditmar, P.; Klees, R.; Farahani, H.H. Analysis of star camera errors in GRACE data and their impact on monthly gravity field models. *J. Geod.* **2015**, *89*, 551–571. [CrossRef]
7. Dahle, C.; Murböck, M.; Flechtner, F.; GFZ GRACE/GRACE-FO SDS Team. Level-2 GRACE RL06 Products from GFZ. Presented at the GRACE/GRACE-FO Science Team Meeting, Potsdam, Germany, 9 October 2018. Available online: https://presentations.copernicus.org/GSTM-2018-90_presentation.pdf (accessed on 19 January 2019).
8. Meyer, U.; Jäggi, A.; Jean, Y.; Beutler, G. AIUB-RL02: An improved time-series of monthly gravity fields from GRACE data. *Geophys. J. Int.* **2016**, *205*, 1196–1207. [CrossRef]
9. Shen, Y.; Chen, Q.; Xu, H. Monthly gravity field solution from GRACE range measurements using modified short arc approach. *Geod. Geodyn.* **2015**, *6*, 261–266. [CrossRef]
10. Chen, Q.; Shen, Y.; Chen, W.; Zhang, X.; Hsu, H. An improved GRACE monthly gravity field solution by modelling the non-conservative acceleration and attitude observation errors. *J. Geod.* **2016**, *90*, 503–523. [CrossRef]
11. Chen, Q.; Shen, Y.; Francis, O.; Chen, W.; Zhang, X.; Hsu, H. Tongji-Grace02s and Tongji-Grace02k: High-Precision Static GRACE-Only Global Earth's Gravity Field Models Derived by Refined Data Processing Strategies. *J. Geophys. Res. Solid Earth* **2018**, *123*, 6111–6137. [CrossRef]
12. Chen, J.L.; Tapley, B.D.; Save, H.; Tamisiea, M.E.; Bettadpur, S.; Ries, J. Quantification of ocean mass change using GRACE gravity, satellite altimeter and Argo floats observations. *J. Geophys. Res. Solid Earth* **2018**. [CrossRef]
13. WCRP Global Sea Level Budget Group. Global sea-level budget 1993–present. *Earth Syst. Sci. Data* **2018**, *10*, 1551–1590. [CrossRef]
14. Feng, W.; Zhong, M.; Lemoine, J.M.; Biancale, R.; Hsu, H.T.; Xia, J. Evaluation of groundwater depletion in North China using the Gravity Recovery and Climate Experiment (GRACE) data and ground-based measurements. *Water Resour. Res.* **2013**, *49*, 2110–2118. [CrossRef]
15. Feng, W.; Shum, C.K.; Zhong, M.; Pan, Y. Groundwater Storage Changes in China from Satellite Gravity: An Overview. *Remote Sens.* **2018**, *10*, 674. [CrossRef]
16. Chen, T.; Shen, Y.; Chen, Q. Mass flux solution in the Tibetan Plateau using mascon modeling. *Remote Sens.* **2016**, *8*, 439. [CrossRef]
17. Chen, J.L.; Wilson, C.R.; Tapley, B.D.; Blankenship, D.; Young, D. Antarctic regional ice loss rates from GRACE. *Earth Planet. Sci. Lett.* **2008**, *266*, 140–148. [CrossRef]
18. Flechtner, F.; Neumayer, K.H.; Dahle, C.; Dobslaw, H.; Fagiolini, E.; Raimondo, J.C.; Guntner, A. What Can be Expected from the GRACE-FO Laser Ranging Interferometer for Earth Science Applications? *Surv. Geophys.* **2016**, *37*, 453–470. [CrossRef]
19. Frank, W. GRACE-FO Mission Status and Further Plans. Presented at the GRACE/GRACE-FO Science Team Meeting, Potsdam, Germany, 9 October 2018.
20. Sheard, B.S.; Heinzel, G.; Danzmann, K.; Shaddock, D.A.; Klipstein, W.M.; Folkner, W.M. Intersatellite laser ranging instrument for the GRACE follow-on mission. *J. Geod.* **2012**, *86*, 1083–1095. [CrossRef]
21. Loomis, B.D.; Nerem, R.S.; Luthcke, S.B. Simulation study of a follow-on gravity mission to GRACE. *J. Geod.* **2012**, *86*, 319–335. [CrossRef]
22. Oppenheim, A.V.; Willsky, A.S.; Nawab, S.H. *Signals and Systems*, 2nd ed.; Prentice Hall: Upper Saddle River, New Jersey, USA, 1996.

23. Han, S.C.; Jekeli, C.; Shum, C.K. Time-variable aliasing effects of ocean tides, atmosphere, and continental water mass on monthly mean GRACE gravity field. *J. Geophys. Res. Solid Earth* **2004**, *109*. [CrossRef]

24. Dobslaw, H.; Bergmann-Wolf, I.; Dill, R.; Forootan, E.; Klemann, V.; Kusche, J.; Sasgen, I. The updated ESA Earth System Model for future gravity mission simulation studies. *J. Geod.* **2015**, *89*, 505–513. [CrossRef]

25. Dobslaw, H.; Bergmann-Wolf, I.; Dill, R.; Poropat, L.; Thomas, M.; Dahle, C.; Esselborn, S.; Konig, R.; Flechtner, F. A new high-resolution model of non-tidal atmosphere and ocean mass variability for de-aliasing of satellite gravity observations: AOD1B RL06. *Geophys. J. Int.* **2017**, *211*, 263–269. [CrossRef]

26. Seo, K.W.; Wilson, C.R.; Han, S.C.; Waliser, D.E. Gravity Recovery and Climate Experiment (GRACE) alias error from ocean tides. *J. Geophys. Res. Solid Earth* **2008**, *113*. [CrossRef]

27. Dobslaw, H.; Bergmann-Wolf, I.; Forootan, E.; Dahle, C.; Mayer-Gürr, T.; Kusche, J.; Flechtner, F. Modeling of present-day atmosphere and ocean non-tidal de-aliasing errors for future gravity mission simulations. *J. Geod.* **2016**, *90*, 423–436. [CrossRef]

28. Wahr, J.; Molenaar, M.; Bryan, F. Time variability of the Earth's gravity field: Hydrological and oceanic effects and their possible detection using GRACE. *J. Geophys. Res. Solid Earth* **1998**, *103*, 30205–30229. [CrossRef]

29. Swenson, S.; Wahr, J. Post-processing removal of correlated errors in GRACE data. *Geophys. Res. Lett.* **2006**, *33*. [CrossRef]

30. Kusche, J.; Schmidt, R.; Petrovic, S.; Rietbroek, R. Decorrelated GRACE time-variable gravity solutions by GFZ, and their validation using a hydrological model. *J. Geod.* **2009**, *83*, 903–913. [CrossRef]

31. Visser, P.N.A.M.; Sneeuw, N.; Reubelt, T.; Losch, M.; van Dam, T. Space-borne gravimetric satellite constellations and ocean tides: Aliasing effects. *Geophys. J. Int.* **2010**, *181*, 789–805. [CrossRef]

32. Wiese, D.N.; Visser, P.; Nerem, R.S. Estimating low resolution gravity fields at short time intervals to reduce temporal aliasing errors. *Adv. Space Res.* **2011**, *48*, 1094–1107. [CrossRef]

33. Daras, I.; Pail, R. Treatment of temporal aliasing effects in the context of next generation satellite gravimetry missions. *J. Geophys. Res. Solid Earth* **2017**, *122*, 7343–7362. [CrossRef]

34. Hauk, M.; Pail, R. Treatment of ocean tide aliasing in the context of a next generation gravity field mission. *Geophys. J. Int.* **2018**, *214*, 345–365. [CrossRef]

35. Wiese, D.N.; Folkner, W.M.; Nerem, R.S. Alternative mission architectures for a gravity recovery satellite mission. *J. Geod.* **2009**, *83*, 569–581. [CrossRef]

36. Bender, P.; Wiese, D.; Nerem, R. A possible dual-GRACE mission with 90 degree and 63 degree inclination orbits. In Proceedings of the International Symposium on Formation Flying, Noordwijk, The Netherlands, 23–25 April 2008; Eur. Space Agency: Noordwijk, The Netherlands, 2008.

37. Wiese, D.N.; Nerem, R.S.; Lemoine, F.G. Design considerations for a dedicated gravity recovery satellite mission consisting of two pairs of satellites. *J. Geod.* **2012**, *86*, 81–98. [CrossRef]

38. Sneeuw, N.; Sharifi, M.; Keller, W. Gravity Recovery from Formation Flight Missions. In *VI Hotine-Marussi Symposium on Theoretical and Computational Geodesy*; Xu, P., Liu, J., Dermanis, A., Eds.; Springer: Berlin/Heidelberg, Germany, 2008; Volume 132, pp. 29–34.

39. Elsaka, B.; Raimondo, J.C.; Brieden, P.; Reubelt, T.; Kusche, J.; Flechtner, F.; Pour, S.I.; Sneeuw, N.; Muller, J. Comparing seven candidate mission configurations for temporal gravity field retrieval through full-scale numerical simulation. *J. Geod.* **2014**, *88*, 31–43. [CrossRef]

40. Cesare, S.; Sechi, G. Next Generation Gravity Mission. In *Distributed Space Missions for Earth System Monitoring*; D'Errico, M., Ed.; Springer: New York, NY, USA, 2013; Volume 31, pp. 575–598.

41. Pail, R.; Gruber, T.; Abrykosov, P.; Hauk, M.; Purkhauser, A. Assessment of NGGM Concepts for Sustained Observation of Mass Transport in the Earth System. Presented at the GRACE/GRACE-FO Science Team Meeting, Potsdam, Germany, 10 October 2018. Available online: https://meetingorganizer.copernicus.org/GSTM-2018/GSTM-2018-10.pdf (accessed on 19 January 2019).

42. Hsu, H.; Zhong, M.; Feng, W.; Wang, C. Overwiew of activities of Chinese gravity mission. Presented at the International Symposium on Geodesy and Geodynamics (ISGG2018)-Tectonics, Earthquake and Geohazards, Kunming, China, 31 July 2018.

43. Reigber, C. Gravity field recovery from satellite tracking data. In *Theory of Satellite Geodesy and Gravity Field Determination*; Sansò, F., Rummel, R., Eds.; Springer: Berlin/Heidelberg, Germany, 1989; Volume 25, pp. 197–234.

44. Zhou, H.; Luo, Z.; Zhou, Z.; Li, Q.; Zhong, B.; Lu, B.; Hsu, H. Impact of Different Kinematic Empirical Parameters Processing Strategies on Temporal Gravity Field Model Determination. *J. Geophys. Res. Solid Earth* **2018**. [CrossRef]
45. Guo, X.; Zhao, Q.; Ditmar, P.; Sun, Y.; Liu, J. Improvements in the Monthly Gravity Field Solutions Through Modeling the Colored Noise in the GRACE Data. *J. Geophys. Res. Solid Earth* **2018**, *123*, 7040–7054. [CrossRef]
46. Wang, C.; Xu, H.; Zhong, M.; Feng, W. Monthly gravity field recovery from GRACE orbits and K-band measurements using variational equations approach. *Geod. Geodyn.* **2015**, *6*, 253–260. [CrossRef]
47. Kaula, W.M. *Theory of Satellite Geodesy: Applications of Satellites to Geodesy*; Dover Publications: New York, NY, USA, 2000.
48. Chen, Q.; Shen, Y.; Zhang, X.; Hsu, H.; Chen, W.; Ju, X.; Lou, L. Monthly gravity field models derived from GRACE Level 1B data using a modified short-arc approach. *J. Geophys. Res. Solid Earth* **2015**, *120*, 1804–1819. [CrossRef]
49. Shen, Y. Algorithm Characteristics of Dynamic Approach-based Satellite Gravimetry and Its Improvement Proposals. *Acta Geod. Cartogr. Sin.* **2017**, *46*, 1308–1315. (In Chinese)
50. Montenbruck, O.; Gill, E. *Satellite Orbits-Models, Methods and Applications*; Springer: Berlin/Heidelberg, Germany, 2000.
51. Kim, J. Simulation Study of a Low-Low Satellite-to-Satellite Tracking Mission. Ph.D. Thesis, The University of Texas, Austin, TX, USA, 2000.
52. Daras, I. Gravity Field Processing towards Future LL-SST Satellite Missions. Ph.D. Thesis, Technische Universität München, München, Germany, 2016.
53. Vallado, D. *Fundamentals of Astrodynamics and Applications*, 4th ed.; Microcosm Press: El Segundo, CA, USA, 2013.
54. Reigber, C.; Lühr, H.; Schwintzer, P. CHAMP mission status. *Adv. Space Res.* **2002**, *30*, 129–134. [CrossRef]
55. Rummel, R.; Yi, W.; Stummer, C. GOCE gravitational gradiometry. *J. Geod.* **2011**, *85*, 777–790. [CrossRef]
56. Förste, C.; Bruinsma, S.L.; Abrikosov, O.; Lemoine, J.; Marty, J.C.; Flechtner, F.; Balmino, G.; Barthelmes, F.; Biancale, R. EIGEN-6C4 The Latest Combined Global Gravity Field Model Including GOCE Data up to Degree and Order 2190 of GFZ Potsdam and GRGS Toulouse. GFZ Data Services. 2014. Available online: http://icgem.gfz-potsdam.de/getmodel/zip/7fd8fe44aa1518cd79ca84300aef4b41ddb2364aef9e82b7cdaabdb60a9053f1 (accessed on 19 January 2019).
57. Savcenko, R.; Bosch, W. *EOT11a-Empirical Ocean Tide Model from Multi-Mission Satellite Altimetry*; Deutsches Geodätisches Forschungsinstitut (DGFI): München, Germany, 2012.
58. Darbeheshti, N.; Wegener, H.; Müller, V.; Naeimi, M.; Heinzel, G.; Hewitson, M. Instrument data simulations for GRACE Follow-on: Observation and noise models. *Earth Syst. Sci. Data* **2017**, *9*, 833–848. [CrossRef]
59. Lyard, F.; Lefevre, F.; Letellier, T.; Francis, O. Modelling the global ocean tides: Modern insights from FES2004. *Ocean Dyn.* **2006**, *56*, 394–415. [CrossRef]
60. Wiese, D.N. Optimizing Two Pairs of GRACE-like Satellites for Recovering Temporal Gravity Variations. Ph.D. Theses, University of Colorado, Boulder, CO, USA, 2011.
61. Meyer, U.; Jean, Y.; Arnold, D.; Jäggi, A.; Kvas, A.; Mayer-Gürr, T.; Lemoine, J.M.; Bourgogne, S.; Dahle, C.; Flechtner, F. The EGSIEM combination service: Final results and further plans. In Proceedings of the EGU General Assembly 2018, Vienna, Austria, 8–13 April 2018; Available online: https://boris.unibe.ch/116268/1/EGU18_EGSIEM-1.pdf (accessed on 19 January 2019).
62. Feng, W. GRAMAT: A comprehensive Matlab toolbox for estimating global mass variations from GRACE satellite data. *Earth Sci. Inform.* **2018**, 1–16. [CrossRef]

 *remote sensing*

Article

Gravity Field Recovery Using High-Precision, High–Low Inter-Satellite Links

Markus Hauk * and Roland Pail

Institute of Astronomical and Physical Geodesy, Technical University of Munich, Arcisstrasse 21,
80333 Munich, Germany; roland.pail@tum.de
* Correspondence: markus.hauk@tum.de

Received: 5 February 2019; Accepted: 27 February 2019; Published: 5 March 2019

Abstract: Past temporal gravity field solutions from the Gravity Recovery and Climate Experiment (GRACE), as well as current solutions from GRACE Follow-On, suffer from temporal aliasing errors due to undersampling of the signal to be recovered (e.g., hydrology), which arise in terms of stripes caused by the north–south observation direction. In this paper, we investigate the potential of the proposed mass variation observing system by high–low inter-satellite links (MOBILE) mission. We quantify the impact of instrument errors of the main sensors (inter-satellite link and accelerometer) and high-frequency tidal and non-tidal gravity signals on achievable performance of the temporal gravity field retrieval. The multi-directional observation geometry of the MOBILE concept with a strong dominance of the radial component result in a close-to-isotropic error behavior, and the retrieved gravity field solutions show reduced temporal aliasing errors of at least 30% for non-tidal, as well as tidal, mass variation signals compared to a low–low satellite pair configuration. The quality of the MOBILE range observations enables the application of extended alternative processing methods leading to further reduction of temporal aliasing errors. The results demonstrate that such a mission can help to get an improved understanding of different components of the Earth system.

Keywords: mass transport in the Earth system; GRACE and GRACE follow-on mission; current and future observation concepts and instruments

---

1. Introduction

In times of a changing climate the need for innovative observation techniques for capturing geophysical processes in the Earth system becomes increasingly urgent. In this context, the observation of the temporal gravity field by satellites from space play an important role when investigating, e.g., rapid changes in the cryosphere, oceans, water cycle, and solid Earth processes on a global scale. For the determination of temporal gravity fields, in the last decade satellite missions such as GRACE [1] or Challenging Minisatellite Payload (CHAMP) [2,3] orbited around the globe and helped to get a better understanding of the Earth's mass flux signals. CHAMP was based on high–low satellite-to-satellite tracking (SST) exploiting the Global Positioning System (GPS) [4] over a time span of 10 years. The accuracy of the CHAMP orbit information of 2–3 cm [5] derived from GPS allowed for resolving only the long wave range of the time varying gravity field, with spatial scales of ≈1000 km, e.g., Baur 2013 [6]. Analyzing the perturbed orbit of other Low Earth Orbiters (LEO), such as the Swarm satellites, allows for a similar performance [7]. The GRACE mission reached spatial scales of the temporal gravity field of ≈300 km and below due to a combination of K-band microwave low–low inter-satellite ranging between two identical satellites following each other in the same orbit at a distance of about 220 km with micrometer precision, and high–low GPS satellite-to-satellite tracking plus accelerometer observations. These missions improved our knowledge of water mass variations on the continents, in the oceans, and the atmosphere to a great extent. Additionally, the static gravity field retrieved from the Gravity field and steady-state Ocean Circulation Explorer (GOCE) [8] mission has improved our knowledge of the long-term static mass distribution, and has provided the physical reference surface of the geoid with centimeter precision with a spatial resolution down to 70–80 km.

The observation of the Earth's gravity field will be continued by the GRACE Follow-On mission [9], which was successfully launched in May 2018. The instruments have been slightly modified compared to those used in GRACE, and additionally the GRACE Follow-On mission includes an inter-satellite laser ranging interferometer as a demonstrator [10], resulting in an increased accuracy of the SST observations to a few nanometers.

One of the main error contributions when observing the time variable gravity field results from geophysical signals with periods shorter than the temporal resolution of the satellite mission, as they alias into the solutions. Traditionally, the temporal resolution of the GRACE data products has been 30 days [1], leading to temporal aliasing effects due to non-tidal mass variations, especially atmospheric, oceanic, and hydrological signals having large amplitudes in the high-frequency range [11], as well as tidal signals with mainly semi-diurnal and diurnal periods. The temporal aliasing errors in GRACE appear in terms of striping effects, which are caused by the anisotropic error behavior resulting from along-track inter-satellite ranging. This type of error pattern also remains for the GRACE Follow-On mission because temporal aliasing errors clearly dominate the error budget of gravity field retrieval [12–15], while instrument errors of the laser interferometer only play a minor role in the total error budget.

In general, there are different approaches of how to deal with temporal aliasing errors: The gravity field solutions retrieved by GRACE are typically treated with de-striping and filtering techniques (e.g., References [16–19]), which are applied a posteriori in order to reduce the striping effects. A further de-aliasing method is proposed by Watkins et al. 2015 [20], where observations from GRACE are being processed using spherical cap mascons resulting in greater resolutions for smaller spatial regions. In the context of a Next Generation Gravity Field Mission (NGGM), various satellite constellations enabling a self-de-aliasing of high-frequency signals were investigated, such as the dual-tandem Bender-type mission [21], consisting of one near-polar pair and one inclined pair. Such a constellation allows the combination of two anisotropic measurements taken in different directions, which increases the isotropy of the combined system. A further concept is the pendulum formation where two satellites are on slightly shifted orbit planes in such a way that the line of sight between the satellites does not only contain along-track components, but also cross-track components [22]. Such innovative satellite constellations offer the application of improved gravity field processing methodologies in order to exploit the full potential of gravity field solutions with enhanced spatial and temporal resolution. Wiese et al. 2011 [23] proposed a method where low-resolution gravity field solutions are co-parameterized for short periods (e.g., daily) together with the long-term solutions (e.g., monthly) in order to mitigate the non-tidal high-frequency signals (especially atmosphere, ocean, and to some extent, also hydrology). A method to reduce tidal aliasing errors proposed by Hauk & Pail 2018 [24] aims at a co-parameterization of ocean tide parameters over time spans of several years, where the estimated tide model is used for de-aliasing during gravity field retrieval in a second processing step.

In this paper we pick up the point of new satellite constellations using an innovative observation concept of high-precision high–low inter-satellite ranging, which was proposed as the (MOBILE) mission [25] in response to European Space Agency (ESA)'s Earth Explorer 10 call. The observation geometry and the measuring principle of this satellite constellation are based on the Geodesy and Time Reference in Space (GETRIS) [26] concept describing a global space-borne infrastructure for data transfer, clock synchronization, and ranging, where gravity field recovery can be one of the first beneficiary applications of such an advanced geodetic space infrastructure.

In this study, we analyze the potential of the MOBILE mission qualitatively and quantitatively, and compare gravity field solutions with GRACE Follow-On-like solutions by means of full-scale numerical simulations. Furthermore, the potential of an extended processing method for the reduction of temporal aliasing is investigated for the MOBILE concept. In the following Section 2, the MOBILE observation configuration is described, while Section 3 gives an overview of the simulation environment including observation equations and stochastic modelling. In Section 4 the estimated gravity field solutions are analyzed and assessed. The main conclusions are summarized in Section 5, and in Section 6, a short outlook is given.

2. Mission Concept

2.1. Observation Geometry

In contrast to the past GRACE and the current GRACE Follow-On missions, which are mainly based on LEO satellites (several hundred km), the MOBILE minimum configuration consists of a constellation of two high and one low orbiting satellites. As done for GRACE and GRACE Follow-On, the main observable is the gravity-induced inter-satellite distance change, which is in case of MOBILE measured between medium orbiting satellites (MEO; several thousand km) and LEO satellites though. As a second gravity observation type, high-precision orbit positions based on Global Navigation Satellite System (GNSS) orbit determination are used. This idea of high-precision high–low tracking was first investigated by Hauk et al. 2017 [26] using the inter-satellite link technique as part of the payload on-board Galileo satellites of future generation in connection with LEO satellites, where the main error sources and the corresponding achievable performance were analyzed. Due to the fact that the MOBILE constellation presents a stand-alone concept without the need to place an additional payload on another space infrastructure, and the very large distance between the high- and the low orbiting satellites, which plays a crucial role in the framework of high-precision high-low tracking, dedicated MEO satellites were included in the concept. It should be emphasized that alternatively, a constellation of one MEO and two LEO satellites could be envisaged, which turns out to provide nearly the same performance as the proposed one, but might be more expensive due to the need to build and maintain at least two LEO satellites in orbit.

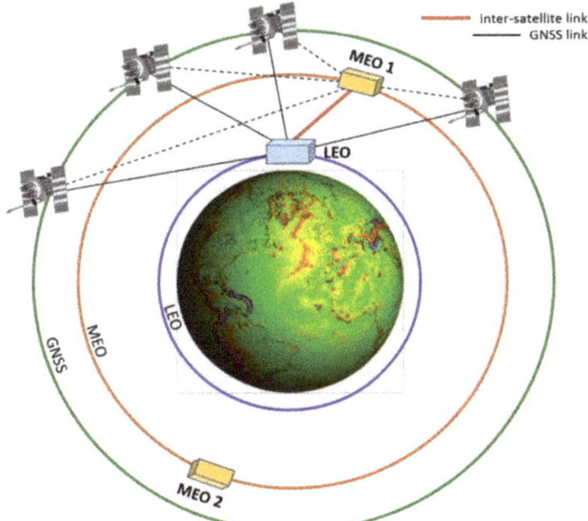

Figure 1. MOBILE mission constellation: high-precision inter-satellite links between LEO and MEOs (red), micro-wave links from GNSS satellites to MEOs and LEO (black).

Figure 1 shows a schematic overview of the MOBILE satellite formation. The orbit parameters, which are used in the simulation environment to build up different satellite constellations, are listed in Table 1. The MEO satellites orbit at an altitude of about 10,150 km in the same orbital plane, separated by an 180-degree mean anomaly as alternating targets of the LEO satellites, in order to maximize the visibility and thus observation time. The LEO satellite is orbiting in an altitude of about 360 km. Both LEO and MEO satellites are flying in polar orbits in order to maintain a long-term stable formation (no relative drifts of the orbit planes). Additionally, two LEO satellites with near-polar orbits flying in an altitude of about 470 km with an inter-satellite distance of 200 km are set up in order to perform comparability studies between the MOBILE constellation and a GRACE Follow-On-like mission. All orbits have certain repeat cycles, after which the satellites reach the same position on Earth again in order to maintain a stable ground track pattern and related stable gravity model quality. The choice of the orbit height of the MEO satellites underlies three major constraints: (1) A high altitude of several thousand kilometers is necessary in order to ensure long observation periods and preferably measurements of multi-directional distance variations, with a strong dominance of the radial component, resulting in a close to isotropic error behavior of the retrieved gravity field solution (see Section 4). (2) The distance between a MEO–LEO pair must not be too large, because the larger the distance, the more difficult it is to fulfill the 1-µm accuracy requirement for the inter-satellite link established by a laser range interferometer (see Section 2.2). (3) The third constraint is driven by solar radiation belts encircling the Earth in which energetic charged particles are trapped inside the Earth's magnetic field [27], which are of different intensities dependent on the solar cycle, altitude, and inclination of the satellite orbit. As a result, altitude ranges of several thousand kilometers below the chosen orbit height drop out. These conditions connected with the repeat orbit lead to an altitude of about 10,000 km for the MEO satellites.

The high–low tracking concept enables a multi-directional observation geometry with differing elevation angles from 3° (assumed minimum elevation angle of visible MEOs observed by the LEO) up to a near-radial direction. However, due to the observation geometry of the MEO–LEO satellite pairs and the changing satellite links from one to the other MEO, data gaps arise for every satellite pair, leading to a non-continuous measurement time series of these pairs. For the simulated MOBILE constellation, this results in a ranging window maximum of 45 min, and a maximum data gap of 18 min. The separation of the two MEO satellites of 180-degree mean anomaly is chosen to keep the time period of the data gap as small as possible. In Figure 2, the LEO ground track of the MOBILE concept is displayed for 1 day together with the corresponding elevation angles.

Table 1. Orbit parameters for satellite constellations.

Satellite	Altitude (km)	Inclination (Degree)	Revolutions/Nodal Days in One Repeat Orbit	Initial Mean Anomaly (Degree)
LEO 1	358	90	476/30	0
MEO 1	10.149	90	124/30	0
MEO 2	10.149	90	124/30	180
Low–low 1	467	89	412/27	53.21
Low–low 2	467	89	412/27	51.51

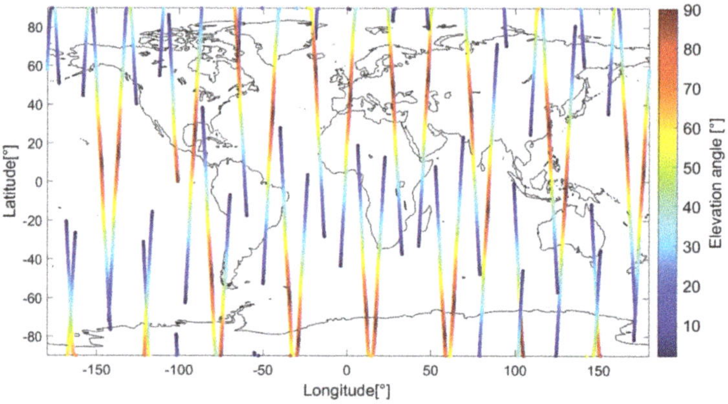

Figure 2. One-day LEO ground track of the MOBILE mission constellation together with the elevation angle of visible MEOs observed by the LEO.

2.2. Instrumentation

The main observable in the MOBILE mission are range measurements from the LEO to the MEOs, where the MEOs are alternating targets. The ranging accuracy is on the micrometer level in order to be sensitive for gravitational forces and its changes on Earth. For distances of several thousand kilometers, a laser-based distance measurement system can reach such an accuracy. The laser range interferometer is placed at the LEO satellite, while the MEOs are equipped with passive reflectors or transponders. In case of the GRACE Follow-On mission, the measurement of inter-satellite ranges by laser range interferometry (LRI) has been successfully established. The link between the two satellites was generated with an active laser on one satellite, and a phase-locked amplifying transponder on the second spacecraft [10]. For the MOBILE concept, the laser ranging instrument needs to be adapted due to the very large distance and the relative motion of the LEO and MEO satellites. In contrast to the GRACE Follow-On, the large distance and the relative speed lead to a range of Doppler shifts of several GHz compared to a few MHz, which causes the need of a reference laser source with a larger range of reference frequencies and a faster phase-tracking capability than implemented for the GRACE Follow-On. The required parameters (<10 GHz range, <10 MHz/s tracking) are within the range of existing, space qualified reference lasers (e.g., the one used for the ATmospheric LIDar (ATLID) instrument on the Earth Clouds Aerosols and Radiation Explorer (CARE) mission) [28], but their compatibility with the needs of an interferometric instrument has to be the subject of further studies. Due to the relative motion of the LEO and MEO satellites, pointing tracking capabilities are required, which requires a modified link implementation. The LEO satellite is selected to play the active part in the tracking mechanism, while the partner satellites (MEOs) are equipped with passive retroreflectors. This type of laser tracking and ranging has been successfully performed for decades with active laser systems on the ground and passive retroreflectors on satellites in orbit (e.g., Laser Geodynamics Satellite (LAGEOS), Ball Lens In The Space (BLITS)) [29,30]. The scientific benefit of deploying a passive payload in space is the significantly increased mission duration when compared to complex active payloads. The main technological challenge in utilizing this setup for an LRI instrument is the need to achieve a sufficiently high level of retrieved power without the need for amplification between the two passes, ideally close to the 80 pW received by the GRACE Follow-On implementation, but at least to levels above $\approx$1 pW in order to allow phase tracking. The main design factors impacting the received power are the initial output power, the size of the retroreflector, and the size of the receiving telescope.

Satellite on-board sensors play an important role in the gravity field retrieval by influencing satellite observations due to correlated noise. In our study, the error assumptions used for the laser ranging instrument in the MOBILE concept are based on the time-series provided by Schäfer et al. 2013 [31], which originated in connection with ESA's GETRIS study, and show micrometer ranging accuracy around 1 MHz. Due to simulation purposes, this time-series was adapted by means of cascaded second order Butterworth auto regressive moving average (ARMA) filter model. The spectral behavior of the LRI is shown in Figure 3 (light green curve) in terms of an amplitude spectral density (ASD). The relative distance measurement errors assumed for the low–low satellite pair are identical to those used in the frame of the ESA-Assessment of Satellite Constellations for Monitoring the Variations in Earth's Gravity Field (SC4MGV) project [32], provided from the consultancy support of Thales Alenia Space Italia, and show a performance of about several 10 nanometers. The corresponding analytical noise model of the used laser interferometer is given by the ASD in terms of range-rates (Figure 3, light blue curve):

$$d_{\text{range-rates}} = 2 \times 10^{-8} \cdot 2\pi f \cdot \sqrt{\left(\frac{10^{-2}\text{Hz}}{f}\right)^2 + 1} \, \frac{\text{m}}{\text{s} \sqrt{\text{Hz}}}. \tag{1}$$

The generation of all noise time-series was done by scaling the spectrum of normally distributed random time-series with their individual spectral model.

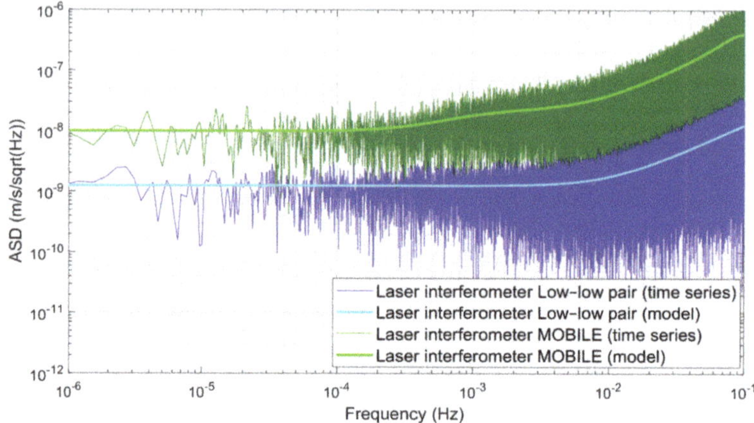

Figure 3. Amplitude spectral density (ASD) of the relative distant measurement errors in terms of range-rates for the low–low pair formation (dark blue) and for MOBILE (dark green). The generation of the noise time-series was done by scaling the spectrum of normally distributed random time-series with their individual spectral model (light blue and light green curves).

The non-gravitational forces are typically sensed by the on-board accelerometers located in the center-of-mass of the satellite. In case of the LEO satellites, the implementation of an accelerometer is absolutely necessary due to air drag as the main contributor. For the low–low pair a GRACE-like electrostatic accelerometer is assumed with two highly sensitive axes oriented in the flight direction (largest signal) and in the radial direction, and one low-sensitive axis in the cross-track direction (see Figure 4, blue and red curves). The accuracy level in terms of accelerations is derived by Iran Pour et al. 2015 [32], and is expressed by:

$$d_{\text{acc. }x} = d_{\text{acc. }z} = 10^{-11} \sqrt{\left(\frac{10^{-3}\text{Hz}}{f}\right)^4 \Big/ \left(\left(\frac{10^{-5}\text{Hz}}{f}\right)^4 + 1\right) + 1 + \left(\frac{f}{10^{-1}\text{Hz}}\right)^4} \frac{\text{m}}{\text{s}^2\sqrt{\text{Hz}}} \tag{2}$$

$$d_{\text{acc. }y} = 10 \cdot d_{\text{acc. }z} \tag{3}$$

with x denoting along-track, y across-track, and z (close to) the radial direction. Based on the heritage of previous gravity missions for MOBILE, we seek a resolution on the level of 10^{-11} m/s^2, which is the same as assumed for the along-track and radial axes of the accelerometer on-board the low–low satellite pair, but ideally with the same performance in all three directions. Furthermore, the slope at frequencies from 10^{-3} Hz and lower is pressed down from $1/f^2$ for the low–low pair to $1/f$ for MOBILE. The performance of the relative acceleration measurement error is displayed in Figure 4 (green curve). While an accelerometer is mandatory for the MOBILE LEO satellite, for the MEOs, less stringent requirements might apply because of the substantially smaller amplitude of the signal and the fact that non-conservative forces can be modelled much more accurately in high altitudes. Also, the design of the MEO could be optimized for the high predictability of non-gravitational forces, e.g., by implementing very simple geometrical surfaces wherever radiative pressure is relevant. In spite of these facts, in the MOBILE concept, the implementation of accelerometers is proposed.

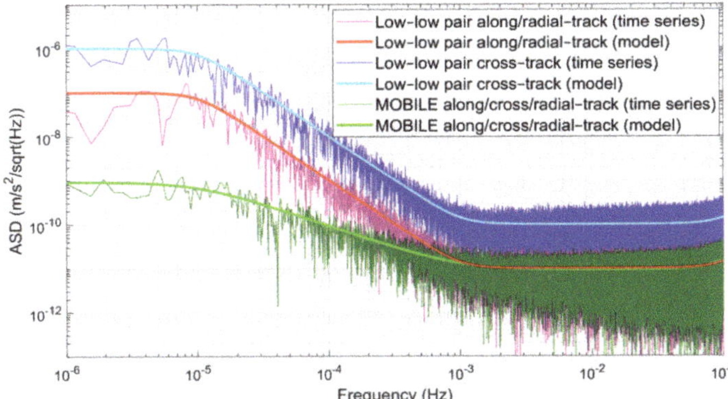

Figure 4. Amplitude spectral density (ASD) of the relative acceleration measurement error in terms of accelerations for the low–low pair formation (dark blue and magenta) and for MOBILE (dark green). The generation of the noise time-series was done by scaling the spectrum of normally distributed random time-series with their individual spectral model (light blue, red, and light green curves).

Geo-location of satellite observations, as well as gravity retrieval, require highly accurate continuous orbit determination, making GNSS space receivers on all satellites obligatory. In our simulations, we assume an absolute kinematic positioning on a cm level. Using a laser ranging instrument as the main measurement system requires exact pointing of the tracking antenna in the order of 10 μrad or less, and therefore the implementation of systems for attitude determination and control. We assume star camera sensor errors for all satellites represented as rotation angles around the along-track (roll), cross-track (pitch), and radial (yaw) axes, expressed by the ASD of the following analytical noise models [32]:

$$d_{\mathrm{roll}} = 10^{-5} \sqrt{\left(\frac{10^{-3}\mathrm{Hz}}{f}\right)^4 \bigg/ \left(\left(\frac{10^{-5}\mathrm{Hz}}{f}\right)^4 + 1\right) + 1} \, \frac{\mathrm{rad}}{\sqrt{\mathrm{Hz}}}, \tag{4}$$

$$d_{\mathrm{pitch}} = d_{\mathrm{yaw}} = 2 \times 10^{-6} \sqrt{\left(\frac{10^{-2}\mathrm{Hz}}{f}\right)^2 \bigg/ \left(\left(\frac{10^{-}\mathrm{Hz}}{f}\right)^2 + 1\right) + 1} \, \frac{\mathrm{rad}}{\sqrt{\mathrm{Hz}}}. \tag{5}$$

In addition, for the MOBILE LEO satellite, a drag-reduction system needs to be implemented in order to maintain the orbit, and not to saturate the accelerometers due to non-gravitational accelerations. The MEO satellites will very likely require an electrical propulsion system to move to their target orbit from the lower separation altitude achievable with a low-cost launcher.

3. Simulation Environment

All simulations were executed with a full numerical mission simulator [33,34], which has already been successfully applied to recover satellite-only gravitational field models from GOCE data [35]. The simulation environment is based on numerical orbit integration, following a multistep method for the numerical integration according to Shampine & Gordon 1976 [36], which applies a modified divided difference form of the Adams predict-evaluate-correct-evaluate (PECE) formulas and local extrapolation. According to this method, the order and the step size are adjusted to control the local error per unit step in a generalized sense. The generation of "true" dynamic orbits and, subsequently, the "true" GNSS high–low SST and low–low laser ranging SST observations, is done by adding different force models according to the "true" world of Table 2. The impact of orbit errors on the gravity field processing is taken into account as well by propagating 1 cm white noise of the integrated orbit positions of each satellite. The resulting erroneous dynamic orbits serve as computational points for the reference values of the observations and enable the computation of the GNSS high–low SST observations in three directions. In order to ensure the accuracy of the inter-satellite link, error-free dynamic orbits are used for the reference values of the low–low SST observations from the laser interferometer system, which are expressed in terms of range-rates.

Table 2. Force and noise models of the "true" and "reference" world used in the simulations.

Model	"True" World	"Reference" World
Static gravity field model	GOCO03s	GOCO03s
Non-tidal gravity field model	ESA AOHIS	-
Ocean tide model	GOT4.7	EOT11a
Noise model	Laser interferometer	-
Noise model	Accelerometer	-
Noise model	Star camera	-
Noise model	GNSS (orbit accuracy)	-

The adopted gravity field approach is based on a modification of the integral equation approach from Schneider 1969 [37] where the orbit is divided into continuous short arcs of 6 h length, and the position vectors at the arc node points are set up as unknown parameters, which are estimated together with the gravity field coefficients. This technique has already been successfully applied in real data applications to recover satellite-only gravitational field models for CHAMP and GRACE [38] (ITSG-Grace2016) [39]. The functional model follows the typical formulation used for low–low SST missions like GRACE, which comprises a high–low SST and a low–low SST component. Position differences between two satellites are used for the computation of the reference values for the high–low SST part of the observation system, whereas the reference values for the low–low SST part are derived by projecting position and velocity differences between two satellites onto the line-of-sight, leading to the computation of inter-satellite range-rates. Table 2 gives an overview of the force and noise models used in the processing for the "true" and "reference" world. The static gravity field model is represented by the GOCO03s model, which is a satellite-only gravity field model based on GRACE, GOCE, and LAGEOS [40]. In order to simulate geophysical signals, ESA's updated Earth system model [41] has been used, which contains the five main geophysical signal components atmosphere (A), ocean (O), hydrology (H), ice (I), and solid Earth (S) with a time resolution of six hours, linearly interpolated to the epochs. The Earth system model covers the time period 1995–2006, and contains plausible variability and trends in both low-degree coefficients and the global mean eustatic sea level. It depicts reasonable mass variability all over the globe at a wide range of frequencies including multi-year trends, year-to-year variability, and seasonal variability, even at very fine spatial scales, which is important for a realistic representation of spatial aliasing and leakage. The impact of ocean tide model errors is assessed by taking the difference of two tide models, EOT11a [42], and GOT4.7 [43].

The total stochastic model for the observations is approximated individually for both satellite formations by means of a cascade of digital Butterworth ARMA filters [44,45]. Filter coefficients are chosen in such a way that the cascade's frequency response optimally matches the inverse of the amplitude spectrum of the previously generated pre-fit residuals. They are estimated as a result of the computation of the linearized normal equations, which include differences between the "true" (only the static GOCO03s gravity field model and sensor noise are included) and the reference observations (only the static GOCO03s gravity field model is included), such that the error sources from the sensors are considered exclusively. Assuming uncorrelated high–low and low–low SST observations, weighting matrices are set up for all observation components separately.

The goal is the retrieval of all spherical-harmonic (SH) coefficients up to a maximum SH degree of 100 from observations sampled every 5 seconds for the first 30 days of the year 2001. Due to the fact of non-linear observation equations, the "reference" observations are reduced from the "true" observations as a result of the linearization process. The gravity field parameters are estimated by solving full normal equations of a least squares system based on a standard Gauss–Markov model using weighted least squares with stochastic models in accordance with the simulated instrument noise levels. The resulting gravity field coefficients are analyzed and compared regarding quality and performance in terms of retrieval errors by removing a monthly average of the true mass transport model from the recovered signal.

4. Results

4.1. Gravity Field Retrieval Performance Due to Instrument Errors

At first the impact of the instrument errors on the gravity field retrieval were quantified. For this task, we performed simulations where each error source according to the assumptions described in Section 2.2 was treated individually. Figure 5 shows the gravity field retrieval performance in terms of equivalent water height (EWH) errors per SH degree per coefficient for the low–low pair constellation and the MOBILE concept. Furthermore, the results were quantified using global RMS values of the errors in the recovered signal expressed in terms of cm of EWH, listed in Table 3 (see part: instrument errors). If only white-noise positioning errors were considered (Figure 5, green curves), the gravity field retrieval performance mainly depended on the observation geometry. The comparison between both satellite concepts revealed strongly reduced retrieval errors for MOBILE, which benefited from multi-directional observations. In the case of accelerometer noise in combination with star camera errors (Figure 5, blue curves), the MOBILE constellation showed reduced error behavior compared to the low–low pair as well. This was mainly caused by the observation geometry, but also by the improved accelerometers ($\approx$23%) with 3D capabilities to certain parts in the case of MOBILE. In contrast, the retrieval performance of the low–low pair benefits from the nanometer accuracy of the laser interferometer compared to the micrometer accuracy of MOBILE's laser link sensor for the most part of the spectrum (>SH degree 10). This became evident when only laser interferometer noise was considered (Figure 5, red curves). However, in the very low degrees, the MOBILE concept performed better than GRACE, which was again owed to the fact of an improved observation geometry. When considering all instrument error sources together, the retrieval errors of the low–low satellite pair are dominated by the accelerometer plus star camera sensor performance, while for the MOBILE constellation, the laser link error was the dominating error source for SH degrees higher than 20, and the accelerometer plus star camera noise only dominated the spectrum in the lower degrees. These results led to the conclusion that the gravity field retrieval based on instrument error sources showed smaller errors below SH degree 40 for the MOBILE concept compared to the low–low pair, but increased errors in the higher frequency spectrum due to the lower accuracy of the laser interferometer.

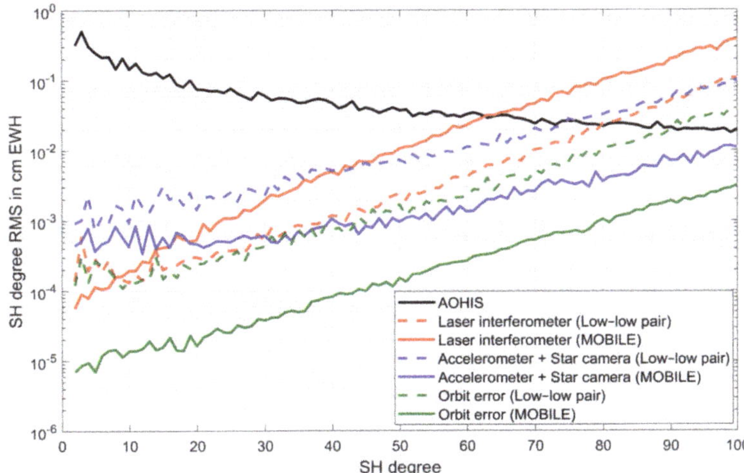

Figure 5. Degree (error) standard deviations after 1 month of full AOHIS signal (black), low–low pair formation (dashed curves), and MOBILE constellation (continuous curves), when including only distance measurement errors (red), acceleration plus star camera errors (blue), and orbit errors (green).

Table 3. Global RMS values in cm EWH for the low–low pair and MOBILE constellation considering different isolated error sources. RMS values are computed up to SH degree 100 for instrument errors, and up to SH degree 50 for temporal aliasing errors. Additionally, the global RMS value of the monthly averaged AOHIS and HIS signal is given (computed up to SH degree 50 and 100).

Errors Included in Simulation	RMS Low–Low Pair (cm EWH)	RMS MOBILE (cm EWH)
Instrument errors up to SH degree 100		
Orbit error (GNSS)	1.58	0.13
Accelerometer + star camera	4.18	0.51
Laser interferometer	4.12	14.80
Temporal aliasing errors up to SH degree 50		
AOHIS aliasing + instruments	28.27	15.76
Ocean tide aliasing + instruments	5.36	3.84
HIS aliasing + instruments	5.13	2.12
HIS aliasing + instruments (ext. processing)	-	1.29
Averaged AOHIS signal up to SH degree 50	4.06	
Averaged AOHIS signal up to SH degree 100	4.63	
Averaged HIS signal up to SH degree 50	3.35	
Averaged HIS signal up to SH degree 100	3.96	

Next to the estimation of SH coefficients, we estimated their formal errors as well, shown in Figure 6. The noise of the different sensors in combination with the observation geometry reveal the performance of a specific satellite concept. In our case, they demonstrated the impact of the MOBILE high–low tracking concept by showing an almost uniform (isotropic) error spectrum and a high sensitivity in the sectorial coefficients (SH degree equal to SH order). In case of the low–low pair configuration especially, the sectorial coefficients were less well-determined than the zonal coefficients (SH order equal to zero). Figure 6b,d gave a closer view of the formal errors located in the long wavelength (low-degree) spectrum. The comparison between MOBILE and the low–low pair led to the assumption that the determination of the very low SH coefficients could be accomplished with a higher sensitivity through the MOBILE concept. In contrast to the observations in the along-track direction of the low–low pearl-string configuration, the multi-directional observations of the high–low tracking concept with a strong dominance of the radial component enabled an improved estimation of the very low SH coefficients. The close to radial observation geometry of MOBILE was comparable to satellite laser ranging (SLR) observations, showing superior performance in observing the very long wavelength gravity field variations, in particular the zonal SH coefficient of degree 2, which physically represented the Earth's dynamic oblateness [46].

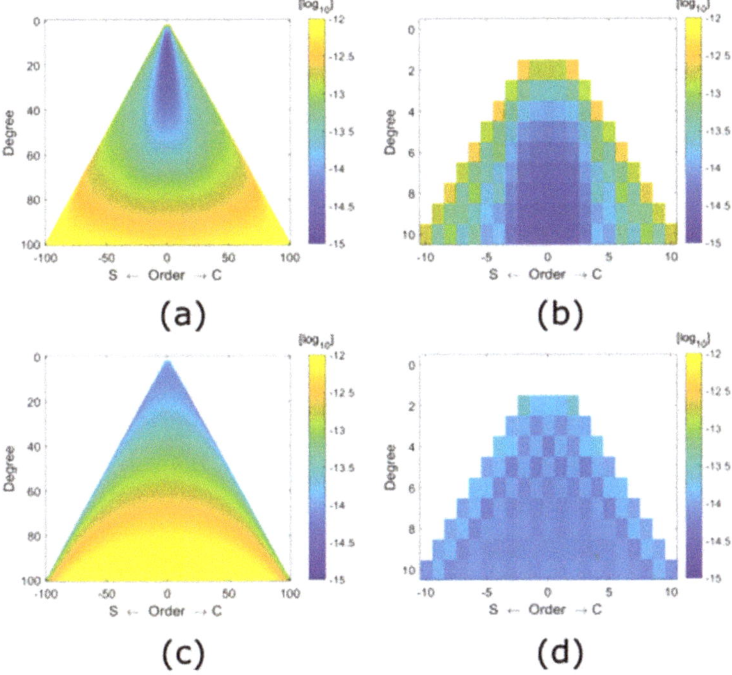

Figure 6. Formal error triangle plots in $\log_{10}$ up to SH degree and order 100 (**a,c**), and up to SH degree and order 10 (**b,d**), for the low–low pair formation (**a,b**), and the MOBILE constellation (**c,d**).

In order to make the effect of the different error spectra of both satellite concepts even more visible, spatial covariance functions were computed for a position at the equator and at 45° latitude (see Figure 7). They describe the correlation of the computation point with its neighborhood in the normal equation system due to the used stochastic model and the observation geometry. The spatial characteristics and the pattern of the covariances provide information about the spatial behavior of the retrieved signals. In our case, the figures show the typical stripes for the low–low tracking concept caused by the north–south observation direction that are known from the GRACE temporal gravity models, while the MOBILE concept exhibited an isotropic error structure at both latitudes.

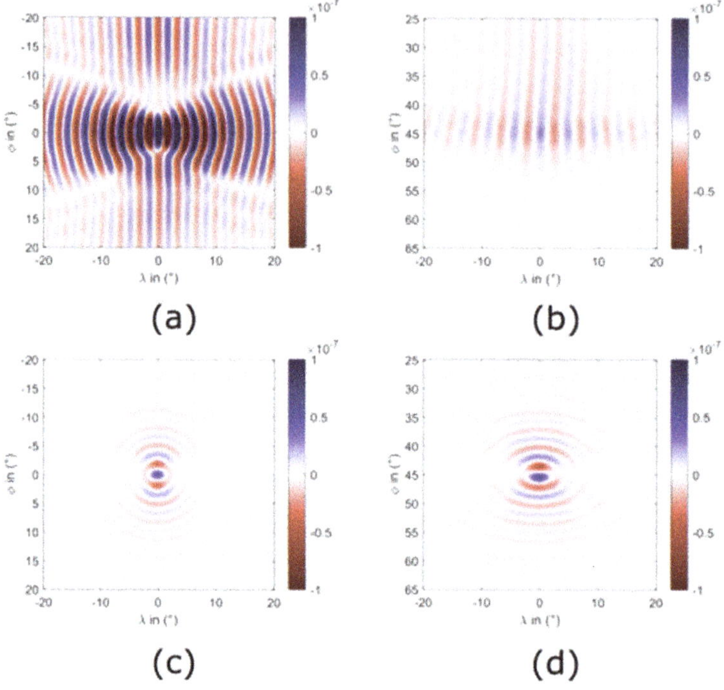

Figure 7. Spatial covariance functions in m^2 for the equator (**a,c**), and for 45° latitude (**b,d**), for the low–low pair formation (**a,b**), and the MOBILE constellation (**c,d**).

4.2. Temporal Gravity Field Retrieval

The retrieval of the temporal gravity field is dominated by temporal aliasing errors due to the undersampling of high frequency geophysical signals and imperfect de-aliasing models, which has already been shown by, e.g., References [47–49] for the GRACE mission. In order to analyze the impact of different time-varying mass signals on gravity field retrieval, we performed simulations by using signals that were subdivided into non-tidal AOHIS, HIS, and tidal signals, including the instrument errors described in Section 2.2. Figure 8 displays the corresponding retrieval errors for both satellite concepts. The results indicate that the errors with the highest signal amplitudes were related to AOHIS signals (Figure 8, red curves), and in particular to atmospheric and oceanic signals. Tidal aliasing effects played a key role in the total error budget as well (Figure 8, blue curves) by representing the highest aliasing errors next to non-tidal atmosphere and ocean aliasing errors. In this context it is important to mention that errors in ocean tide models are considered as one of the major sources of error in the determination of temporal gravity field models from GRACE data [50,51]. Our simulations show that the MOBILE configuration can reduce non-tidal aliasing errors ($\approx$45%) as well as tidal aliasing errors ($\approx$30%) over the whole spectrum significantly (see also Table 3, part: temporal aliasing errors). Despite the fact that for the high–low tracking concept the assumed arrangement of satellites causes incomplete data time series, the multi-directional observation geometry enables the sampling of time varying signals with reduced aliasing errors compared to the low–low pair configuration.

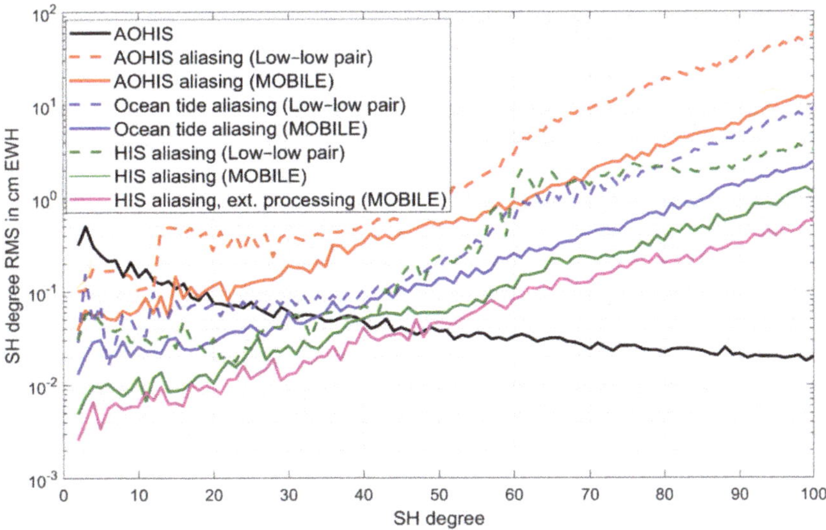

Figure 8. Degree (error) standard deviations after 1 month of full AOHIS signal (black), low–low pair formation (dashed curves), and MOBILE constellation (continuous curves), when including only AOHIS signals + instrument errors (red), ocean tide signals + instrument errors (blue), HIS signals + instrument errors (green), and HIS signals + instrument errors using the extended processing method (magenta).

Usually high-frequency mass signals are a priori reduced based on atmosphere and ocean de-aliasing (AOD) products [52], and ocean tide de-aliasing models. The resulting temporal gravity field models thus contain mainly information on sub-seasonal, seasonal, and secular continental hydrological mass variations and ice mass variations on Earth [53,54], and solid Earth signals related to glacial isostatic adjustment (GIA), and co- and post-seismic gravity changes of big earthquakes. For the analysis of such mass flux signals we performed simulations by using only the HIS signal. The resulting gravity field retrieval errors (Figure 8, green curves) again revealed smaller aliasing effects for the MOBILE concept (≈60%), which is even better visible when looking at the spatial domain, shown in Figure 9. As already suggested by Figure 7, the retrieved HIS fields demonstrate, that the error pattern of MOBILE was much more homogeneous, and the typical striping of a low–low along-track ranging system is significantly reduced, particularly in the equatorial regions where the orbit ground tracks were less dense. This resulted in a clearly improved free representation of hydrological and ice mass signals for the MOBILE concept.

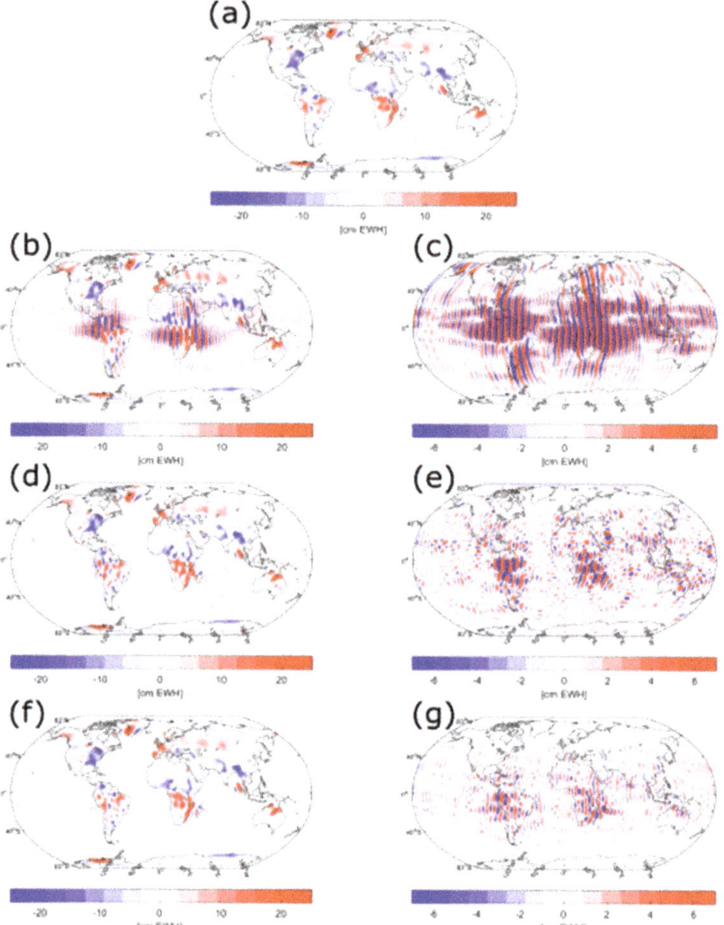

Figure 9. Global grids of EWH (cm) up to SH degree and order 50 after 1 month. The grids show the true HIS signal (**a**), the recovered signal (**b,d,f**), and the differences of the true HIS signal and the recovered signal (**c,e,g**), for the low–low pair configuration (**b,c**), for the MOBILE concept (**d,e**), and for the MOBILE concept using the extended processing method (**f,g**).

The high quality of multi-directional observations of the high–low tracking concept allows the application of an extended alternative processing method first proposed by Wiese et al. 2011 [23], which enables the mitigation of temporal aliasing effects due to non-tidal time varying signals, as it was stated in Section 1. Wiese et al. 2011 [23] demonstrated the benefit of the co-parameterization of additional daily low degree and order gravity field coefficients for Bender-type satellite constellations. We investigated the potential of this methodology regarding the MOBILE concept by simulating a monthly solution while co-estimating daily gravity fields up to SH degree and order 10, including HIS signal plus instrument errors. The resulting retrieval errors are displayed in Figure 8 (magenta curve). They revealed an error reduction of about 40% compared to the nominal solution (green curve), which led to an increased spatial resolution of about SH degree 50 ($\approx$400 km) instead of 40 ($\approx$500 km). The corresponding spatial plot (Figure 9, f and g) shows a global reduced aliasing pattern, especially in higher latitudes. The comparison between the true HIS signal and the MOBILE recovered signal (nominal and extended processed) displayed in Figure 9 shows that the quality of the solutions could be improved to such a level that de-striping and smoothing the solutions was no longer necessary when examining signals to degree and order 50. Therefore, a possible loss of signal by a posteriori filtering of the gravity field solutions recovered by MOBILE could be avoided.

5. Conclusions

In this study, we investigated the gravity field retrieval performance of the novel and innovative MOBILE high–low satellite tracking concept and compared it with a low–low GRACE Follow-On-like configuration qualitatively and quantitatively. Based on full numerical simulations, gravity field parameters were estimated in terms of SH coefficients by solving a least-squares system by inverting full normal equations over a time span of 1 month. The most important error sources affecting the gravity field retrieval performance, key instruments on-board the satellites, as well as time varying mass flux signals, were included in order to assess their impact on gravity field retrieval for both mission concepts.

The results regarding the instrumental impact on the gravity field solution show that the performance of the MOBILE configuration was mainly limited by the assumed micrometer accuracy of the laser interferometer, especially in the short wavelength spectrum, while the performance in the lower wavelengths of the gravity field benefited from the multi-directional observation geometry and optimized 3D accelerometer. In contrast, the gravity field retrieval of the low–low pair constellation was limited mainly by the accelerometer, which predominated the nanometer accuracy of the assumed laser interferometer. The multi-directional observations of MOBILE mentioned above included a strong radial component and led to an almost uniform (isotropic) error spectrum, while the low–low tracking concept showed the typical stripes caused by the north–south observation direction. However, the high accuracy of the low–low satellite pair's inter-satellite link led to an improved gravity field performance from SH degree 40 and higher compared to MOBILE, which performed better in the long wavelength spectrum where the largest amplitudes of time varying gravity field signals occurred.

The benefit of MOBILE's multi-directional observation geometry arose when including tidal and non-tidal mass variation signals into the simulation process. The results revealed significantly reduced temporal aliasing errors in the recovered gravity field signal compared to the low–low tracking concept over the whole spectrum. In the case of the separate treatment of the HIS signal, the resulting gravity field error performance of MOBILE improved even by about 60%, and the application of an extended processing method to reduce temporal aliasing errors by co-estimation of daily gravity field parameters, led to a further reduction of retrieval errors of about 40%. Furthermore, the results show that the quality of recovered MOBILE gravity field solutions could make a treatment of such solutions using a posteriori filtering techniques obsolete.

The gravity field solutions retrieved using MOBILE can contribute to an improved understanding of different components of the Earth system, such as the estimation of continental water storage and freshwater fluxes, the quantification of large-scale flood and drought events and their monitoring and forecasting, or understanding the mass balance of ice sheets and larger glacier systems, just to name a few. The application of the extended processing method implied the co-parameterized gravity field parameters (which, in our case, are daily gravity fields) as a side product. These daily solutions with low spatial resolution could aid in improving atmospheric models, and possibly be beneficial to the oceanography community as well, as many of these short-term signals have large spatial scales.

6. Outlook

The gravity field solutions of the high–low tracking concept presented in this paper were based on the minimal configuration of MOBILE. On top of this scenario, the optional implementation of a third or fourth MEO satellite, but also a second LEO, could be considered to further increase the mission performance, but also significantly improve the temporal resolution. Due to the largely passive instrumentation of this mass transport mission, the function of the MEO satellites could be implemented as a backpack application of other MEO missions, such as the Galileo next-generation satellites, in order to extend and maintain the infrastructure for laser ranging payloads. Aside, one of the most important fields of research is the mitigation of temporal aliasing errors. In this context it is important to mention that aliasing effects due to imperfect ocean tide models represent one of the largest error sources in temporal gravity field retrieval. The capability of the multi-directional observations of MOBILE of co-parameterize tidal parameters over long time spans, as proposed in Reference [24], in order to improve current ocean tide models will be the subject of further study.

Author Contributions: Conceptualization, M.H. and R.P.; methodology, M.H.; software, M.H.; validation, M.H. and R.P.; formal analysis, M.H.; investigation, M.H.; resources, M.H. and R.P.; data curation, M.H.; writing—original draft preparation, M.H.; writing—review and editing, R.P.; visualization, M.H.; supervision, R.P.; project administration, R.P.; funding acquisition, R.P.

Funding: This research was funded by *Deutsche Forschungsgemeinschaft (DFG)*, grant number, *PA 1543/8-2* and the APC was funded by *Institutional Open Access Program of the Technichal University of Munich (TUM)*.

Acknowledgments: A big part of the investigations presented in this paper was performed in the framework of the study "Two-way satellite tracking to provide a basis for gravity field mission scenarios—a simulation study with detailed error analysis II, " Deutsche Forschungsgemeinschaft (DFG), Contract No. *PA 1543/8-2* funded by DFG.

Conflicts of Interest: The authors declare no conflict of interest. The funders had no role in the design of the study; in the collection, analyses, or interpretation of data; in the writing of the manuscript, or in the decision to publish the results.

References

1. Tapley, B.D.; Bettadpur, S.; Watkins, M.; Reigber, C. The gravity recovery and climate experiment experiment, mission overview and early results. *Geophys. Res. Lett.* **2004**, *31*, L09607. [CrossRef]
2. Reigber, C.; Schwintzer, P.; Lühr, H. The CHAMP geopotential mission, in bollettino di geofisica teoretica ed applicata, 40/3-4, September–December 1999. In Proceedings of the Second Joint Meeting of the International Gravity and the International Geoid Commission, Trieste, Italy, 7–12 September 1998; pp. 285–289.
3. Reigber, C.; Schwintzer, P.; Neumayer, K.H.; Barthelmes, F.; König, R.; Förste, C.; Balmino, G.; Biancale, R.; Lemoine, J.M.; Loyer, S.; et al. Earth gravity field model EIGEN-2. *Adv. Space Res.* **2003**, *31*, 1883–1888. [CrossRef]
4. Xu, G. *GPS. Theory, Algorithms and Applications*; Springer: Berlin, Germany, 2003; ISBN 3-540-67812-3.
5. Van den Ijssel, J.; Visser, P.; Patino Rodriguez, E. Champ precise orbit determination using GPS data. *Adv. Space Res.* **2003**, *31*, 1889–1895. [CrossRef]
6. Baur, O. Greenland mass variation from time-variable gravity in the absence of GRACE. *Geophys. Res. Lett.* **2013**, *40*, 4289–4293. [CrossRef]
7. Encarnação, J.T.; Arnold, D.; Bezděk, A.; Dahle, C.; Doornbos, E.; van den Ijssel, J.; Jäggi, A.; Mayer-Gürr, T.; Sebera, J.; Visser, P.; et al. Gravity field models derived from swarm GPS data. *Earth Planets Space* **2016**, *68*, 127. [CrossRef]
8. Drinkwater, M.R.; Floberghagen, R.; Haagmans, R.; Muzi, D.; Popescu, A. GOCE: ESA's first Earth explorer core mission. In *Earth Gravity Field from Space—From Sensors to Earth Science, Space Sciences Series of ISSI*; Kluwer Academic Publishers: Dordrecht, The Netherlands, 2003; Volume 18, pp. 419–432. ISBN 1-4020-1408-2.
9. Flechtner, F.; Neumayer, K.H.; Dahle, C.; Dobslaw, H.; Fagiolini, E.; Raimondo, J.C.; Günter, A. What can be expected from the GRACE-FO laser ranging interferometer for Earth science applications? *Surv. Geophys.* **2016**, *37*, 453–470. [CrossRef]
10. Sheard, B.S.; Heinzel, G.; Danzmann, K.; Shaddock, D.; Kilpstein, W.; Folkner, W. Intersatelite laser ranging instrument for the GRACE follow-on mission. *J. Geod.* **2012**, *86*, 1083–1095. [CrossRef]
11. Murböck, M.; Pail, R. Reducing non-tidal aliasing effects by future gravity satellite formations. In *Earth on the Edge: Science for a Sustainable Planet*; Rizos, C., Willis, P., Eds.; International Association of Geodesy Symposia; Springer: Berlin/Heidelberg, Germany, 2014; Volume 139, pp. 407–412. ISBN 978-3-642-37222-3.

12. Visser, P. Designing Earth gravity field missions for the future: A case study. In *Gravity, Geoid and Earth Observation, International Association of Geodesy Symposia*; Mertikas, S., Ed.; Springer-Verlag: Berlin/Heidelberg, Germany, 2010; Volume 135, pp. 131–138.

13. Visser, P.; Sneeuw, N.; Reubelt, T.; Losch, M.; van Dam, T. Space-borne gravimetric satellite constellations and ocean tides: Aliasing effects. *Geophys. J. Int.* **2010**, *181*, 789–805. [CrossRef]

14. Wiese, D.N.; Nerem, R.S.; Han, S.C. Expected improvements in determining continental hydrology, ice mass variations, ocean bottom pressure signals, and earthquakes using two pairs of dedicated satellites for temporal gravity recovery. *J. Geophys. Res.* **2011**. [CrossRef]

15. Daras, I.; Pail, R. Treatment of temporal aliasing effects in the context of next generation satellite gravimetry missions. *J. Geophys. Res.* **2017**, *122*, 7343–7362. [CrossRef]

16. Swenson, S.; Wahr, J. Post-processing removal of correlated errors in GRACE data. *Geophys. Res. Lett.* **2006**. [CrossRef]

17. Kusche, J. Approximate decorrelation and non-isotropic smoothing of time-variable GRACE-type gravity field models. *J. Geod.* **2007**, *81*, 733–749. [CrossRef]

18. Werth, S.; Güntner, A.; Schmidt, R.; Kusche, J. Evaluation of GRACE filter tools from a hydrological perspective. *Geophys. J. Int.* **2009**, *179*, 1499–1515. [CrossRef]

19. Horvath, A. Retrieving Geophysical Signals from Current and Future Satellite Gravity Missions. Ph.D. Thesis, Technische Universität München, Munich, Germany, 2017.

20. Watkins, M.; Wiese, D.N.; Yuan, D.N.; Boening, C.; Landerer, F.W. Improved methods for observing Earth's time variable mass distribution with GRACE using spherical cap mascons. *J. Geophys. Res. Solid Earth* **2015**, *120*, 2648–2671. [CrossRef]

21. Bender, P.L.; Wiese, D.N.; Nerem, R.S. A possible dual-GRACE mission with 90 degree and 63 degree inclination orbits. In Proceedings of the Third International Symposium on Formation Flying, Missions and Technologies, Noordwijk, The Netherlands, 23–25 April 2008; pp. 1–6.

22. Sharifi, M.A.; Sneeuw, N.; Keller, W. Gravity recovery capability of four generic satellite formations. In Proceedings of the 1st International Symposium of the International Gravity Field Sevice "Gravity Field of the Earth", Istanbul, Turkey, 28 August–1 September 2007.

23. Wiese, D.N.; Visser, P.; Nerem, R.S. Estimating low resolution gravity fields at short time intervals to reduce temporal aliasing errors. *Adv. Space Res.* **2011**, *48*, 1094–1107. [CrossRef]

24. Hauk, M.; Pail, R. Treatment of ocean tide aliasing in the context of a next generation gravity field mission. *Geophys. J. Int.* **2018**, *214*, 345–365. [CrossRef]

25. Pail, R.; Bamber, J.; Biancale, R.; Bingham, R.; Braitenberg, C.; Cazenave, A.; Eicker, A.; Flechtner, F.; Gruber, T.; Güntner, A.; et al. Mass variation observing system by high low inter-satellite links (MOBILE)—A mission proposal for ESA Earth Explorer 10. In Proceedings of the 20th EGU General Assembly, EGU2018, Vienna, Austria, 4–13 April 2018.

26. Hauk, M.; Schlicht, A.; Pail, R.; Murböck, M. Gravity field recovery in the framework of a Geodesy and Time Reference in Space (GETRIS). *Adv. Space Res.* **2017**, *59*, 2032–2047. [CrossRef]

27. Van Allen, J.A. Discovery of the magnetosphere. In *History of Geophysics*; Gillmor, C.S., Spreiter, J.R., Eds.; American Geophysical Union: Washington, DC, USA, 1997; Vol. 7, pp. 235–251. ISBN 978-0875902883.

28. Illingworth, A.J.; Barker, H.W.; Beljaars, A.; Ceccaldi, M.; Chepfer, H.; Cole, J.; Delanoë, J.; Domenech, C.; Donovan, D.P.; Fukuda, S.; et al. The Earthcare satellite: The next step forward in global measurements of clouds, aerosols, precipitation and radiation. *Bull. Am. Meteorol. Soc.* **2015**, *96*, 1311–1332. [CrossRef]

29. Spencer, R.L. Lageos—A geodynamics tool in the making. *J. Geol. Educ.* **1977**, *25*, 38–42. [CrossRef]

30. Kucharski, D.; Kirchner, G.; Lim, H.C.; Koidl, F. Optical response of nanosatellite BLITS measured by the Graz 2 kHz SLR system. *Adv. Space Res.* **2011**, *48*, 1335–1340. [CrossRef]

31. Schäfer, W.; Flechtner, F.; Nothnagel, A.; Bauch, A.; Hugentobler, U. *Geodetic Time Reference in Space (GETRIS)*; ESA Study AO/1-6311/2010/F/WE Final Report GETRIS-TIM-FR-0001; 2013.

32. Iran Pour, S.; Reubelt, T.; Sneeuw, N.; Daras, I.; Murböck, M.; Gruber, T.; Pail, R.; Weigelt, M.; van Dam, T.; Visser, P.; et al. *Assessment of Satellite Constellations for Monitoring the Variations in Earth Gravity Field—SC4MGV*; ESA—ESTEC Contract No. AO/1-7317/12/NL/AF, Final Report; 2015.

33. Daras, I.; Pail, R.; Murböck, M.; Yi, W. Gravity field processing with enhanced numerical precision for LL-SST missions. *J. Geod.* **2015**, *89*, 99–110. [CrossRef]

34. Daras, I. Gravity Field Processing Towards Future LL-SST Satellite Missions. Ph.D. Thesis, Technische Universität München, Munich, Germany, 2016; pp. 23–39.
35. Yi, W. The Earth's Gravitational Field from GOCE. Ph.D. Thesis, Centre of Geodetic Earth System Research, Munich, Germany, 2012.
36. Shampine, L.F.; Gordon, M.K. *Computer Solution of Ordinary Differential Equations: The Initial Value Problem*; W.H. Freeman: San Francisco, CA, USA, 1975.
37. Schneider, M. Outline of a general orbit determination method. In *Space Research IX, Proceedings of the Open Meetings of Working Groups (OMWG) on Physical Sciences of the 11th Plenary Meeting of the Committee on Space Research (COSPAR), Tokyo, Japan*; Mitteilungen aus dem Institut für Astronomische und Physikalische Geodäsie, Nr. 51; Champion, K.S.W., Smith, P.A., Smith-Rose, R.L., Eds.; North Holland Publ. Company: Amsterdam, The Netherlands, 1969; pp. 37–40.
38. Mayer-Gürr, T. Gravitationsfeldbestimmung aus der Analyse Kurzer Bahnbögen am Beispiel der Satellitenmissionen CHAMP und GRACE. Ph.D. Thesis, Rheinische Friedrich-Wilhelms-Universität zu Bonn, Bonn, Germany, 2006.
39. Mayer-Gürr, T.; Jäggi, A.; Meyer, U.; Yoomin, J.; Susnik, A.; Weigelt, M.; van Dam, T.; Flechtner, F.; Gruber, C.; Günter, A.; et al. European gravity service for improved emergency management—Status and project highlights. In Proceedings of the EGU General Assembly, EGU 2016, Vienna, Austria, 17–22 April 2016.
40. Mayer-Gürr, T.; Rieser, D.; Höck, E.; Brockmann, J.M.; Schuh, W.D.; Krasbutter, I.; Kusche, J.; Maier, S.; Krauss, S.; Hausleitner, W.; et al. The new combined satellite only model GOCO03s. In Proceedings of the IAG Symposium Gravity, Geoid and Height Systems, Venice, Italy, 9–12 October 2012.
41. Dobslaw, H.; Bergmann-Wolf, I.; Dill, R.; Forootan, E.; Klemann, V.; Kusche, J.; Sasgen, I. The updated ESA Earth system model for future gravity mission simulation studies. *J. Geod.* **2015**, *89*, 505–513. [CrossRef]
42. Savcenko, R.; Bosch, W. *EOT11a-Empirical Ocean Tide Model from Multi-Mission Satellite Altimetry*; DGFI Report No. 89; Deutsches Geodätisches Forschungsinstitut: München, Germany, 2012.
43. Ray, R. *A Global Ocean Tide Model from Topex/Poseidon Altimetry: Got99.2*; NASA Technical Memorandum 209478; Goddard Space Flight Center: Greenbelt, Maryland, 1999.
44. Siemes, C. Digital Filtering Algorithms for Decorrelation within Large Least Square Problems. Ph.D. Thesis, University of Bonn, Bonn, Germany, 2008.
45. Pail, R.; Bruinsma, S.; Migliaccio, F.; Förste, C.; Goiginger, H.; Schuh, W.D.; Höck, E.; Reguzzoni, M.; Brockmann, J.M.; Abrikosov, O.; et al. First GOCE gravity field models derived by three different approaches. *J. Geod.* **2011**, *85*, 819–843. [CrossRef]
46. Cheng, M.K.; Tapley, B.D. Seasonal variations in low degree zonal harmonics of the Earth's gravity field from satellite laser ranging observation. *J. Geophys. Res.* **1999**, *104*, 2667–2681. [CrossRef]
47. Zenner, L.; Gruber, T.; Jaeggi, A.; Beutler, G. Propagation of atmospheric model errors to gravity potential harmonics-impact on GRACE de-aliasing. *Geophys. J. Int.* **2010**, *182*, 797–807. [CrossRef]
48. Han, S.; Jekeli, C.; Shum, C. Time-variable aliasing effects of ocean tides, atmosphere, and continental water mass on monthly mean GRACE gravity field. *J. Geophys. Res.* **2004**, *109*, B04403. [CrossRef]
49. Thompson, P.; Bettadpur, S.; Tapley, B.D. Impact of short period, non-tidal, temporal mass variability on GRACE gravity estimates. *Geophys. Res. Lett.* **2004**, *31*, L06619. [CrossRef]
50. Knudsen, P.; Andersen, O. Correcting GRACE gravity fields for ocean tide effects. *Geophys. Res. Lett.* **2002**, *29*, 1178. [CrossRef]
51. Seo, K.W.; Wilson, C.R.; Han, S.C.; Waliser, D.E. Gravity recovery and climate experiment (GRACE) alias error from ocean tides. *J. Geophys. Res.* **2008**, *113*, B03405. [CrossRef]
52. Flechtner, F. *AOD1b Product Description Document for Product Releases 01 to 04*; GRACE 327-750; GeoForschungsZentrum Potsdam: Potsdam, Germany, 2007.
53. Tiwari, V.M.; Wahr, J.; Swenson, S. Dwindling groundwater resources in northern India, from satellite gravity observations. *Geophys. Res. Lett.* **2009**, *36*, L18401. [CrossRef]
54. Velicogna, I.; Sutterley, T.C.; van den Broeke, M.R. Regional acceleration in ice mass loss from Greenland and Antarctica using GRACE time-variable gravity data. *Geophys. Res. Lett.* **2014**, *41*, 8130–8137. [CrossRef]

Article

High-Resolution Mass Trends of the Antarctic Ice Sheet through a Spectral Combination of Satellite Gravimetry and Radar Altimetry Observations

Ingo Sasgen [1],*, Hannes Konrad [1,2], Veit Helm [1] and Klaus Grosfeld [1]

[1]   Alfred Wegener Institute, Helmholtz Centre for Polar and Marine Research, 27570 Bremerhaven, Germany; hannes.konrad@dwd.de (H.K.); veit.helm@awi.de (V.H.); klaus.grosfeld@awi.de (K.G.)

[2]   Deutscher Wetterdienst, 63067 Offenbach, Germany

*   Correspondence: ingo.sasgen@awi.de; Tel.: +49-(0)471-4831-2468

Received: 5 November 2018; Accepted: 9 January 2019; Published: 14 January 2019

Abstract: Time-variable gravity measurements from the Gravity Recovery and Climate Experiment (GRACE) and GRACE-Follow On (GRACE-FO) missions and satellite altimetry measurements from CryoSat-2 enable independent mass balance estimates of the Earth's glaciers and ice sheets. Both approaches vary in terms of their retrieval principles and signal-to-noise characteristics. GRACE/GRACE-FO recovers the gravity disturbance caused by changes in the mass of the entire ice sheet with a spatial resolution of 300 to 400 km. In contrast, CryoSat-2measures travel times of a radar signal reflected close to the ice sheet surface, allowing changes of the surface topography to be determined with about 5 km spatial resolution. Here, we present a method to combine observations from the both sensors, taking into account the different signal and noise characteristics of each satellite observation that are dependent on the spatial wavelength. We include uncertainties introduced by the processing and corrections, such as the choice of the re-tracking algorithm and the snow/ice volume density model for CryoSat-2, or the filtering of correlated errors and the correction for glacial-isostatic adjustment (GIA) for GRACE. We apply our method to the Antarctic ice sheet and the time period 2011–2017, in which GRACE and CryoSat-2 were simultaneously operational, obtaining a total ice mass loss of 178 ± 23 Gt yr^{-1}. We present a map of the rate of mass change with a spatial resolution of 40 km that is evaluable across all spatial scales, and more precise than estimates based on a single satellite mission.

Keywords: Mass balance; Ice Sheets; Sea-level Rise; Antarctica; GRACE; CryoSat-2; GRACE-Follow On; GRACE-FO; downward continuation; spectral methods

---

1. Introduction

The Antarctic ice sheet is the largest reservoir of non-oceanic water mass. Projections estimate the ice sheet's potential to raise sea levels by 15 m by the year 2500 for scenarios of unabated climate change [1]. It is currently in a state of overall decline [2,3], but regional differences are very prominent. Most of East Antarctica is relatively stable [4], while many glaciers in West Antarctica and the Antarctic Peninsula are losing mass due to enhanced discharge [5,6] and retreat [7] following the basal melt or structural collapse of ice shelves [8]. Some of these West Antarctic glaciers represent a serious threat to global sea levels as they could collapse within a few centuries, raising global mean sea level by centimeters to meters [9–11].

Satellite technology enabled much better coverage of mass balance estimates of the Antarctic and Greenland ice sheets than could be achieved with in situ or airborne measurements [2]. Three different techniques have been commonly employed to assess ice sheet mass balance: (1) The mass budget method subtracts ice discharge into the oceans and ice shelves from the net mass flux from the atmosphere onto the ice sheet's upper surface [12]. (2) Satellite gravimetry observations (mostly by the Gravity Recovery and Climate Experiment, GRACE [13]) measure the perturbations in the gravitational potential of the Earth caused by redistribution of Earth's mass. (3) Satellite altimetry enables repeat measurements of surface elevation, which can be used to quantify the total ice volume change and estimates of total mass change if the density at which volume changes occur can be inferred [14].

The gravimetric and altimetric approaches can both quantify the mass changes as spatial fields. However, their characteristics are very different. For example, the CryoSat-2 radar altimetry mission offers highly resolved maps of height changes [15], however, it is influenced by uncertainties from the backscattering properties of snow and firn, as well as limited resolvability of terrain with steep gradients in coastal or mountainous areas [16,17]. In addition, the conversion of volume to mass changes requires knowledge of the contributions to height changes by depositional processes which occur at the density of snow, by ice-dynamical imbalance which occurs at the density of ice, and by firn densification, which implies a height change without a mass change [18].

GRACE, on the other hand, measures mass change of the whole ice column. Although it is not affected by the presence of different processes in the snow, firn, and ice column, GRACE has a significantly lower effective spatial resolution and needs to be corrected for all other mass change processes not related to the present-day ice mass change (for example caused by short-term mass variability in the atmosphere and the ocean). GRACE mass trends need to be corrected for the Earth's viscoelastic response to past ice changes (glacial isostatic adjustment, GIA), which is a prominent trend signal in polar regions and superimposes with the present-day mass balance (for example [19]).

Here we present a novel method to derive spatial maps of ice sheet mass change that exploits the advantages of these two techniques, while minimizing the uncertainties that are associated with each type of observation. Satellite gravimetry and laser altimetry observations of ice change have been combined before in order to separate the signals of difference processes (for example [20–23]) and also to increase the spatial resolution [24]. Here, we perform for the first time, to our knowledge, a combination in the spectral domain (i.e., in terms of the representation of the spatial fields as spherical harmonic coefficients), the common representation in which the GRACE gravity fields are estimated and distributed (Level 2 data). The combination approach can be regarded as a downward-continuation of GRACE gravity measurements to mass changes on the Earth's surface, using mass change fields derived from CryoSat-2 data. We show that our approach allows overcoming the limited resolution of the GRACE data, producing a field of ice mass trends that is evaluable across all spatial scales, and more precise than that recovered by a single sensor. The method is applicable to other regions and components of the Earth system and may be useful to join GRACE/GRACE-FO data and additional measurements into new combined Level 4 data products.

2. Data and Methods

2.1. GRACE Satellite Gravimetry

GRACE observations of the gravitational potential are typically released by the respective processing centers as monthly sets of spherical harmonics coefficients for integer degree j and order $-j \leq m \leq j$. Coefficients of degree $j = 1$ have to be obtained from different observations (see below), the coefficient of $j = 0$ is constant by definition (conservation of total mass). The spatial representation of the GRACE geoid height observation at colatitude θ, longitude φ, and time t can be expressed as

$$g(\theta, \varphi, t) = \sum_{j=0}^{j_{max}} \sum_{m=-j}^{j} \left(C_{jm}(t) \, Y_{jm}(\theta, \varphi) \right) \tag{1}$$

with $Y_{jm}(\theta, \varphi) = \begin{cases} P_{jm}(\cos \theta) \cos(m\varphi) & for \ m \geq 0 \\ P_{j|m|}(\cos \theta) \sin(|m|\varphi) & for \ m < 0 \end{cases}$ and $C_{j0} \equiv 0$ for $m < 0$, where P_{jm} is the normalized associated Legendre polynomial of degree j and order m [25,26]. We utilize monthly coefficients $C_{jm}(t)$ of release 5 by the Center for Space Research, The University of Texas at Austin from $j = 2$ up to $j_{\max} = 90$ (CSR RL05; [27]). The GRACE coefficients $(j, m) = (2, 0)$ are replaced by values from satellite laser ranging [28]. The GRACE spectrum is completed with coefficients of $j = 1$ estimated by the approach of Swenson et al. [29]. A pole drift correction has not been applied.

We decompose the time series of each coefficient by adjusting, in a least squares sense, a temporal model consisting of the most pronounced oscillations (an annual cycle at 1 year period, the S2 tide with a 171 days repeat cycle, and the P1 tide at 161 days), as well as a linear and quadratic trend in time. For consistency with CryoSat-2 data, the adjustment period is from February 2011 to June 2017, which is the last available GRACE monthly solution. We create an ensemble of the linear trend $\frac{\partial g}{\partial t}$ by coefficient-wise random perturbation of the fitted linear trend according to the propagated standard deviation of the trend estimate (30 realizations). We subtract GIA contributions according to the three different estimates IJ05r2 [30], AGE1a [31] (this is the version independent of GRACE observations), and ICE6G [32]. Thus, we obtain 90 ensemble members representing both the uncertainties of the GRACE coefficients, as well as uncertainties of the GIA correction. Figure 1 shows the mean and standard deviation of the ensemble, as well as the uncertainty components of each ensemble as a degree-power spectrum. It is visible that the GIA-induced uncertainty peaks in the range of degree 5 to 15, while the measurement uncertainty of the coefficients gradually increases.

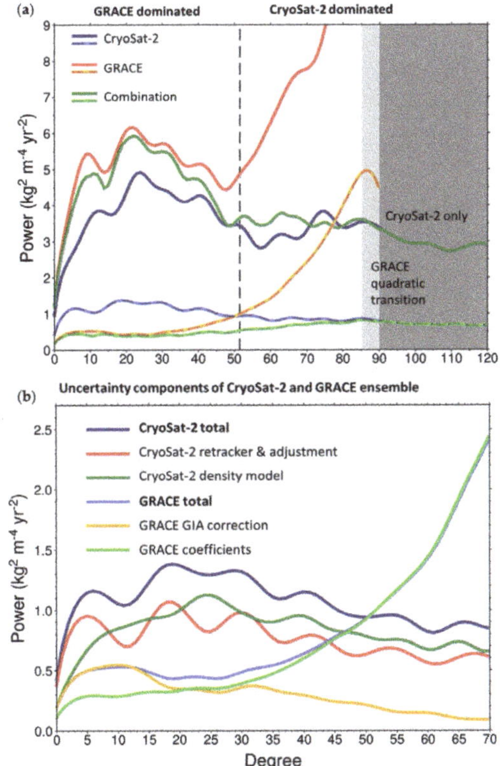

Figure 1. Degree-power spectrum of the (**a**) rate of mass change (kg^2 m^{-4} yr^{-2}) for the ensemble mean (solid) and ensemble standard deviation (dashed) of CryoSat-2 (blue), GRACE (red), and the combined solution (green). The vertical dashed line indicates the degree at which the power of the GRACE and

CryoSat-2 uncertainties are equal, separating the spectrum into a GRACE and CryoSat-2 dominated part. Light shading shows the ad hoc quadratic transition to the GRACE related weights of zero for $j \geq 90$, meaning that degrees and order 90 to 512 are only supplied by CryoSat-2. (**b**) Degree-power spectrum of the ensemble standard deviation for CryoSat-2 (dark blue) and GRACE (light blue), along with the uncertainty components re-tracker and adjustment method (red), snow/ice density model (dark green) for CryoSat-2, and the glacial-isostatic adjustment (GIA) correction (orange) and uncertainty of the GRACE trend coefficients (light green). Note that a different range of degrees is plotted in (**a**) and (**b**).

Note that regional optimized models exist for the Amundsen Sea Embayment [33], the Antarctic Peninsula [34], and the Siple Coast [35], yielding GIA-induced apparent mass changes of 17 Gt yr^{-1}, 3 Gt yr^{-1}, and a range of $\pm$ 6–8 Gt yr^{-1}, respectively. Using the model for the Amundsen Sea Embayment will increase the mass loss from GRACE and the discrepancy to the CryoSat-2 estimate shown later. However, these models are computed with Earth structure models optimized for these regions and should not be simply superimposed with continent-wide GIA simulations adopting an average Earth structure for Antarctica.

Each member of the geoid height trend $\frac{\partial g}{\partial t}$ ensemble is then converted to mass on the Earth surface $\frac{\partial \sigma}{\partial t}$ according to Wahr et al. [36], using load Love numbers corresponding to an elastic Earth represented by the Preliminary Reference Earth Model [37]. For clarity, we refer from now on to this surface-mass density change simply as mass change or surface load. Then, a mask is applied to the spherical harmonic spectrum using the transform method of Martinec [38] in order to reduce the far-field signal and obtain a spectral representation of the mass changes over Antarctica only. The mask is initially designed in space, $M(\theta, \varphi)$, and then analysed in terms of spherical harmonics coefficients up to $j_{\max} = 90$,

$$M_{jm} = \int_0^{2\pi} d\varphi \int_0^{\pi} d\theta \, \sin\theta \, M(\theta, \varphi) \, Y_{jm}(\theta, \varphi) \tag{2}$$

According to the relation between coefficients and the spatial function that they represent, as in Equation (1); $M(\theta, \varphi)$ is zero at any point that is more than 300 km away from the Antarctic grounding line (from Rignot et al. [39]), one at any point that is less than 200 km away from it, and linearly interpolated in between. The spectral multiplication increases the necessary degrees for the representation of the masked field to $j_{\max} = 180$ [38].

Later, in the spectral combination, we will refer to the respective spherical harmonic coefficients of the $\frac{\partial \sigma}{\partial t}$ ensemble member as $C_{jm}^{\mathrm{GRACE},k}$, where k specifies the ensemble member. From $C_{jm}^{\mathrm{GRACE},k}$, we calculate ensemble mean C_{jm}^{GRACE} (indicated by the dropped superscript k) and ensemble standard deviation $\Delta C_{jm}^{\mathrm{GRACE}}$. For each degree j, the respective degree powers (Figure 1) are proportional to $\sum_{m=-j}^{j} \left(C_{jm}^{\mathrm{GRACE}} \right)^2$ and $\sum_{m=-j}^{j} \left(\Delta C_{jm}^{\mathrm{GRACE}} \right)^2$. The magnitudes of means and standard deviations of the individual coefficients are shown in Figure 2. The spatial representations are calculated after the synthesis of the spherical harmonic spectrum of each ensemble member, as per-grid cell rates of mass change $\frac{\partial \sigma}{\partial t}$ and their respective standard deviation $\Delta \frac{\partial \sigma}{\partial t}$. Unless stated otherwise, we express the mass changes in the unit kg m^{-2}yr^{-1}, or equivalently in mm of the water column of density 1000 kg m^{-3} referred to as mm water equivalent, per year (mm we yr^{-1}).

Note that after isotropic filtering of the ensemble mean spatial field, the typical North-South oriented pattern of uncertainties [40] are still present in the ensemble mean, but are successfully removed by our spectral combination with CryoSat-2 observations (see Section 2.3). Our analysis shows that de-striping the gravity field observation before the combination using the filter of Swenson and Wahr [40] does not markedly change our combination results and can be omitted. Furthermore, we find that accounting for co-variances in the monthly GRACE coefficients' using the m-block approximation [41] does not significantly alter the linear-trend estimate. We therefore adopt the variances of GRACE coefficients estimated from the post-fit residual.

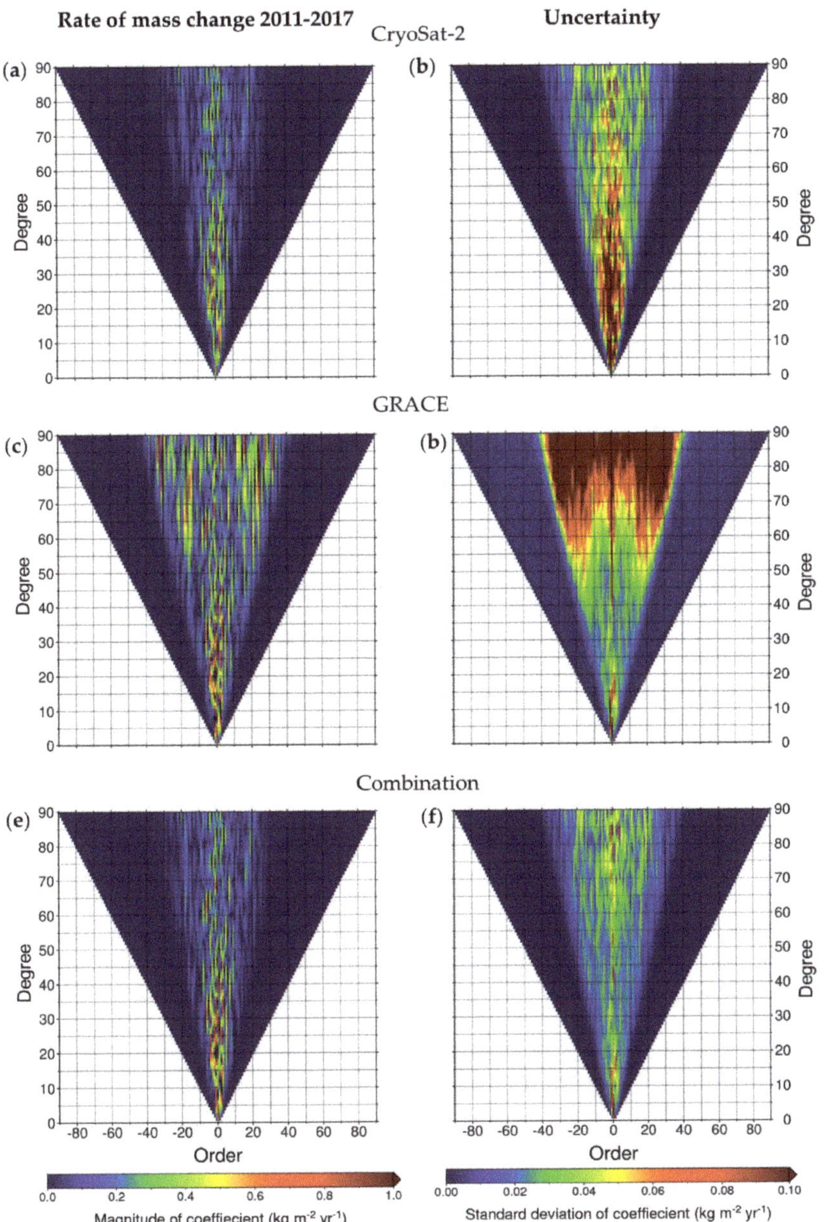

Figure 2. Spectral magnitude of the mass change coefficients ($kg\ m^{-2}\ yr^{-1}$) for the ensemble mean of (**a**) CryoSat-2, (**c**) GRACE, and the (**e**) combined solution and the respective ensemble standard deviation in (**b**), (**d**), and (**f**). Shown are coefficients up to degree and order 90. The complimentary uncertainty characteristics of GRACE and CryoSat-2, and the overall reduced uncertainty of the combined fields are visible. Time period is February 2011 to June 2017.

2.2. CryoSat-2 Satellite Radar Altimetry

The initial measurements of the radar altimeter SIRAL on board of CryoSat-2 are radar echoes, or waveforms, from prominent reflectors in the uppermost part of the firn body, from which we eventually derive elevation rates over the period 2011–2017, following the processing scheme of Helm et al. [42], with some modifications (see Appendix A). We create an ensemble of CryoSat-2 $\frac{\partial h}{\partial t}$ estimates with the aim of representing uncertainties arising from methodical differences, as well as uncertainties from the influence of volume scatter to the elevation estimates. The ensemble consists of seven re-tracker solutions, i.e., algorithms to detect the timing of the incoming wave from the reflector, and thus elevation, as well as four least-square space-time fitting methods for aggregating the elevation measurements into linear rates of elevation change. Permutation of these processing choices gives us in total, 28 independent ensemble members. Formal measurement uncertainties are not considered in the ensemble, as our analysis indicated that uncertainties from the re-tracker and plane fitting choices are dominant. Details on the processing of the CryoSat-2 data, the re-trackers and fitting schemes are provided in the Appendix A.

For the conversion of mass changes $\frac{\partial \sigma}{\partial t}$ in the spatial domain into the spherical harmonic spectrum, all gaps in the respective height change field $\frac{\partial h}{\partial t}$ must be filled, due to the global integration over the spherical harmonic base functions (for example Equation (2)). We fill the gaps in the $\frac{\partial h}{\partial t}$ fields by interpolating using inverse distance weighting. Other, more elaborate techniques, such as kriging, are rejected here, because they are unlikely to improve the filling of the gaps, which often occur in the Antarctic Peninsula and the Transantarctic Mountains as the complex terrain distorts waveforms, and thus lets the conventional re-tracking algorithms fail [42]. Yet it is worth keeping in mind that the CryoSat-2 field is consequently less accurate in these areas. Surface elevation change outside the grounding line are set to zero, as respective floating ice changes are not visible directly in the gravity field, respectively, or the derived mass change $\frac{\partial \sigma}{\partial t}$. We do not correct for GIA in the CryoSat-2 data, as bedrock topography changes are well below 5 mm yr^{-1} in most places [22], which is small in comparison with the measured elevation rates. Higher values have been reported in the Amundsen Sea Embayment [33], but this is also where the surface elevation rates are highest, and so the relative uncertainty remains very low. We note that the possible bias and uncertainty induced by neglecting GIA-induced crustal displacements in the CryoSat-2 only is around 9 ± 6 Gt yr^{-1} [23].

We convert each of the above 28 realizations of CryoSat-2 surface elevation trends $\frac{\partial h}{\partial t}$ into trends of mass change $\frac{\partial \sigma}{\partial t}$, using four different assumptions on the significance of snow/ice processes, which generates a total of 112 ensemble members. Three of these methods are based on grid-based multiplying $\frac{\partial h}{\partial t}$, with the density associated with the assumed most dominant process [15,43]. The fourth method is based on the output of a regional climate model [44]. Figure 1 shows the uncertainty associated with the density models, as wells as with the re-tracker and adjustment method as a degree-power spectrum.

The CryoSat-2 fields, being available at a resolution of 5 km here, would, in principle, allow a spherical harmonic expansion to degree $j_{\max} \approx \frac{20000\ km}{5\ km} = 4000$. For the sake of efficiency, we opt for a maximum spherical harmonic degree of $j_{\max} = 512$, equivalent to about 40 km spatial resolution in latitudinal direction, which is sufficient to demonstrate the feasibility of our approach, however, is somewhat coarser than the spatial resolutions < 32 km, adopted by many continent-wide ice sheet models [45]. We transfer the $\frac{\partial \sigma}{\partial t}$ from the grid equidistant in polar stereographic coordinates to a grid equidistant in latitude and longitude ($0.2°$) by bilinear interpolation to the polar stereographic projected latitude and longitude nodes. The spherical harmonics spectra of the 112 CryoSat-2 ensemble members are then generated from these re-gridded fields, according to the relation given in Equation (2). In accordance with our nomenclature for spherical harmonic coefficients of the mass change field $\frac{\partial \sigma}{\partial t}$ from GRACE, we will refer to the respective ensemble member l, represented by its coefficients as $C_{jm}^{CS2,l}$, where superscript 'CS2' stands for CryoSat-2. Likewise, the ensemble mean and standard deviations are C_{jm}^{CS2} and ΔC_{jm}^{CS2}, respectively.

2.3. Spectral Combination

The combination of GRACE and CryoSat-2 exploits the different noise characteristics of each satellite observation that is dependent on the spatial wavelength of the mass fields. While GRACE uncertainties are known to increase with spatial resolution, due to the ill-posed and unstable nature of the gravimetric inversion problem (for example [46]), uncertainties of CryoSat-2 are expected to be sensitive to large-scale offsets or regional uncertainties in the snow/ice density necessary for the conversion from volume rates to mass rates. The GRACE coefficient at degree and order larger than 50 (spatial resolution of ca. 400 km) are typically dominated by noise, coefficients beyond degree and order ca. 90 (spatial resolution of 220 km) are often not provided in the GRACE gravity field solutions. In this sense, our combination can be interpreted as augmenting the low-frequency GRACE spherical-harmonic spectrum with higher frequencies provided by CryoSat-2, using an uncertainty weighted, optimal blending of both data sets in the spectral range $j \leq 90$. From a geophysical point of view, our combination is a downward-continuation of the GRACE-measured gravitational perturbation at satellite altitude to the sources of mass change on the Earth's surface with CryoSat-2.

For the combined spectrum of GRACE and CryoSat-2, we find weights for GRACE (w_{jm}^{GRACE}) and CryoSat-2 (w_{jm}^{CS2}) spherical harmonic coefficients according to the standard deviation of the ensemble for each mission in each coefficient

$$w_{jm}^{GRACE|CS2} = N_{jm}^i \left(\Delta C_{jm}^{GRACE|CS2} \right)^{-2},$$ (3)

with the normalization factor

$$N_{jm}^i = \left(\left(\Delta C_{jm}^{GRACE} \right)^{-2} + \left(\Delta C_{jm}^{CS2} \right)^{-2} \right)^{-1}.$$ (4)

Due to the decreasing signal-to-noise ratio in the GRACE observations with increasing degrees (for example [36]), we set the GRACE weights to zero beyond degree 90, and ensure a smooth transition within 5 degrees by the ad hoc multiplication of ΔC_{jm}^{GRACE} by $\left(1 - \left(\frac{i-85}{5} \right)^2 \right)$ for $85 \leq j \leq 90$ in Equations (3) and (4). For GRACE ensemble member k and CryoSat-2 ensemble member l, the resulting coefficients of the combined field (superscript 'Comb.') are then given as

$$C_{jm}^{Comb.,k,l} = C_{jm}^{GRACE,k} \, w_{jm}^{GRACE} + C_{jm}^{CS2,l} \, w_{jm}^{CS2}.$$ (5)

This means that the weighting factors calculated from (3) are the same for all ensemble members k and l in the combination (5), respectively. The resulting full spectrum up to degree and order 512 (not shown) is transferred into the spatial domain (see Section 3.2) for respective ensemble mean and standard deviation. Note that due to the noise characteristics of GRACE and CryoSat-2, w_{jm}^{GRACE} represents a low-pass filter, while w_{jm}^{CS2} represents a high-pass filter, as seen by the uncertainty characteristics shown in Figures 1 and 2.

2.4. Limitations of the Spectral Combination

The spectral combination makes use of complementary wavelength-dependent noise characteristics and resolution capabilities of GRACE and CryoSat-2. However, as a downside of the spectral combination, artefacts may appear if both spectral parts are not fully consistent, which is likely, as they are obtained from two observing systems sensitive to different processes. For example, Figure 1 shows lower degree-power for CryoSat-2 in spectral range j < 50 compared to GRACE, which, in our case, translates into a lower magnitude of total mass balance (see Section 3.3). This spectral difference will, in combination with GRACE, lead to an inconsistent spherical-harmonic representation of the true (unknown) mass field. Therefore, signal artefacts may appear, visible in our combined field as minor mass changes over ocean areas that were previously set to zero by masking (Sections 2.1

and 2.2). For example, a slightly negative signal is visible in the ocean part of the Amundsen Sea Sector (see Section 3); here, the high-frequency supplied by CryoSat-2 does not cancel ocean leakage from GRACE completely. Note that similar—however, less obvious—issues arise when inconsistent gridded fields are averaged.

2.5. GRACE and CryoSat-2 Contributions

We quantify how much signal power GRACE and CryoSat-2 contribute to the combined field at each spatial scale, i.e., up to each harmonic degree j. For this, we quantify the cumulative sum of the degree power according to $p_j^{\text{GRACE|CS2}} \propto \sum_{j'=0}^{j} \sum_{m=-j'}^{j'} \left(C_{j'm}^{\text{GRACE|CS2}} \, w_{j'm}^{\text{GRACE|CS2}} \right)^2$ and evaluate this quantity relative to the cumulative degree power of the mean combined spectrum:

$$P_j^{\text{GRACE|CS2}} = p_j^{\text{GRACE|CS2}} \Big/ \left(p_j^{\text{GRACE}} + p_j^{\text{CS2}} \right) \tag{6}$$

This choice of relating $P_j^{\text{GRACE|CS2}}$ to the overall cumulative power of the combined field has the advantage that $P_j^{\text{GRACE}} + P_j^{\text{CS2}} = 100\%$, and the disadvantage that the term $p_{j'}^{\text{GRACE}} + p_j^{\text{CS2}}$ does not fully represent the cumulative power of the mean combined field $p_j^{\text{Comb.}}$, due to the quadratic term in the computation of degree power. However, $p_j^{\text{GRACE}} + p_j^{\text{CS2}}$ and the actual cumulative degree power of the mean combined spectrum $p_j^{\text{Comb.}}$ differ by a maximum of $\sim 6\%$ of $p_j^{\text{Comb.}}$ (reached at $j = 90$), indicating that our approach is valid. In turn, this means that we can determine the optimum mix of the two sensors' observations for a targeted spatial resolution.

2.6. Basin Averages and Transects

For comparison with studies providing GRACE only estimates (for example [3]), we provide integrated mass balance for 25 commonly used Antarctic drainage basins, shown in Figure 3 (after Rignot et al. [12]; used in Sasgen et al. [23]). We quantify the mass balance within the basin based on the spatial representation for each k (GRACE), l (CryoSat-2), and (k, l) (combined) ensemble member according to $m_N = \sum_n \frac{\partial \sigma_n}{\partial t} A_n$, where n indicates the running index of grid elements within a certain basins N, and A_n is the associated area of this grid element. Based on these ensembles, we compute mean and standard deviations of the integrated basin mass balances for GRACE, CryoSat-2, and the combined solution. In the following, the uncertainties provided represent one standard deviation.

In addition, we assess the spatial resolution and decorrelation effects of our combined estimate by evaluating $\frac{\partial \sigma}{\partial t}$ and $\Delta \frac{\partial \sigma}{\partial t}$ field along three 2000 km long transects, which run approximately parallel to the grounding line, however shifted inland by approximately 200 km (Figure 3). We have selected three regions of particular interest: (1) Wilkes Land, East Antarctica (transect AA'), where very localized ice dynamic imbalance has been noted at the Totten glacier system (for example [47]). (2) Dronning, Maud, and Enderby Land (transect BB'), where strong accumulation variations are observed [48]. And (3) the Amundsen Sea and Bellingshausen Sea Sectors (transect CC'), where the largest ice dynamic losses for Antarctica are recorded (for example [49]). The transects are chosen approximately across the ice-dynamic flow line to assess whether glacial entities smaller than the typical basin scale can be resolved. In addition, we assess the mass rate fields locally perpendicular to the transects described above, namely along Totten Glacier (aa'), Shirase Glacier (bb'), and Pine Island Glacier (cc'). Crossing the division between the Antarctica continent and the surrounding ocean or ice shelf areas allows us to assess the signal leakage beyond the coastline into the open ocean. Note that we do not attempt to adopt the exact grounding and flow line positions of the ice streams, which is beyond the capabilities of CryoSat-2, and thus, the combined product.

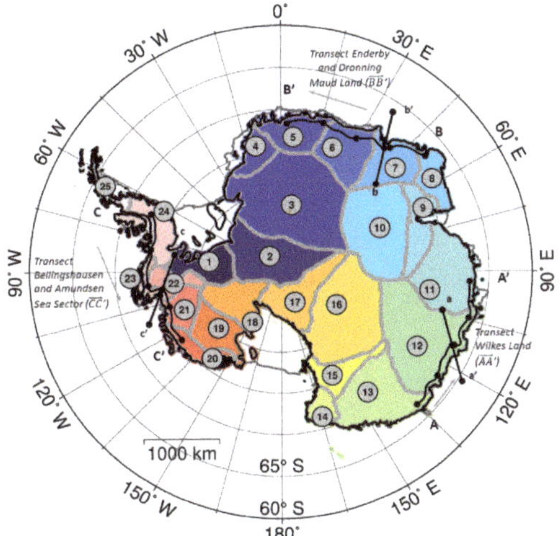

Figure 3. Definition of Antarctic drainage basins and transects used for the evaluation of the mass balance fields. We adopt 25 drainage basins merged from the basin division. In addition, we assess our results along three 2000 km long transects AA′, BB′, and CC′, approximately parallel and 200 km inland of the grounding line [34], as well as three 1000 km long transect locally perpendicular to the grounding line (aa′, bb′, and cc′). The projection is Polar Stereographic centered at 90°S and 0°E, with the true latitude of 71°S (applies to scale) and WGS84 (EPSG:3031).

3. Results

3.1. Spectral Representation

The magnitude of the coefficients of the ensemble mean and the ensemble standard deviations of CryoSat-2, GRACE, and the combined solution are shown in Figure 2 for $j \leq 90$. The magnitude per coefficient is centered close to the zonal coefficients (m = 0), with lower values in the sectorial coefficients (j = m). This is a result of Antarctica's geographical position approximately centered on the South Pole, with data coverage only south of 60°S. Note that the spectrum of CryoSat-2 is more focused towards the zonal coefficients than GRACE, owing to the mask buffer zone of 300 km adopted for GRACE (Section 2.1) and the North-South oriented noise pattern. The overall magnitude of GRACE coefficients is larger in the low degrees and orders, and noise is starting to dominate around degree and order 50.

The ensemble per-coefficient standard deviations (Figure 2, right panels) confirm the noise structure shown in Figure 1; the variability caused by the choice of the re-tracker and density model in CryoSat-2 creates uncertainties in the lower degree part of the spectrum, similar to the characteristics signal spectrum itself. For GRACE, the well-known increase of the noise with degree and order is visible, suggesting an onset of the noise dominated regime at about degree and order 60. The combined solution retains the spectral magnitude in the low degrees and orders, while reducing noise in the high degrees and orders.

3.2. Spatial Representation

Figure 4 shows the spatial representation of the CryoSat-2, GRACE and the combined field together with their respective uncertainties. For reference, we have labeled prominent features of mass change in the spatial representation of the ensemble mean CryoSat-2 field (Figure 4). These include the well-known hotspots of ice dynamics losses in the Amundsen Sea Embayment (Figure 4, Label 1; for example [49]), the slowing of Ice Stream C (Label 2; for example [50]), and the Totten glacier system (Label 3; for example [47]). In addition, CryoSat-2 shows a large-scale, low-magnitude mass loss signal in the interior part of Wilkes Land (Label 4) and prominent accumulation driven mass gain is shown along the Antarctic Peninsula (Label 5; [6]).

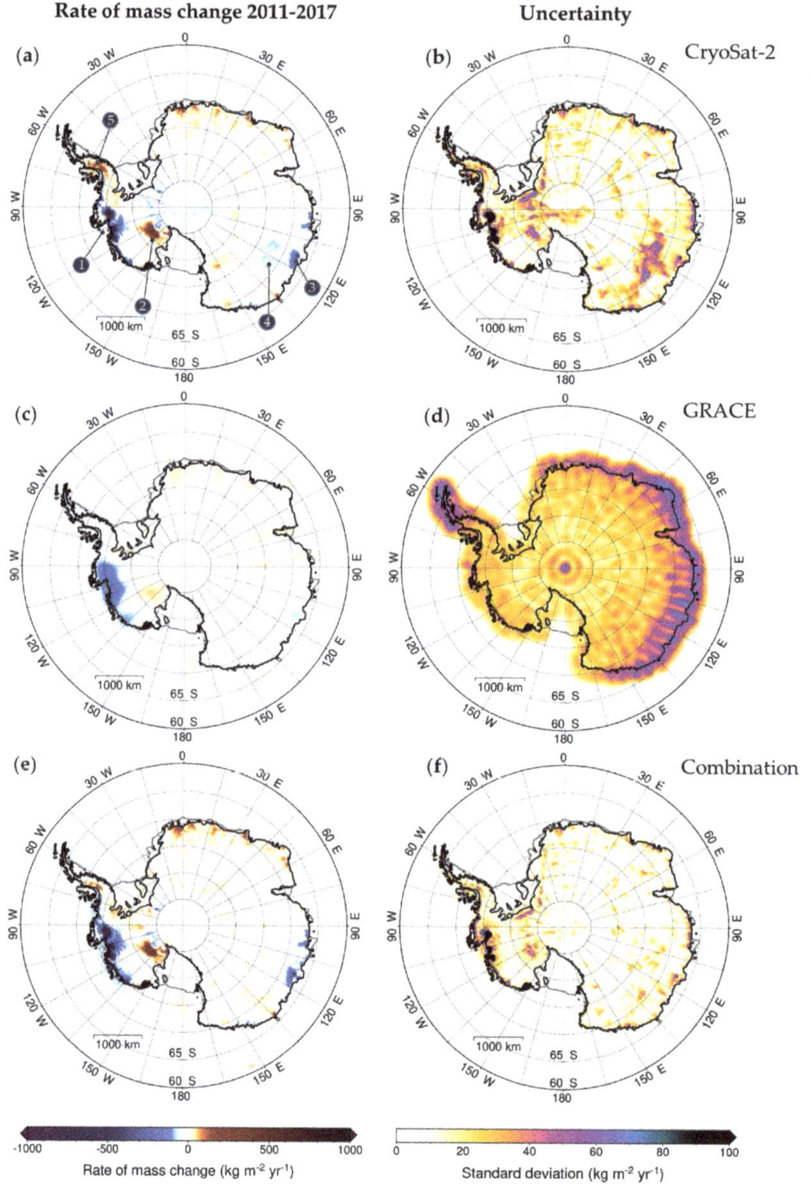

Figure 4. Spatial rate of mass change ($\mathrm{kg\ m^{-2}\ yr^{-1}}$) for the ensemble mean of (**a**) CryoSat-2, (**c**) GRACE, and the (**e**) combined solution and the respective ensemble standard deviation in (**b**), (**d**), and (**f**). Note that the saturation of the color bar in (**a**), (**c**), and (**f**) enhances signals of relatively low magnitude. Note that the GRACE trend in (**c**) is filtered with a Gaussian filter of half-width 1.3°, to reduce noise for visualization. The time period is February 2011 to June 2017. The projection is Polar Stereographic centered at 90°S and 0°E, with the true latitude of 71°S (applies to scale) and WGS84 (EPSG:3031).

The uncertainty of the CryoSat-2 ensemble mostly correlates with the signal structure, as the variability of the density models has the largest effects where the CryoSat-2 $\frac{\partial h}{\partial t}$ signal. However, in the interior of Wilkes Land (Label 4), uncertainties are induced by differences in the re-trackers due to varying backscattering properties of the snow and ice (for example [42]). Another remarkable feature in the CryoSat-2 uncertainty is CryoSat-2's mode mask boundary south of the Filchner Ice Shelf. Note that also some Gibb artefacts (for example [51]) are present, mostly beyond the ice sheet boundaries, for example in the Amundsen Sea, caused by the representation of the $\partial\sigma/\partial t$ field by a finite spherical harmonic expansion and synthesis.

The GRACE $\partial\sigma/\partial t$ field (Figure 4) shows the prominent mass loss in the Amundsen Sea Embayment. For visualization, we present the GRACE trends after smoothing with a Gaussian filter of $1.3°$ (for example [36]), reducing most of the noise and revealing the mass change anomalies. The GRACE noise field is not filtered for correlated north-south striping errors. The uncertainty represented by the ensemble standard deviation shows a striped pattern, caused by the correlation of uncertainties in the GRACE coefficients, the uncertainty due to the Polar gap of $\pm\,0.5°$, as well as an uncertainty increasing with latitude towards the equator, which is due to decreasing ground track density of the GRACE near-polar orbits (for example [52]). It is worth noting that CryoSat-2 and GRACE show very different, and, to some extent, complementary patterns of the uncertainty, for example in the Amundsen Sea Embayment.

In the combined field, some important differences are visible; first, the magnitude of mass losses in the Amundsen Sea Embayment is increased with respect to CryoSat-2 only (Figure 4, Label 1), as the signal magnitude is adjusted towards GRACE based on each sensor's uncertainty. In contrast, the magnitudes for Ice Stream C (Label 2) and Totten (Label 3) are unchanged, suggesting initial consistency between GRACE and CryoSat-2. However, the mass loss signal and its uncertainty in the interior of Wilkes Land (Label 4) is strongly reduced, mitigating the artefacts caused by the re-trackers, and possibly an overestimation of mass loss by CryoSat-2 caused by a depletion in snow. Similarly, the mass increase along the Antarctic Peninsula (Label 5) visible in CryoSat-2 is reduced by combining with GRACE. The height change in CryoSat-2 is likely caused by snow accumulation (more than is assumed in the density models), and thus detected only at lower magnitudes by GRACE. Compared to an individual sensor, the uncertainty of the combined solution (Figure 4) is considerably reduced, removing the sensor-specific patterns of regional or zonal uncertainty.

3.3. Basin Averages

Next, we evaluate the GRACE, CryoSat-2, and combined fields, as integrated over the 25 Antarctic drainage basins shown in Figure 3, which are considered independently resolvable by GRACE (for example [53,54]). Table 1 (color enhanced) lists the mass balances along with the respective uncertainties (see Section 2.6). Note that the total value for the Antarctic ice sheet of GRACE was estimated, including the buffer zone of the mask described above. The values for individual basins do not include any correction for leakage to the ocean.

Table 1. Mass balance of 25 Antarctic drainage basins shown in Figure 3. Listed are basin area, mass balance, uncertainty of mass balance as a result of this study, as well as the mass balance from GRACE Level 3 mascon product of Center for Space Research, University of Texas (CSR RL05 M) and the gridded product of the Climate Change Initiative (CCI) of the European Space Agency (ESA). Color coding is the same for the mass balances and the uncertainties, respectively. Red colors denote rates of mass gain, blue colors rates of mass loss. The color range is mapped to the value range from zero to the largest negative and positive values, respectively.

Basin No. N	Area (10^4km^2)	Mass Rates (Gt yr^{-1}), This Study			Uncertainty (Gt yr^{-1}), This Study			Other Products (Gt yr^{-1})	
		m_N^{Comb}	m_N^{CS2}	m_N^{GRACE}	Δm_N^{Comb}	Δm_N^{CS2}	Δm_N^{GRACE}	CSR RL05 M †	ESA CCI ‡
1	31.8	5.8	−0.9	−1.3	1.5	5.2	1.6	1.6	1.8
2	71.8	−9.8	−14.0	−4.1	2.0	7.5	2.4	−12.4	−0.4
3	154.7	−6.0	−5.5	0.8	2.5	7.6	3.2	−0.4	11.2
4	19.5	0.9	1.0	0.5	1.0	1.8	1.6	4.6	4.7
5	34.9	13.8	13.3	12.6	1.3	4.4	1.5	12.7	16.9
6	45.6	6.7	9.3	5.9	1.0	3.1	2.0	6.4	9.7
7	40.6	4.9	6.2	5.6	1.7	4.6	1.8	8	4.6
8	23.4	4.6	2.7	5.1	0.9	1.5	2.3	7.3	12
9	94.3	−10.4	−6.5	3.5	2.3	3.5	2.5	−2.9	2.2
10	31.6	−1.9	−0.3	−5.2	1.0	1.1	2.1	−1.3	−1.1
11	68.2	−12.9	−11.9	−14.0	1.9	6.3	2.0	−9.2	−11.2
12	115.7	−32.4	−44.6	−25.1	3.7	22.2	3.9	−23.8	−20.4
13	73.0	−0.5	−3.2	4.7	2.0	10.2	2.2	0.9	−1.3
14	14.6	−1.3	0.9	−8.3	0.7	1.5	1.4	−1.5	−0.8
15	26.1	−2.0	2.5	−0.2	0.8	0.8	1.7	−0.6	0.8
16	111.3	−3.7	7.7	0.1	2.4	2.7	3.2	−4.2	6.2
17	47.6	−3.5	−7.3	0.7	1.0	4.1	1.7	−4.5	1.9
18	36.4	15.9	14.4	14.6	2.4	4.5	3.2	9.4	16.5
19	35.5	−4.8	1.5	−3.8	1.6	0.7	2.0	−3.9	2
20	17.7	−21.0	−11.6	−18.3	1.0	2.2	1.5	−22.1	−33
21	22.1	−61.7	−52.1	−52.6	2.6	8.0	1.3	−52.5	−61.5
22	16.8	−45.8	−37.4	−44.1	2.4	6.3	1.7	−41	-47.4
23	7.5	−7.4	−2.6	−10.0	1.5	1.8	0.9	−7.6	−15.6
24	33.4	−1.3	10.5	−1.2	2.4	3.2	2.9	−4	−13.6
25	8.0	−3.8	−1.8	−1.7	0.9	1.1	1.3	−6.4	−12.7
Total	1182.1	−177.6	−129.9	−182.4	22.6	58.2	24.9	−147.4	−128.4

† January 2011 to June 2017 (same as this study); ‡ January 2011 to June 2016.

Overall, the characteristics of positive and negative mass balances are similar for GRACE and CryoSat-2, and this is preserved in the combination. However, it is apparent that the combination is not merely a weighted average of the individual inputs, as GRACE recovers signals at or below the spatial extent of the basins, while CryoSat-2 de-correlates signals for higher resolutions. This is visible in the relative cumulative contribution of GRACE and CryoSat-2 calculated according to Equation (6) and shown in Figure 5. The combined solution (Figure 4) features the high-resolution characteristics of the CryoSat-2 input field, because short wavelength features are dominated by CryoSat-2. Figure 5 shows that about 74% of the signal power in the combined field, integrated up until degree and order 512, is contributed by the CryoSat-2 data. However, GRACE remains the dominating source for our combined product (88%) if limited to the spatial scale of 500 km (degree and order 40).

For example, as a consequence, mass balance for basin 1 turns positive in the combination, even though GRACE and CryoSat-2 inputs are both negative in sign. Another effect is the localization of the signal (reduction of leakage), visible for basins 21. Here, the combined solution shows higher mass loss (−61.7 Gt yr^{-1}) compared to CryoSat-2 (−52.1 Gt yr^{-1}), which tends to be less negative in the entire Amundsen Sea Embayment, but also as GRACE (−52.6 Gt yr^{-1}), for which some signal is lost due to leakage. Another example is the southern Antarctic Peninsula (basin 24), where strong mass gains inferred from CryoSat-2 (10.5 Gt yr^{-1}) are entirely suppressed by the GRACE contribution in the combination (−1.3 Gt yr^{-1}), resulting in a combined estimate close to GRACE.

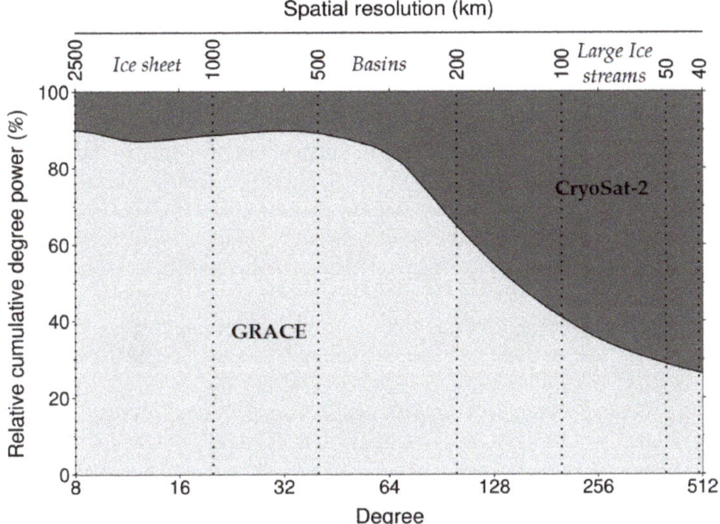

Figure 5. Relative cumulative contribution of the GRACE and CryoSat-2 data to the degree power of the combined mass balance field. GRACE poses the dominant contribution for scales up to 500 km (typical for the spatial extent of the 25 basins (Figure 3) or j $\leq$ 40. Equal contributions are obtained at j $\approx$ 150 or 133 km, while for features at the spatial scale of the nominal resolution corresponding to our maximum degree j = 512 (40 km), GRACE and CryoSat-2 contribute ca. 26% and 74% of the power, respectively. Time period is January 2011 to June 2017. Note that due to non-zero weights, CryoSat-2 contributes about 10% of the power also in the low degrees j < 32.

In terms of the uncertainty, the combined solution is similar or lower compared to GRACE only, on average about 27 % (basins 20, 21, and 22 excluded). Note that the GRACE uncertainty is underestimated, since signal (and noise) leakage between basins is not accounted for in our uncertainty estimate at basin scale. As a consequence, the combined estimates for basins 20, 21, and 22 show an increased uncertainty compared to the GRACE only estimates, but also significant changes in the mass loss values, particularly basin 23. Also, note that the combined results feature the full resolution of 40 km, meaning that the averages are calculated using the full spectrum up to degree and order 512. Compared to CryoSat-2, the uncertainties reduce for all basins (except basin 19), typically by more than half. The dominant uncertainties for basins 11, 12, and 13, caused by sensitivity of the re-tracker (see Figure 4: uncertainty in Wilkes Land, Label 4), are drastically reduced. The uncertainty of the ice mass change for the entire Antarctic ice sheet is reduced from $\pm$ 58.2 Gt yr^{-1} for CryoSat-2 only to $\pm$ 22.6 Gt yr^{-1} for the combined solution (GRACE only is $\pm$ 24.9 Gt yr^{-1}). This again exemplifies that up to the basin level, GRACE improves CryoSat-2 estimates, and for higher resolutions, it is the other way around.

3.4. Transects

The evaluation of the spatial fields along the transects shown in Figure 6 demonstrates that our combined field of mass changes provide a much finer resolution than GRACE on its own. For example, the combined field resolves highly localized hotspots of mass change at the Totten Glacier snout (around 1000 km of the tangential profile AA' and around 500 km along the perpendicular profile aa') or along the various Amundsen Sea glaciers (several very prominent spots in CC', and the strong signal along Pine Island Glacier in cc'). The magnitude of the associated peak signals is clearly underestimated by GRACE on its own. However, even smaller and less prominent spots, like the ~75 kg m^{-2} yr^{-1} mass gain around 500 km of the orthogonal transect at Shirase Glacier (bb'), far beyond the detection capabilities of GRACE, are clearly resolved by the combination, due to the high-resolution CryoSat-2 input. The transect along Pine Island Glacier (cc') highlights how the combination of GRACE with CryoSat-2 rectifies one of the major shortcomings of GRACE. Early truncation of the GRACE spherical harmonic series results in a smooth decline of the signal close to the grounding-line position (around 500 km in cc') into the ice shelf and open ocean (around 800–900 km in cc'). This leakage is not only problematic at the ice/ocean boundary (see aa', bb' and cc' in Figure 6), but also transverse to the flow (see AA', BB' and CC' in Figure 6), because the mass imbalance should be focused within the shear margins of the ice streams (for example [55]). This is for example visible in transect CC', where between 1000 km and 2000 km, the highly resolved CryoSat-2 only and combined mass losses only peak where the ice flows relatively fast, whereas the GRACE-only signal leaks over a larger part of the section.

The advantage of supporting GRACE with CryoSat-2 for enhancing the level of spatial detail unresolvable with GRACE on its own is complemented by the ability of GRACE to mitigate large-scale ambiguities in the CryoSat-2 data and a smaller uncertainty than a single CryoSat-2 or GRACE solution in most locations. The long-wavelength contribution that comes mostly from GRACE adjusts the regional mean of the CryoSat-2 only solution. For example, the addition of GRACE to the CryoSat-2 signal increases mass loss rates along transect CC' (Amundsen Sea Embayment) from less than $100 \text{ kg m}^{-2} \text{ yr}^{-1}$ around 0 km up to additional $-250 \text{ kg m}^{-2} \text{ yr}^{-1}$ at the Haynes, Pope, Smith, and Kohler Glaciers glacier systems (HSK) (around 1650 km). The long wavelength offset is also visible in the mass balance of the entire region (basins 20 through 23), increasing the mass balance from $101 \pm 10 \text{ Gt yr}^{-1}$ for CryoSat-2 to $129 \pm 4 \text{ Gt yr}^{-1}$ for the combined field (Table 1). In addition, systematic noise in the GRACE data (striping) carried by individual coefficients is efficiently suppressed by the combination, as seen in transect BB'.

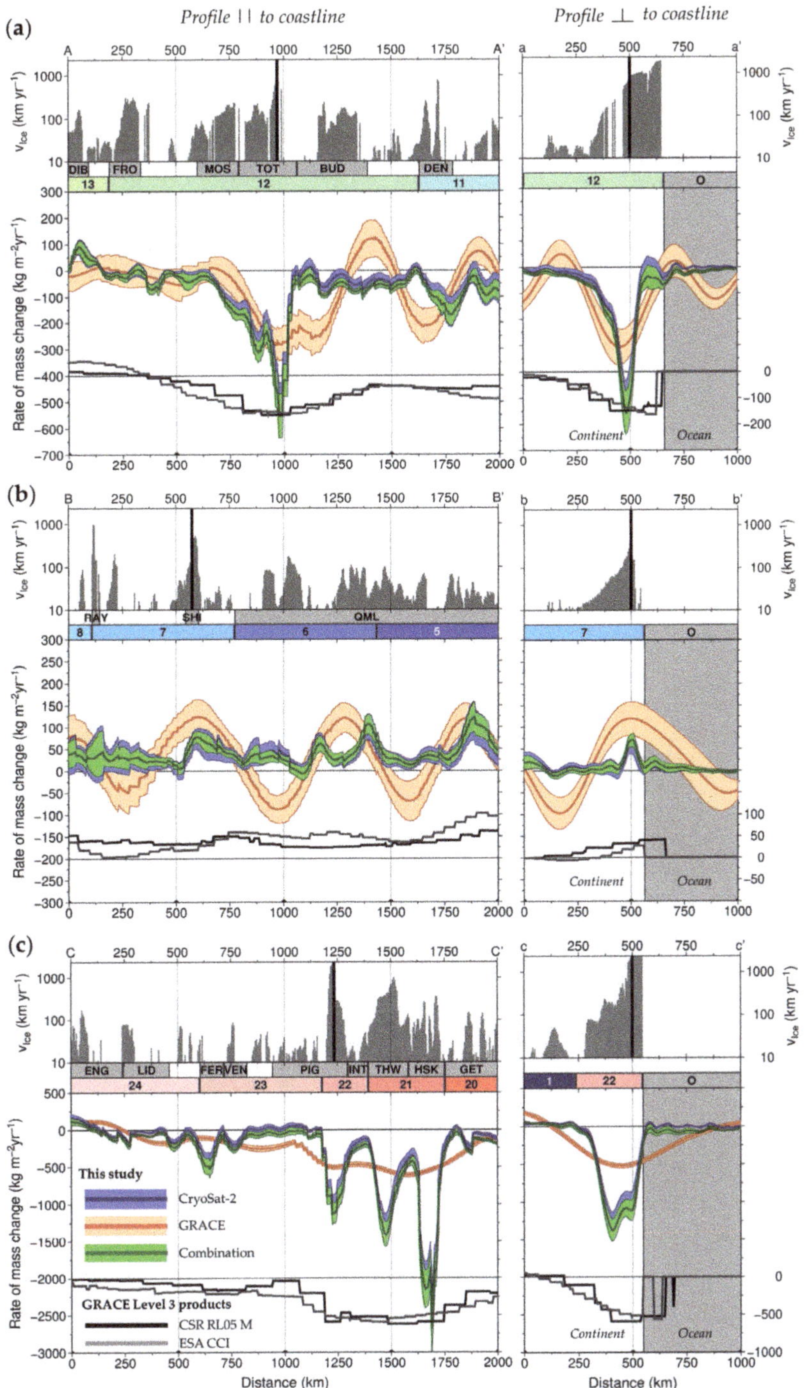

Figure 6. Profiles along (left) and across (right) the coastline in (**a**) Wilkes Land, (**b**) Dronning, Maud, and Enderby Land, and the (**c**) Bellingshausen Sea and Amundsen Sea Sector. Shown rates of mass

change (kg m^{-2} yr^{-1}) along transects indicated in the map shown in Figure 3, GRACE (unfiltered), CryoSat-2, and the combined field (middle panel). The basin attribution and the surface-ice velocity (m yr^{-1}) are shown in the top panel. The bottom inset of the lower panel shows the same transect evaluated for the GRACE Level 3 data products of the CSR and the CCI of ESA (see main text). Note that the Level 3 curves are offset (right scale applies), the scale however is unchanged, and thus directly comparable to our combined solution. The glacier and ice steams labels refer to: Budd Coast (BUD), Denman Glacier (DEN), Dibble Ice Stream (DIB), English Coast (ENG), Ferringo Ice Stream (FER), Frost Glacier (FRO), Glaciers flowing into Getz Ice Shelf (GET), Haynes Pope Smith and Kohler Glaciers (HSK), Inter-stream ridge (INT), Lidke and other glaciers (LID), Moscow University Ice Shelf (MOS), Pine Island Glacier (PIG), Queen Maud Land (QML), Raynor and Thyer Glacier (RAY), Shirase Glacier (SHI), Thwaites Glacier (THW), Totten Glacier (TOT) and Glaciers flowing into Venable Ice Shelf (VEN). The projection is Polar Stereographic centered at 90°S and 0°E, with the true latitude of 71°S and WGS84 (EPSG:3031).

4. Discussions

4.1. Comparison with GRACE Level 3 Data

We compare our combined estimate to the gridded mass rate fields from the Level 3 mascon product of Center for Space Research, University of Texas (CSR RL05 M; [56]) and the gridded product of the Climate Change Initiative (CCI) of the European Space Agency (ESA) [57]. Note that both gridded mass balance products rely on GRACE data only (Level 3 data), with some assumptions on geographic boundaries, as well as signal and noise characteristics. In the logic of the product hierarchy, our combined solution should be considered Level 4 data, as it involves ancillary data compared to GRACE-only mass balance grids. Note that the data sets differ in the underlying GRACE data-CSR solutions for CSR RL05 M and ITSG-Grace2016 [58] for ESA CCI Antarctica, and adopt different corrections of GIA, both of which are part of our ensemble (ICE6G [32] computed by A et al. [59] and IJ05r2 [30], respectively). These post-processing choices may cause considerable differences in the total mass change, but are less important for the decorrelation of basin-scale and local mass rates assessed here. Also, the ESA CCI data set is based on the time span February 2011 to June 2016, while CSR RL05 M is based on the same interval as our data (February 2011 to June 2017). Note that our combined estimate of -178 $\pm$ 23 Gt yr^{-1} lies well within the range of estimates obtained in the inter-comparison exercise presented in Shepherd et al. [3]; for comparison, we state that the mean and standard of multiple GRACE analysts are -179 ± 43 Gt yr^{-1}.

4.1.1. Basin Averages

Figure 7 and Table 1 present basin-average mass rates for the combined, the GRACE-only, and CryoSat-2 only estimates. The estimates of mass change at basin level of our combined estimate (also GRACE and CryoSat-2 only) and CSR RL05 M and ESA CCI data products are generally in agreement (Table 1). However, the variation of mass change from basin-to-basin is greater in our combined field compared to the GRACE-only estimates (i.e., our GRACE estimates, CSR RL05M, and ESA CCI), suggesting a higher level of decorrelation already at basin-scale level. Differences between GRACE, CryoSat-2, and the combined estimates arise from a stronger decorrelation (for example basin 9 for GRACE), a suppression of uncertainties (for example basin 12 for CryoSat-2), and presumably, a particularly high sensitivity to snow accumulation (for example CryoSat-2 in basin 24). The total mass balance of the Antarctic Ice Sheet of our combined estimate is with -178 ± 23 Gt yr^{-1}, about 31 Gt yr^{-1} more negative than that obtained from CSR RL05M for the same time span (February 2011 to June 2017). The lower mass loss estimates from ESA CCI (-129 Gt yr^{-1}) are most likely a consequence of the shorter time span covered by ESA CCI. Similarly, the distribution of mass change among the drainage basins of our combination is more similar to the CSR RL05 M than to ESA CCI product. However, combined mass loss estimates of the drainage basin 21 (Thwaites glacier; $-62 \pm$ 3 Gt yr^{-1}), which shows the strongest mass loss in Antarctica, is in better agreement with ESA CCI.

In contrast, our GRACE only estimate of -53 ± 1 Gt yr^{-1} for the same basin is in agreement with CSR RL05M. A similar pattern is observed for basin 18 (Ice Stream C), for which CSR RL05M matches our GRACE only estimate (no leakage correction applied), but is lower in magnitude than both the ESA CCI and our combined estimate. A likely cause for this difference could be signal loss due to leakage, but a more standardized inter-comparison at different processing levels is needed to provide a definitive answer. Note again that ESA CCI adopts the GIA correction IJ05r2 [30], while CSRL RL05 M adopts ICE6G [32] computed by A et al. [59], which produces a 7 to 17 Gt yr^{-1} greater apparent mass change [3]. Both GIA corrections are part of our GRACE ensemble. Nevertheless, the comparison shows that basin estimates of our combined field are consistent with other GRACE data sets, with some improvement in the decorrelation of basin-scale estimates and the reduction of signal leakage.

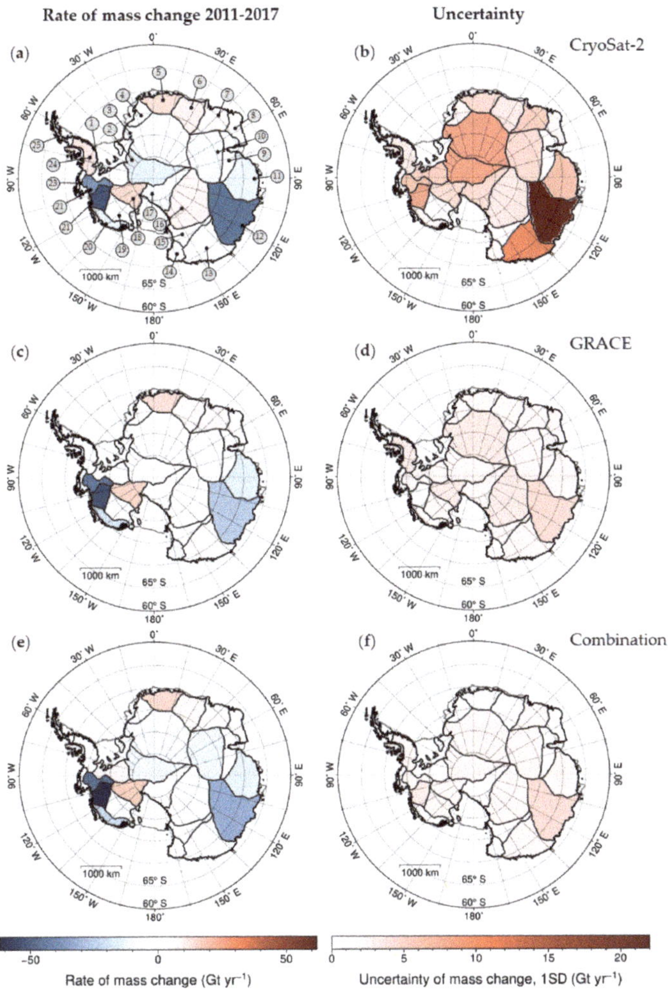

Figure 7. Basin integrated rate of mass change (Gt yr^{-1}) for the ensemble mean of (**a**) CryoSat-2, (**c**) GRACE, and the (**e**) combined solution, and the respective ensemble standard deviation in (**b**), (**d**), and (**f**). Numbered labels refer to the basins shown in Figure 3. Note that the saturation of the color bar in (**a**), (**c**), and (**f**) enhances signals of relatively low magnitude. The time period is February 2011 to June 2017. The projection is Polar Stereographic centered at 90°S and 0°E, with the true latitude of 71°S (applies to scale) and WGS84 (EPSG:3031).

4.1.2. Transects

The lower panels of Figure 6 show evaluation of the Level 3 data products along the transect in Wilkes Land, Dronning Maud, Enderby Land, and the Amundsen Sea Embayment. It is visible that the noise present in the GRACE data is successfully reduced and the geographic boundaries (continent/ocean) are implemented, even though the exact location of the coastline differs between the data products. The wavelength of the resolved patterns of the Level 3 products are similar and corresponds to 200 to 400 km, corresponding to our profile based on GRACE only. However, the magnitude of the mass changes is considerably lower than that of our combined solution, and the signals remain highly correlated between independent glacial entities. For example, while the combined solution is able to resolve individual signals for Thwaites (THW) and Haynes, Pope, Smith, and Kohler Glaciers (HSK) in Figure 6 (CC′), these signals are merged into one anomaly for the Level 3 data. This inter-basin leakage is an unresolved problem of the GRACE-only gridded data sets, limiting their use for basins integrals or for assimilation into glaciological models. Even though local mass rates may be well enough recovered with CryoSat-2 data alone, the combination with GRACE leads to reduced uncertainties across all spatial scales.

4.2. Remaining Inconsistencies

To achieve optimum results, we determine the difference between the combined mass balance, and the uncertainty-weighted mean field for GRACE (including a buffer zone) and CryoSat-2. We apply this value as an ad hoc correction term, distributing the mismatch evenly over the Antarctic ice sheet (-3.4 kg m^{-2} yr^{-1}), i.e., well below the mean uncertainty of the combined solution of 9.9 kg m^{-2} yr^{-1} (peak uncertainty is 318.5 kg m^{-2} yr^{-1}).

In additional analysis, we tried to make use of the spectral mismatch between GRACE and CryoSat-2 to identify the ensemble member (k, l) of the combination that minimizes the artefacts. Within the range of the GRACE and CryoSat-2 ensemble spread, the artefact could be reduced to some extent (by ca. 10 to 20 %), but could not be removed completely. We infer that the ensemble spread still underestimates the true uncertainty in one or both data sets. Possible candidates for underestimated uncertainties are the influence of far-field signals in the Northern Hemisphere on the GRACE signal over Antarctica, or the range of schemes for converting CryoSat-2 elevation rates to mass rates. For the time being, however, we accept the inconsistency between both data sets and resolve it as stated above. However, improvements to our approach could be made by introducing an a posteriori weighting of the ensemble members, for example according to the inverse of the signal artefacts that are created, instead of adopting an unweighted ensemble mean as we did in this study.

5. Conclusions

We have presented an approach for combining GRACE and CryoSat-2 data in the spectral domain, resulting in a spatially highly resolved mass balance of the Antarctica ice sheet. We treat the combination as a downward continuation of the GRACE coefficients with CryoSat-2 data, accounting for the respective wavelength-dependent noise characteristics. We obtain a total ice mass balance for Antarctica of -178 ± 23 Gt yr^{-1} for the time period of February 2011 to June 2017; basin-averaged mass rates are presented in Figure 7. Based on the analysis of statistical ensembles, we have shown that GRACE and CryoSat-2 have complementary characteristics regarding the noise power at different spatial wavelengths. Thus, up to degree and order j = 40 (500 km), GRACE contributes about 88 % (CryoSat-2 is 12 %) to the power of the mass rate field, while at the maximum cut-off degree of 512 (40 km), the cumulative GRACE contribution is reduced to about 26 % (CryoSat-2 is 74 %; see Figure 5). The combined mass rate field has the resolution of the CryoSat-2 field (here, 40 km, due to cut-off degree $j = 512$), and therefore provides independent mass rate estimates beyond the typical basin scale (25 Antarctic drainage basins). The combined field exhibits smaller uncertainties compared to estimates based on single sensors for all spatial scales and successfully reduces systematic noise patterns in GRACE and CryoSat-2. Compared to alternative gridded mass products from GRACE data alone (Level 3 data from CSR and ESA CCI), our combined GRACE and CryoSat-2 estimate is higher resolved, more accurate, and largely suppresses leakage to the ocean and between basins. Further developments will help identifying the optimal scheme for converting elevation rates to surface load rates based on the mismatch of the GRACE and CryoSat-2 spectra. Beyond improving ice sheet mass balances, the spectral combination method may offer the possibility to merge GRACE/GRACE-Follow On data and other water storage measurements into a combined Level 4 data product.

Author Contributions: I.S. conceived the study with contributions by H.K. and K.G. H.K. programmed most of the spherical harmonic combination algorithm and pre-processed the input satellite data with contributions and supervision by I.S. V.H. calculated the ensemble of CryoSat-2 for different re-trackers and fitting methods. I.S. provided the graphical displays. All authors discussed the results and contributed to the writing of the manuscript.

Funding: This research was funded through, and is a contribution to, the Regional Climate Change Initiative REKLIM of the Helmholtz Association.

Acknowledgments: We are grateful for F. Simons providing scripts for handling spherical harmonics at http://geoweb.princeton.edu/people/simons/software.html. We would like to thank the German Space Operations Center (GSOC) of the German Aerospace Center (DLR) for providing continuously and nearly 100% of the raw telemetry data of the twin GRACE satellites, and ESA for providing Level-1B/2I Baseline_C CryoSat-2 data. In addition, we wish to thank those who have made the following ancillary data available: GRACE Mascons CSR RL05M (http://www2.csr.utexas.edu/grace/RL05_mascons.html), ESA CCI Antarctica ice mass product (http://esa-icesheets-antarctica-cci.org/), and RACMO2/ANT (https://www.projects.science.uu.nl/iceclimate/publications/data/2018/index.php).

Conflicts of Interest: The authors declare no conflict of interest.

Appendix A

Processing of CryoSat-2 Level 1 and 2 data

We use the Level-1B/2I Baseline_C CryoSat-2 data provided by the ESA as initial data product, from which we eventually derive elevation rates. For each waveform (i.e., the radar echo detected by the satellite), the range is estimated using four re-tracker algorithms implemented in our processing scheme: TFMRA, AWI_OCOG, AWI_ICE2, and EC_TFMRA, respectively. The three remaining re-tracker solutions are extracted from the Level_2I (LRM) product ESA_OCOG, ESA_ICE, ESA_OCEAN. In the SARIn case, only one ESA Re-tracker solution is provided, which we used in combination with the three different ESA LRM products.

The AWI_ICE2 follows the Ice2 re-tracker designed by LEGOS to process ERS1 data over continental ice sheets [60,61], whereas AWI_OCOG uses a modified version of the algorithm developed by Wingham et al. [62]. The leading edge of the waveform is re-tracked at the first intersection of 30% of the OCOG amplitude. EC_TFMRA uses the TFMRA solution corrected by the leading-edge width. The leading-edge width of the waveform is estimated as linear regression between 15% and 80% of the leading edge of the first maximum. In general, EC_TFMRA and TFMRA are less sensitive to volume scattering, followed by AWI_OCOG and ESA_OCOG. The three-model based re-tracker solutions AWI_ICE2, ESA_OCEAN, and ESA_ICE are more sensitive to contributions of volume scatter, as they generally re-track at 50% or more of the maximum power. This has been demonstrated for Antarctica by Helm et al. [42] and for Greenland by Nilsson et al. [16].

All seven LRM solutions are further corrected for slope, estimating the point of closest approach (POCA) using the refined relocation method [42,63]. For SARIn, the POCA is determined using the interferometric phase at the re-tracked position (only for the four AWI solutions). For the SARIn ESA solution, we use the POCA given in the L2I product. Each of those seven independent elevation products are finally used to obtain Antarctic-wide $\partial h/\partial t$ estimates, using four different least square fitting methods (M1 to M4). In all cases, we apply the fit to all data points of the 2011 to 2017 time series falling within a pixel size of 2 km. This intermediate $\partial h/\partial t$ raster is interpolated using inverse distance weighting with a radius of 25 km to obtain the final $\partial h/\partial t$ grid with a 5 km pixel spacing. The differences between M1 to M4 are due to the unknown topography within a 2 km pixel, which needs to be considered, as the elevation trend is estimated at the center of each pixel. In case M1, we use an external DEM to estimate the subpixel topography. To estimate $\partial h/\partial t$, the topography is subtracted using bilinear interpolation of a DEM [42], before a linear regression on the elevation residuals is applied. For M2, M3, and M4 the topography is estimated in combination with the elevation trend as a polynomial, linear, and quadratic surface fit, respectively.

Conversion of CryoSat-2 elevation rates into rates of surface loading

A simple method of converting elevation rates, $\frac{\partial h}{\partial t}$, into rates of surface loading, $\frac{\partial \sigma}{\partial t}$, is to assume that one can identify the main process causing surface elevation change (ice dynamical imbalance vs. snow fall anomalies) and multiply surface height trends with the respective density (ice vs. snow, for example [15,43]). To account for some of the uncertainties introduced by the assumptions, we utilize this approach in three different realizations: (i) All surface elevation change is due to ice dynamical imbalance (i.e., $\frac{\partial \sigma}{\partial t} = 910 \frac{\text{mm we}}{\text{m}} \frac{\partial h}{\partial t}$); (ii) surface elevation change is homogenously 50% due to ice dynamical imbalance and 50% due to snow fall anomaly, and the two processes act in the same direction (in- or deflation of the surface; factor $650 \frac{\text{mm we}}{\text{m}}$ instead of $910 \frac{\text{mm we}}{\text{m}}$); (iii) areas of ice dynamical imbalance are associated with fast-flowing or fast lowering regions (factor $910 \frac{\text{mm we}}{\text{m}}$), while in remaining areas the density of snow is assumed, modulated by the changes in firn densification (factor $300 \frac{\text{mm we}}{\text{m}}$ to $500 \frac{\text{mm we}}{\text{m}}$; [23], Section 3.4.1 and Figure 7 therein). Each of these assumptions will be valid in some places and wrong in others, but the ensemble of all of them together probably contains the actual situation in most places. Note that opposing anomalies of ice-dynamical imbalance and accumulation may lead to apparent density changes beyond the physical range of the respective end members of $300 \frac{\text{mm we}}{\text{m}}$ and $910 \frac{\text{mm we}}{\text{m}}$.

Additionally, as a fourth method of converting from $\frac{\partial h}{\partial t}$ to $\frac{\partial \sigma}{\partial t}$, we remove modelled trends in surface height due to surface processes (snow fall variability, firn densification, $\frac{\partial h_S}{\partial t}$) over the same period, leaving supposedly ice-dynamical height changes as the residual. After the conversion from height changes to mass changes (factor $910 \frac{\text{mm we}}{\text{m}}$), the trends in the mass balance (after removal of the long-term average) over the same period, $\frac{\partial \sigma_{\text{SMB}}}{\partial t}$ is added back to restore mass variability due to snowfall. Both fields are available locally based on the Regional Climate Model RACMO2/ANT [44]. In this realization, the volume to mass conversion is

$$\frac{\partial \sigma}{\partial t} = 910 \frac{\text{mm we}}{\text{m}} \left(\frac{\partial h}{\partial t} - \frac{\partial h_S}{\partial t} \right) + \frac{\partial \sigma_{\text{SMB}}}{\partial t} \tag{A1}$$

References

1. DeConto, R.M.; Pollard, D. Contribution of Antarctica to past and future sea-level rise. *Nature* **2016**, *531*, 591. [CrossRef] [PubMed]
2. Shepherd, A.; Ivins, E.R.; Geruo, A.; Barletta, V.R.; Bentley, M.J.; Bettadpur, S.; Briggs, K.H.; Bromwich, D.H.; Forsberg, R.; Galin, N. A reconciled estimate of ice-sheet mass balance. *Science* **2012**, *338*, 1183–1189. [CrossRef] [PubMed]
3. Shepherd, A.; Ivins, E.; Rignot, E.; Smith, B.; van den Broeke, M.; Velicogna, I.; Whitehouse, P.; Briggs, K.; Joughin, I.; Krinner, G. Mass balance of the Antarctic Ice Sheet from 1992 to 2017. *Nature* **2018**, *556*, 219–222.
4. Gardner, A.S.; Moholdt, G.; Scambos, T.; Fahnstock, M.; Ligtenberg, S.; van den Broeke, M.; Nilsson, J. Increased West Antarctic and unchanged East Antarctic ice discharge over the last 7 years. *Cryosphere* **2018**, *12*, 521–547. [CrossRef]
5. Mouginot, J.; Rignot, E.; Scheuchl, B. Sustained increase in ice discharge from the Amundsen Sea Embayment, West Antarctica, from 1973 to 2013. *Geophys. Res. Lett.* **2014**, *41*, 1576–1584. [CrossRef]
6. Hogg, A.E.; Shepherd, A.; Cornford, S.L.; Briggs, K.H.; Gourmelen, N.; Graham, J.A.; Joughin, I.; Mouginot, J.; Nagler, T.; Payne, A.J.; et al. Increased ice flow in Western Palmer Land linked to ocean melting. *Geophys. Res. Lett.* **2017**, *44*, 4159–4167. [CrossRef]
7. Pritchard, H.D.; Ligtenberg, S.R.M.; Fricker, H.A.; Vaughan, D.G.; van den Broeke, M.R.; Padman, L. Antarctic ice-sheet loss driven by basal melting of ice shelves. *Nature* **2012**, *484*, 502–505. [CrossRef] [PubMed]
8. Scambos, T.A. Glacier acceleration and thinning after ice shelf collapse in the Larsen B embayment, Antarctica. *Geophys. Res. Lett.* **2004**, *31*. [CrossRef]

9.   Joughin, I.; Smith, B.E.; Medley, B. Marine ice sheet collapse potentially under way for the Thwaites Glacier Basin, West Antarctica. *Science* **2014**, *344*, 735–738. [CrossRef]

10.   Cornford, S.L.; Martin, D.F.; Payne, A.J.; Ng, E.G.; Le Brocq, A.M.; Gladstone, R.M.; Edwards, T.L.; Shannon, S.R.; Agosta, C.; Van Den Broeke, M.R. Century-scale simulations of the response of the West Antarctic Ice Sheet to a warming climate. *Cryosphere* **2015**. [CrossRef]

11.   Goldberg, D.N.; Heimbach, P.; Joughin, I.; Smith, B. Committed retreat of Smith, Pope, and Kohler Glaciers over the next 30 years inferred by transient model calibration. *Cryosphere* **2015**, *9*, 2429–2446. [CrossRef]

12.   Rignot, E.; Bamber, J.L.; Van Den Broeke, M.R.; Davis, C.; Li, Y.; Van De Berg, W.J.; Van Meijgaard, E. Recent Antarctic ice mass loss from radar interferometry and regional climate modelling. *Nat. Geosci.* **2008**, *1*, 106. [CrossRef]

13.   Velicogna, I.; Wahr, J. Measurements of time-variable gravity show mass loss in Antarctica. *Science* **2006**, *311*, 1754–1756. [CrossRef] [PubMed]

14.   Wingham, D.J.; Shepherd, A.; Muir, A.; Marshall, G.J. Mass balance of the Antarctic ice sheet. *Philos. Trans. R. Soc. Lond. A Math. Phys. Eng. Sci.* **2006**, *364*, 1627–1635. [CrossRef] [PubMed]

15.   McMillan, M.; Shepherd, A.; Sundal, A.; Briggs, K.; Muir, A.; Ridout, A.; Hogg, A.; Wingham, D. Increased ice losses from Antarctica detected by CryoSat-2. *Geophys. Res. Lett.* **2014**, *41*, 3899–3905. [CrossRef]

16.   Nilsson, J.; Gardner, A.; Sandberg Sørensen, L.; Forsberg, R. Improved retrieval of land ice topography from CryoSat-2 data and its impact for volume-change estimation of the Greenland Ice Sheet. *Cryosphere* **2016**, *10*, 2953–2969. [CrossRef]

17.   Simonsen, S.B.; Sørensen, L.S. Implications of changing scattering properties on Greenland ice sheet volume change from Cryosat-2 altimetry. *Remote Sens. Environ.* **2017**, *190*, 207–216. [CrossRef]

18.   Huss, M. Density assumptions for converting geodetic glacier volume change to mass change. *Cryosphere* **2013**, *7*, 877–887. [CrossRef]

19.   van der Wal, W.; Whitehouse, P.L.; Schrama, E.J. Effect of GIA models with 3D composite mantle viscosity on GRACE mass balance estimates for Antarctica. *Earth Planet. Sci. Lett.* **2015**, *414*, 134–143. [CrossRef]

20.   Riva, R.E.; Gunter, B.C.; Urban, T.J.; Vermeersen, B.L.; Lindenbergh, R.C.; Helsen, M.M.; Bamber, J.L.; van de Wal, R.S.; van den Broeke, M.R.; Schutz, B.E. Glacial isostatic adjustment over Antarctica from combined ICESat and GRACE satellite data. *Earth Planet. Sci. Lett.* **2009**, *288*, 516–523. [CrossRef]

21.   Gunter, B.C.; Didova, O.; Riva, R.E.M.; Ligtenberg, S.R.M.; Lenaerts, J.T.M.; King, M.A.; van den Broeke, M.R.; Urban, T. Empirical estimation of present-day Antarctic glacial isostatic adjustment and ice mass change. *Cryosphere* **2014**, *8*, 743–760. [CrossRef]

22.   Martín-Español, A.; King, M.A.; Zammit-Mangion, A.; Andrews, S.B.; Moore, P.; Bamber, J.L. An assessment of forward and inverse GIA solutions for Antarctica. *J. Geophys. Res. Solid Earth* **2016**, *121*, 6947–6965. [CrossRef] [PubMed]

23.   Sasgen, I.; Martín-Español, A.; Horvath, A.; Klemann, V.; Petrie, E.J.; Wouters, B.; Horwath, M.; Pail, R.; Bamber, J.L.; Clarke, P.J.; et al. Joint inversion estimate of regional glacial isostatic adjustment in Antarctica considering a lateral varying Earth structure (ESA STSE Project REGINA). *Geophys. J. Int.* **2017**, *211*, 1534–1553. [CrossRef]

24.   Colgan, W.; Abdalati, W.; Citterio, M.; Csatho, B.; Fettweis, X.; Luthcke, S.; Moholdt, G.; Simonsen, S.B.; Stober, M. Hybrid glacier Inventory, Gravimetry and Altimetry (HIGA) mass balance product for Greenland and the Canadian Arctic. *Remote Sens. Environ.* **2015**, *168*, 24–39. [CrossRef]

25.   Heiskanen, W.A.; Moritz, H. Physical geodesy. *Bull. Géod. (1946–1975)* **1967**, *86*, 491–492. [CrossRef]

26.   Wieczorek, M.A.; Meschede, M. SHTools: Tools for Working with Spherical Harmonics. *Geochem. Geophys. Geosyst.* **2018**, *19*, 2574–2592. [CrossRef]

27.   Bettadpur, S. *Level-2 Gravity Field Product User Handbook*; GRACE 327-734; Revision 4.0; Center for Space Research, The University of Texas at Austin: Austin, TX, USA, 25 April 2018.

28.   Cheng, M.; Tapley, B.D.; Ries, J.C. Deceleration in the Earth's oblateness: J2 VARIATIONS. *J. Geophys. Res. Solid Earth* **2013**, *118*, 740–747. [CrossRef]

29.   Swenson, S.; Chambers, D.; Wahr, J. Estimating geocenter variations from a combination of GRACE and ocean model output: Estimating geocenter variations. *J. Geophys. Res. Solid Earth* **2008**, *113*. [CrossRef]

30.   Ivins, E.R.; James, T.S.; Wahr, J.; Schrama, E.J.O.; Landerer, F.W.; Simon, K.M. Antarctic contribution to sea level rise observed by GRACE with improved GIA correction. *J. Geophys. Res. Solid Earth* **2013**, *118*, 3126–3141. [CrossRef]

31. Sasgen, I.; Konrad, H.; Ivins, E.R.; Van den Broeke, M.R.; Bamber, J.L.; Martinec, Z.; Klemann, V. Antarctic ice-mass balance 2003 to 2012: Regional reanalysis of GRACE satellite gravimetry measurements with improved estimate of glacial-isostatic adjustment based on GPS uplift rates. *Cryosphere* **2013**, *7*, 1499–1512. [CrossRef]

32. Peltier, W.R.; Argus, D.F.; Drummond, R. Space geodesy constrains ice age terminal deglaciation: The global ICE-6G_C (VM5a) model: Global Glacial Isostatic Adjustment. *J. Geophys. Res. Solid Earth* **2015**, *120*, 450–487. [CrossRef]

33. Barletta, V.R.; Bevis, M.; Smith, B.E.; Wilson, T.; Brown, A.; Bordoni, A.; Willis, M.; Khan, S.A.; Rovira-Navarro, M.; Dalziel, I.; et al. Observed rapid bedrock uplift in Amundsen Sea Embayment promotes ice-sheet stability. *Science* **2018**, *360*, 1335–1339. [CrossRef]

34. Nield, G.A.; Barletta, V.R.; Bordoni, A.; King, M.A.; Whitehouse, P.L.; Clarke, P.J.; Domack, E.; Scambos, T.A.; Berthier, E. Rapid bedrock uplift in the Antarctic Peninsula explained by viscoelastic response to recent ice unloading. *Earth Planet. Sci. Lett.* **2014**, *397*, 32–41. [CrossRef]

35. Nield, G.A.; Whitehouse, P.L.; King, M.A.; Clarke, P.J. Glacial isostatic adjustment in response to changing Late Holocene behaviour of ice streams on the Siple Coast, West Antarctica. *Geophys. J. Int.* **2016**, *205*, 1–21. [CrossRef]

36. Wahr, J.; Molenaar, M.; Bryan, F. Time variability of the Earth's gravity field: Hydrological and oceanic effects and their possible detection using GRACE. *J. Geophys. Res. Solid Earth* **1998**, *103*, 30205–30229. [CrossRef]

37. Dziewonski, A.M.; Anderson, D.L. Preliminary reference Earth model. *Phys. Earth Planet. Inter.* **1981**, *25*, 297–356. [CrossRef]

38. Martinec, Z. Program to calculate the spectral harmonic expansion coefficients of the two scalar fields product. *Comput. Phys. Commun.* **1989**, *54*, 177–182. [CrossRef]

39. Rignot, E.; Jacobs, S.; Mouginot, J.; Scheuchl, B. Ice-shelf melting around Antarctica. *Science* **2013**, *341*, 266–270. [CrossRef] [PubMed]

40. Swenson, S.; Wahr, J. Post-processing removal of correlated errors in GRACE data. *Geophys. Res. Lett.* **2006**, *33*. [CrossRef]

41. Gerlach, C.; Fecher, T. Approximations of the GOCE error variance-covariance matrix for least-squares estimation of height datum offsets. *J. Geod. Sci.* **2012**, *2*, 247–256. [CrossRef]

42. Helm, V.; Humbert, A.; Miller, H. Elevation and elevation change of Greenland and Antarctica derived from CryoSat-2. *Cryosphere* **2014**, *8*, 1539–1559. [CrossRef]

43. Zwally, H.J.; Li, J.; Robbins, J.W.; Saba, J.L.; Yi, D.; Brenner, A.C. Mass gains of the Antarctic ice sheet exceed losses. *J. Glaciol.* **2015**, *61*, 1019–1036. [CrossRef]

44. van Wessem, J.M.; van de Berg, W.J.; Noël, B.P.Y.; van Meijgaard, E.; Amory, C.; Birnbaum, G.; Jakobs, C.L.; Krüger, K.; Lenaerts, J.T.M.; Lhermitte, S.; et al. Modelling the climate and surface mass balance of polar ice sheets using RACMO2—Part 2: Antarctica (1979–2016). *Cryosphere* **2018**, *12*, 1479–1498. [CrossRef]

45. Nowicki, S.M.J.; Payne, A.; Larour, E.; Seroussi, H.; Goelzer, H.; Lipscomb, W.; Gregory, J.; Abe-Ouchi, A.; Shepherd, A. Ice Sheet Model Intercomparison Project (ISMIP6) contribution to CMIP6. *Geosci. Model Dev.* **2016**, *9*, 4521–4545. [CrossRef] [PubMed]

46. Sjöberg, L. Solutions to Linear Inverse Problems on the Sphere by Tikhonov Regularization, Wiener filtering and Spectral Smoothing and Combination—A Comparison. *J. Geod. Sci.* **2012**, *2*, 31–37. [CrossRef]

47. Li, X.; Rignot, E.; Morlighem, M.; Mouginot, J.; Scheuchl, B. Grounding line retreat of Totten Glacier, East Antarctica, 1996 to 2013. *Geophys. Res. Lett.* **2015**, *42*, 8049–8056. [CrossRef]

48. Boening, C.; Lebsock, M.; Landerer, F.; Stephens, G. Snowfall-driven mass change on the East Antarctic ice sheet. *Geophys. Res. Lett.* **2012**, *39*. [CrossRef]

49. Sutterley, T.C.; Velicogna, I.; Rignot, E.; Mouginot, J.; Flament, T.; van den Broeke, M.R.; van Wessem, J.M.; Reijmer, C.H. Mass loss of the Amundsen Sea Embayment of West Antarctica from four independent techniques. *Geophys. Res. Lett.* **2014**, *41*, 8421–8428. [CrossRef]

50. Joughin, I.; Tulaczyk, S. Positive mass balance of the Ross ice streams, West Antarctica. *Science* **2002**, *295*, 476–480. [CrossRef]

51. Gottlieb, D.; Shu, C.-W. On the Gibbs phenomenon and its resolution. *SIAM Rev.* **1997**, *39*, 644–668. [CrossRef]

52. Wiese, D.N.; Landerer, F.W.; Watkins, M.M. Quantifying and reducing leakage errors in the JPL RL05M GRACE mascon solution. *Water Resour. Res.* **2016**, *52*, 7490–7502. [CrossRef]

53. Horwath, M.; Dietrich, R. Signal and error in mass change inferences from GRACE: The case of Antarctica. *Geophys. J. Int.* **2009**, *177*, 849–864. [CrossRef]

54. Sasgen, I.; Dobslaw, H.; Martinec, Z.; Thomas, M. Satellite gravimetry observation of Antarctic snow accumulation related to ENSO. *Earth Planet. Sci. Lett.* **2010**, *299*, 352–358. [CrossRef]

55. Bindschadler, R.A.; Scambos, T.A. Satellite-image-derived velocity field of an Antarctic ice stream. *Science* **1991**, *252*, 242–246. [CrossRef]

56. Save, H.; Bettadpur, S.; Tapley, B.D. High-resolution CSR GRACE RL05 mascons. *J. Geophys. Res. Solid Earth* **2016**, *121*, 7547–7569. [CrossRef]

57. Groh, A.; Horwath, M. The method of tailored sensitivity kernels for GRACE mass change estimates. In Proceedings of the EGU General Assembly Conference Abstracts, Vienna Austria, 17–22 April 2016; Volume 18, p. 12065.

58. Mayer-Gürr, T.; Behzadpour, S.; Ellmer, M.; Kvas, A.; Klinger, B.; Zehentner, N. *ITSG-Grace2016—Monthly and Daily Gravity Field Solutions from GRACE*; Graz University of Technology: Graz, Austria, 2016.

59. Wahr, J.; Zhong, S. Computations of the viscoelastic response of a 3-D compressible Earth to surface loading: An application to Glacial Isostatic Adjustment in Antarctica and Canada. *Geophys. J. Int.* **2013**, *192*, 557–572.

60. Legrésy, B. *Etude du retracking des formes d'onde altimétriques au-dessus des calottes polaires*; Report for European Space Agency, CT/ED/TU/UD/96.188, Contract 856/2/95/CNES/006; Centre National D'Études Spatiales: Toulouse, France, 1995.

61. Legresy, B.; Papa, F.; Remy, F.; Vinay, G.; Van den Bosch, M.; Zanife, O.-Z. ENVISAT radar altimeter measurements over continental surfaces and ice caps using the ICE-2 retracking algorithm. *Remote Sens. Environ.* **2005**, *95*, 150–163. [CrossRef]

62. Wingham, D.J.; Rapley, C.G.; Griffiths, H. New techniques in satellite altimeter tracking systems. In Proceedings of the 1986 International Geoscience and Remote Sensing Symposium (IGARSS '86) on Remote Sensing, Zurich, Switzerland, 8–11 September 1986; European Space Agency: Zurich, Switzerland, 1986; Volume 86, pp. 1339–1344.

63. Roemer, S.; Legrésy, B.; Horwath, M.; Dietrich, R. Refined analysis of radar altimetry data applied to the region of the subglacial Lake Vostok/Antarctica. *Remote Sens. Environ.* **2007**, *106*, 269–284. [CrossRef]

Article

The Rapid and Steady Mass Loss of the Patagonian Icefields throughout the GRACE Era: 2002–2017

Andreas Richter [1,2,3,*], **Andreas Groh** [1], **Martin Horwath** [1], **Erik Ivins** [4], **Eric Marderwald** [2,3], **José Luis Hormaechea** [5], **Raúl Perdomo** [2] and **Reinhard Dietrich** [1]

[1]   Technische Universität Dresden, Institut für Planetare Geodäsie, 01062 Dresden, Germany;
andreas.groh@tu-dresden.de (A.G.); Martin.Horwath@tu-dresden.de (M.H.);
reinhard.dietrich@tu-dresden.de (R.D.)

[2]   Laboratorio MAGGIA, Facultad de Ciencias Astronómicas y Geofísicas, Universidad Nacional de La Plata,
La Plata B1900FWA, Argentina; emarderwald@fcaglp.unlp.edu.ar (E.M.); perdomo@presi.unlp.edu.ar (R.P.)

[3]   Consejo Nacional de Investigaciones Científicas y Técnicas CONICET, La Plata B1904CMC, Argentina

[4]   Jet Propulsion Laboratory, California Institute of Technology, Pasadena, CA 91109, USA;
Erik.R.Ivins@jpl.nasa.gov

[5]   Estación Astronómica Río Grande, Río Grande V9420EAR, Argentina; jlhor@fcaglp.unlp.edu.ar

*   Correspondence: andreas.richter@tu-dresden.de

Received: 18 March 2019; Accepted: 10 April 2019; Published: 14 April 2019

Abstract: We use the complete gravity recovery and climate experiment (GRACE) Level-2 monthly time series to derive the ice mass changes of the Patagonian Icefields (Southern Andes). The glacial isostatic adjustment is accounted for by a regional model that is constrained by global navigation satellite systems (GNSS) uplift observations. Further corrections are applied concerning the effect of mass variations in the ocean, in the continental water storage, and of the Antarctic ice sheet. The 161 monthly GRACE gravity field solutions are inverted in the spatial domain through the adjustment of scaling factors applied to a-priori ice mass change patterns based on published remote sensing results for the Southern and Northern Patagonian Icefields, respectively. We infer an ice mass change rate of -24.4 ± 4.7 Gt/a for the Patagonian Icefields between April 2002 and June 2017, which corresponds to a contribution to the eustatic sea level rise of 0.067 ± 0.013 mm/a. Our time series of monthly ice mass changes reveals no indication for an acceleration in ice mass loss. We find indications that the Northern Patagonian Icefield contributes more to the integral ice loss than previously assumed.

Keywords: ice mass; satellite gravimetry; Patagonia; GRACE

1. Introduction

The Southern Patagonian Icefield (SPI; 12,500 km^2) and Northern Patagonian Icefield (NPI; 4000 km^2) constitute, together, the largest temperate ice mass in the Southern Hemisphere. They are aligned in a north–south (N–S) direction along the crest of the Southern Andes (Figure 1). The icefields are strongly influenced by the Westerlies and the transport of wet air mass from the Pacific Ocean, receiving precipitation of up to 10 m/a water equivalent [1]. The quantification of present-day ice mass changes contributes to our understanding of ongoing climate change in the southern mid-latitudes and the southeastern Pacific. An observational assessment of the demise of the two Patagonian Icefields provides robust information for predicting the future of the regional cryosphere and the water cycle, as well as its contribution to the global sea level. Previous works suggest a disproportionally large contribution of the Patagonian Icefields to eustatic sea level rise when compared to their area, for example, 0.067 ± 0.004 mm/a [2], 0.105 ± 0.011 mm/a [3], and 0.059 ± 0.005 mm/a [4]. Bedrock global navigation satellite systems (GNSS) observations adjacent to and within the SPI reveal an uplift as high as 4 cm/a [5–7] related to glacial isostatic adjustment (GIA), which combines the elastic response to present-day ice load change, and the viscoelastic adjustment to the ice mass gain and loss during and following the Little Ice Age (LIA) [8,9]. We make use of recently determined GIA models, which are fitted to the uplift observations [6] in order to compute the solid Earth contribution to the observed mass changes. The contrast between two recent regional GIA models highlights the degree of uncertainty in the regional ice load history, especially for the most recent time slice 1995–2013 [6] (Table 2).

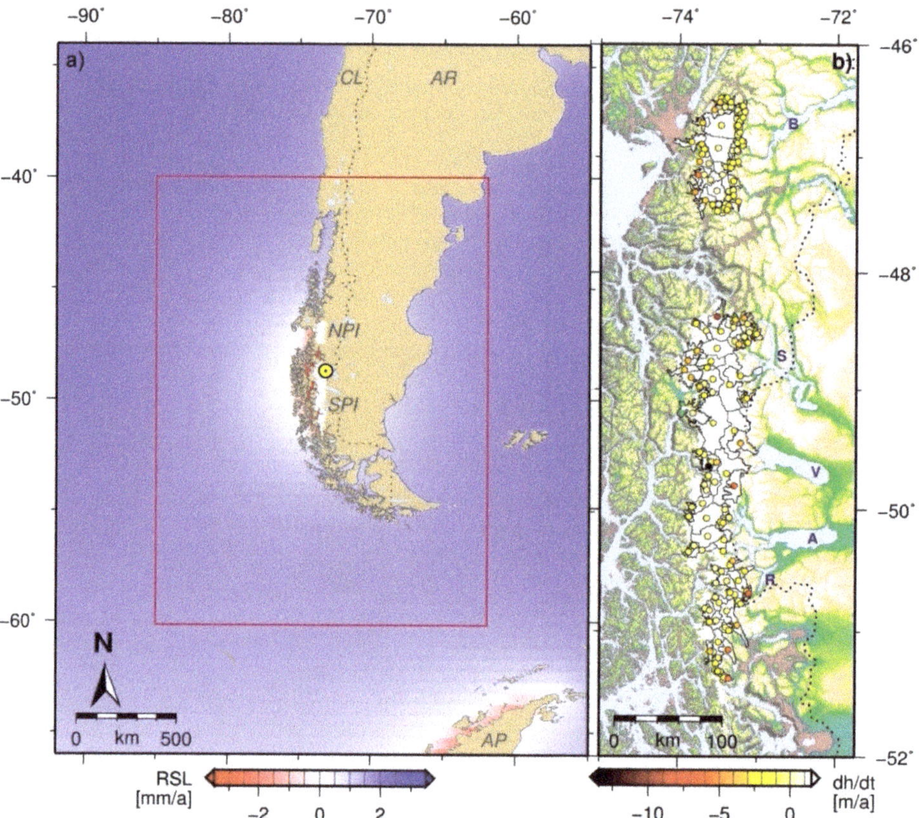

Figure 1. Maps of the region under investigation. (**a**) Location of the Northern (NPI) and Southern Patagonian Icefields (SPI) in southernmost South America. Color scale shows the ocean mass change rates contributing to the relative sea level according to our correction model (see text); yellow circle: data grid node for which time series are shown in Figure 2; red box indicates the extension of the grid used in the computations and corresponds to the map outline of Figure 3. AR—Argentina; CL—Chile; AP—Antarctic Peninsula. (**b**) Detailed map of the NPI and SPI area. Thin polygons on the icefields outline the accumulation and ablation zones of 87 glacier basins. Colored dots show the a-priori mass-rate pattern derived from ice elevation rates [2] at the barycenter coordinates for each zone. Lakes mentioned in the text: B—Lago Buenos Aires/General Carrera; S—Lago San Martín/O'Higgins; V—Lago Viedma; A—Lago Argentino; R—Brazo Rico and Brazo Sur.

Southern Patagonia is characterized by a series of distinct Neoglacial advances and retreats during mid-to-late-Holocene times (0–5 ka before present) [10,11]. The most recent advance corresponds to the regional LIA. This history is inferred from datable moraine materials [11] and is consistent with the regional paleo-climatological record [12,13]. While the climatic origins of these fluctuations are unknown, they are thought to be linked by global teleconnections [14,15]. More importantly, they establish a pattern of natural variability that is now perhaps in sync with anthropogenically driven warm late-summer conditions [16,17]. The latter exacerbates the negative net mass balance of the mountain glacier systems over the past 30–40 years, and that we know to be global in nature [18]. The future of the Patagonian Icefields may well be determined by the climatic conditions generated by the evolving state of the atmosphere and ocean on their Western Flank, as these are, in turn, linked to changes in both the El Niño and Southern Oscillation (ENSO) and the Pacific Decadal Oscillation (PDO), phenomena that we now suspect have been modulated by anthropogenic forcing since the 1990's [19].

Our study of the 15-year mass evolution of the Patagonian Icefields provides new clues about the present-day land-ice trajectory of the higher latitude Southern Hemisphere. We discover that glacier mass loss is dominated by a strong, and relatively uninterrupted, linear trend, possibly lending new insights to some of the standing questions regarding the future of the Southern Ocean and its environs [20].

2. Data and Methods

Natural conditions at the Patagonian Icefields limit the applicability of observational techniques that can be used to determine ice mass changes. Table 1 summarizes the previously published present-day ice mass change rates for the Patagonian Icefields. Conventional satellite radar altimetry is not appropriate because of the large footprint size compared to the narrow, fragmented, and often steep ice surface. Satellite laser altimetry provides insufficient data because of the frequent cloud cover and the unfavorable N–S orientation of the icefields. Recent advances using Cryosat-2 swath radar altimetry [4] are promising, but are limited in time to 2010 onward. The surface mass balance estimates derived from atmospheric modelling [17,21,22], or the changes derived from field glaciological measurements, suffer from a sparse distribution of meteorological stations and sampling sites that do not adequately capture the large variability over very short distances and rugged topography. Remote sensing techniques based on both optical [2,3,23–25] and radar imagery [26–28] provide valuable data for quantifying the ice volume and ice mass changes of the SPI and NPI, in spite of having to deal with cloud cover or uncertain radar signal penetration. All of the remote sensing results (Table 1) have a partially reconcilable agreement on the net ice mass loss to both icefields. However, the published rates can differ by as much as a factor of 2, a discrepancy that cannot be explained by the differing imagery acquisition time spans. Mouginot and Rignot [29] present a comprehensive determination of the ice flow velocities over SPI and NPI based on satellite data covering 30 years (1984–2014). They reveal a heterogeneous behavior over time among the individual glaciers—some of the major glaciers accelerate significantly, others decelerate or show a reversal in acceleration/deceleration over a decade or so.

Time-variable global gravity field solutions provided by the gravity recovery and climate experiment (GRACE; [30]) satellite mission have proven an efficient basis for the quantification of ice mass changes (e.g., [31–35], as well as the references therein). Previous inversions of GRACE data targeted explicitly on SPI and NPI [36,37] had to rely on relatively short time spans compared to the complete mission record through 2017, along with larger uncertainties in the earlier GRACE Level-2 data releases and less refined correction models available at that time (e.g., for correcting the effects of GIA and the surface mass changes, and for improving the low-degree Stokes coefficients, i.e.: C_{10}, C_{11}, S_{11}, C_{20}, C_{20}, and S_{21}). More recent ice mass change rates for southern Patagonia are presented as part of the global inversions that account for all mountain glacier and ice cap contributions to sea level rise [32,38–40].

Here, we derive the ice mass changes of the Patagonian Icefields between April 2002 and June 2017 from the monthly ITSG-Grace2016 [41] gravity field solutions. The monthly SPI and NPI ice mass changes are derived from a GRACE Level-2 data analysis approach similar to the regional point mass inversion presented in Forsberg and Reeh [42] for Greenland. This approach has been refined and modified (e.g., [43–45]), and it has also been used to derive ice-mass change time series for Antarctica and Greenland for a reconciled estimate of ice sheet mass balance [34,35]. In our case, we used a-priori information on the location of the point masses and on their changes with time. This allows for the set-up of a very dense pattern of masses (SPI: 98, NPI: 76, in total 174), and we determined a common scaling parameter for the a-priori mass change rates.

There is no sharp spatial resolution limit of GRACE. The ability to separate the ice mass changes of the SPI and the NPI (with a 320 km distance between their aerial barycenters and a 80 km minimum distance between their areas) depends on a number of aspects, namely: (i) the noise level of the used GRACE products with respect to the specific geographic situation, (ii) the strength of the signal, and (iii) the exploitation of the a-priori information. The a-priori information on the geographic patterns of the mass change have been widely used to separate nearby signals. This has been shown to have great advantages for dealing with coastlines [46]. We explore the separability of the NPI and SPI signals, because each of these three factors may be readily exploited. (i) ITSG-Grace2016 has a reduced noise level w.r.t. to the previous GRACE Level-2 releases [47]. The higher frequency and shorter wavelength noise of the GRACE gravity field solutions is inherently lower for the along-track (N–S) accelerations detected by the K-Band Ranging (KBR) system, relative to the noise in an east–west (E–W) (track-normal) direction, wherein the KBR system does not directly provide gravity information [48]. (ii) The sustained mass change of the Patagonian Icefields is among the largest spatially concentrated long-term signals worldwide. (iii) We can exploit the a-priori information on the distribution of the glacier mass loss of the two icefields along with the extensive a-priori information on the disturbing signals outside the icefields, as explained below and in Appendix A.

Table 1. Summary of ice mass change rates. The "comment" indicates the spatial extent for analyses using gravity recovery and climate experiment (GRACE) and Cryosat-2, and the data sources in the case of the remote sensing studies. SPI—Southern Patagonian Icefield; NPI—Northern Patagonian Icefield; SRTM—Shuttle Radar Topography Mission.

Reference	Comment	Time Span	Ice Mass Rate (Gt/a)	
GRACE satellite gravimetry				
Chen et al. [36]	SPI and NPI	04/2002–12/2006	-25.1 ± 9.9	*
Ivins et al. [37]	SPI and NPI and Ant. Penins.	01/2003–03/2009	-26 ± 6	
Jacob et al. [38]	Global (Patagonia)	01/2003–12/2010	-23 ± 9	
Gardner et al. [32]	Global (Southern Andes)	08/2003–08/2009	-29 ± 10	
Schrama et al. [39]	Global (Patagonia)	02/2003–06/2013	-22.1 ± 7	
Reager et al. [40]	Global (Southern Andes)	04/2002–12/2014	-33.1 ± 12.1	
this study	*SPI and NPI*	*04/2002–06/2017*	-24.4 ± 4.7	
Cryosat-2 swath radar altimetry				
Foresta et al. [4]	SPI andNPI	04/2011–03/2017	-21.3 ± 2.0	
Remote sensing-SPI				
Rignot et al. [3]	SRTM, Photogrammetry	1968/75–2000	-12.2 ± 0.7	*
		1995–2000	-34.8 ± 4.0	*
Willis et al. [2]	ASTER, SRTM	2000–2012	-20.0 ± 1.2	
Jaber [28]	TanDEM-X, SRTM	2000–2011/12	-13.14 ± 0.4	
Malz et al. [27]	TanDEM-X, SRTM	2000–2015/16	-11.84 ± 3.3	
Braun et al. [26]	TanDEM-X, SRTM	2000–2015	-12.64 ± 0.98	
Remote sensing-NPI				
Rignot et al. [3]	SRTM, Photogrammetry	1968/1975–2000	-2.9 ± 0.4	*
Rivera et al. [25]	ASTER, Photogr.	1979–2001	-5.1	*
Willis et al. [23]	ASTER, SRTM	2000–2011	-3.40 ± 0.07	
Willis et al. [2]	ASTER, SRTM	2000–2011	-4.4 ± 0.3	*
Jaber [28]	TanDEM-X, SRTM	2000–2014	-3.96 ± 0.1	
Dussaillant et al. [24]	SPOT5, SRTM	2000–2012	-4.1 ± 0.4	*
	ASTER	2000–2012	-4.2 ± 0.3	*
Braun et al. [26]	TanDEM-X, SRTM	2000–2015	-4.77 ± 0.48	

* Converted to an ice mass change rate from the originally published ice volume change rate, applying an ice density of 0.9 g/cm^3.

In the following section, our inversion approach is described, and further details of the analysis procedure are given in the Appendix A. We used unconstrained monthly gravity field solutions of release ITSG-Grace2016 provided by TU Graz [41]. These are provided as Stokes coefficients to a spherical harmonic degree and order of 120. This time series comprises 161 monthly solutions that cover the period April 2002–June 2017. We added degree one coefficients (C_{10}, C_{11}, and S_{11}) derived from the ITSG-Grace2016 solutions [49]. The C_{20}-term was replaced by the values obtained by satellite laser ranging [50]. The C_{21} and S_{21} coefficients were corrected for the pole tide, as recommended by Wahr et al. [51]. The monthly Stokes coefficients were corrected for GIA by applying the regional model A according to Lange et al. [6] (henceforth referred to as La-A), after removing the elastic contribution to the uplift. The Stokes coefficients were then further corrected for the effect of elastic loading deformation [44] (Equation (2)).

We retrieved GRACE pseudo-observations on a regional grid at the approximate GRACE satellite orbit altitude (500 km) from each monthly set of corrected Stokes coefficients [42–45]. Figure 1a shows the extension of this grid. The spacing is approximately 20 km, with a constant longitude increment and latitude increments that provide for an equal area of all of the grid cells. We chose to generate our pseudo-observations at the orbit altitude in order to avoid the amplification of noise in the downward continuation. No explicit filtering was applied to the GRACE data. For each grid node and month, 3D Cartesian gravity vector components were computed (instead of the scalar gravity disturbances used by Forsberg et al. [44] as pseudo-observations). These monthly sets of 3D gravity vectors were interpreted with respect to the mass redistributions at the Earth's surface. For comparison, an additional mass corresponding to a uniform layer of 1cm water equivalent of global extent at the surface (sea level) implies a change in the length of our gravity vectors at the orbit altitude of +7.22 nm s^{-2}.

The monthly gravity vector grids were corrected explicitly for five surface mass variation processes—the Antarctic ice sheet (AIS) [52], ocean mass change, changes in continental water storage [53,54], ice mass change rates in South America outside Patagonia [26], and water mass changes in the great Patagonian lakes [55]. Our ocean mass correction consisted of a gravitationally self-consistent distribution of the mass discharge from five major ice sheets/ice caps (AIS, Greenland ice sheet, Patagonian Icefields, Alaska, and the Canadian Arctic) and continental water storage over the ocean [56] (using ice mass change estimates [52,57] and continental water storage [53] as input), complemented by an empirical regional ocean mass change rate increment. The mass signals from outside our data grid domain may affect our pseudo-observations through leakage [44]. Our approach, with the global extent of the correction models (ocean mass and continental water storage) plus the additional solved-for ocean mass change rate increment, ensures that possible leakage effects are minimized. The time series of the pseudo-observations prior and after the correction, as well as of the individual mass-variation corrections, are shown for one data grid node in Figure 2. The individual impact of the applied corrections on the SPI and NPI ice mass change rate is given in Table 2. Figure 3 shows maps of the gravity rate components prior (top) and after (center) the application of all corrections.

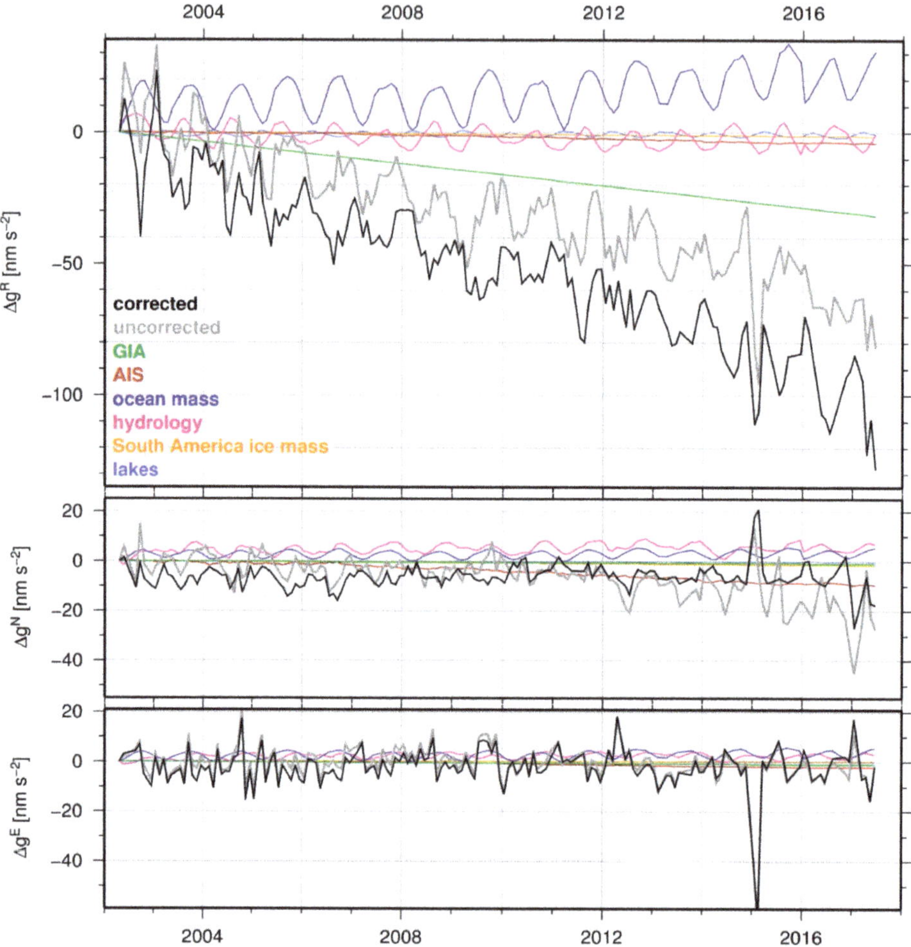

Figure 2. Time series of the gravity signal and the applied corrections for a selected data grid node. The relative change in gravity at the approximate GRACE orbit altitude prior (uncorrected, grey) and after (corrected, black) the application of the corrections described in the text is shown in the radial (top), north (center), and east (bottom) vector components. The applied corrections for the five mass-variation processes are shown in different colors. The location of this grid node is indicated in Figure 1 (yellow circle).

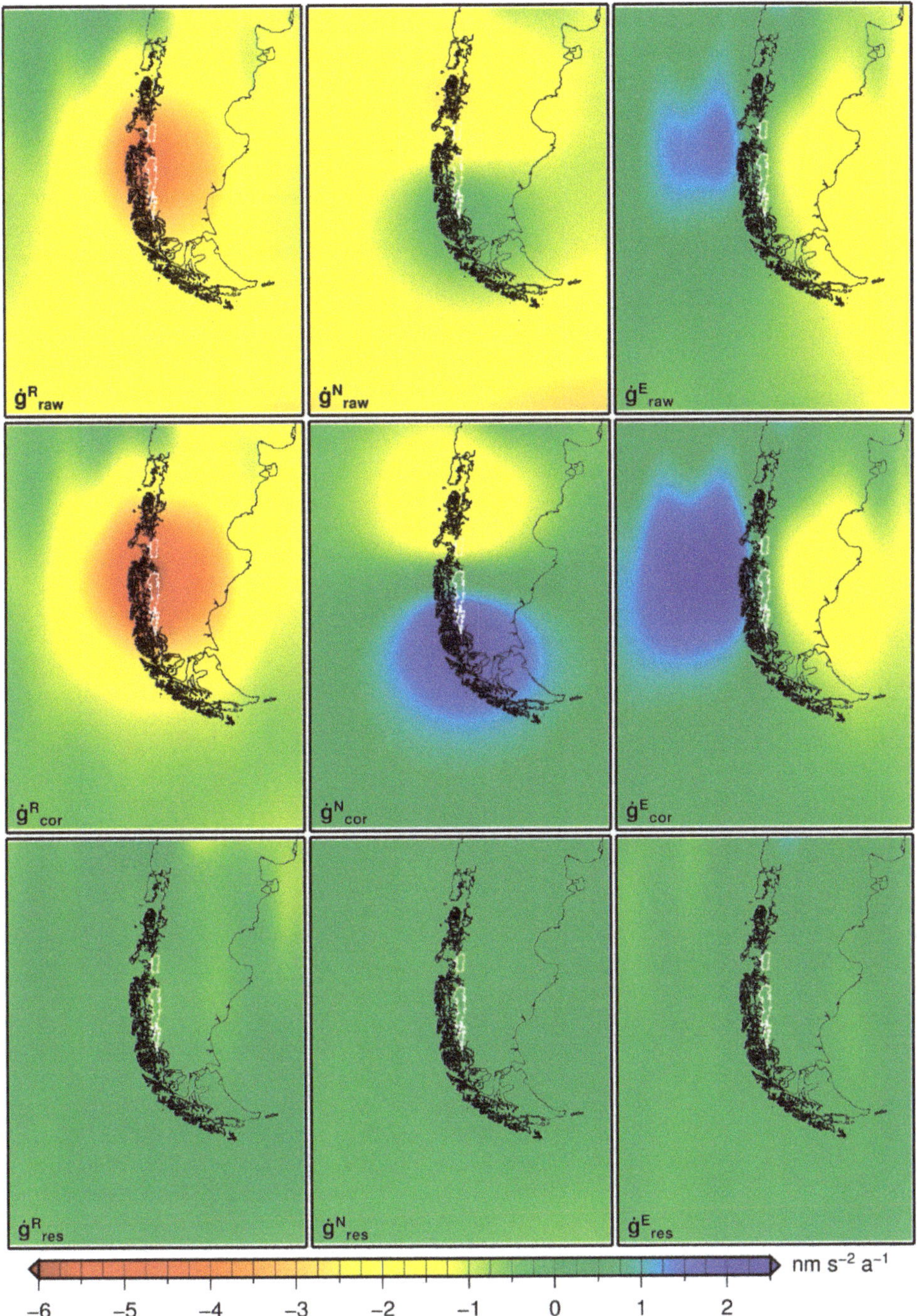

Figure 3. Map representation of gravity rate vector components at the approximate GRACE orbit altitude. Topocentric vector components are shown: radial (left), north (center), and east (right). Top: gravity rate derived from the original monthly GRACE solutions prior to the corrections; center: gravity rate after application of the corrections described in the text; bottom: residuals after adjustment of the ice mass scale factors.

Table 2. Impact of individual corrections applied in the GRACE data analysis scheme on the derived SPI and NPI ice mass change rate. Column 2 is the difference between our solution, including all of the corrections minus a modified solution, including all but the correction (or employing the alternative data set) specified in column 1. Column 3 indicates the percentage of column 2 with respect to the rate including all of the corrections (−24.4 Gt/a). Column 4 shows, for each modified solution, the ratio of the gravity vector RMS over the regional grid between the residuals after the adjustment and the corrected observations (for comparison, this ratio amounts to 0.152 for the solution including all corrections).

Correction	Change in Ice-Mass Rate (Gt/a)	(%)	RMS Ratio
Inclusion of degree one Stokes coefficients	+0.09	+0.37	0.166
Replacement of C20 coefficients from SLR	−0.15	−0.62	0.191
Pole tide correction of C21, S21 coefficients	−0.00	−0.02	0.177
CSR release 06 instead of ITSG2016 Level-2 data	+0.75	+3.17	0.117
Regional GIA: model La-A in Lange et al. [6]	+9.12	+37.38	0.243
using GIA model La-B instead of model La-A	−1.41	−5.77	0.142
Contribution of elastic deformation	+1.42	+5.81	0.151
Global ocean water mass	−2.10	−8.59	0.152
Antarctic ice sheet	+1.14	+4.65	0.315
Global continental water storage	−0.27	−1.09	0.183
Patagonian lakes	+0.01	+0.06	0.152
South American glaciers outside Patagonia	−0.30	−1.21	0.142
Empiric regional ocean mass-rate increment	+0.22	+0.89	0.181
Optimization of the a-priori ice-mass change pattern	+0.63	+2.59	0.176
Downweighting of monthly solutions after March 2016	−1.32	−5.41	0.141

We synthesized an a-priori pattern of the ice mass change at SPI and NPI. A scaling factor for this mass change pattern was derived from each monthly field of the corrected gravity vectors, simultaneously with the parameters of the empirical regional ocean mass change rate increment, by a least-squares adjustment without any regularization. In this way, a time series of the integral ice mass change was obtained (Figure 4). First, an SPI and NPI a-priori pattern consisting of 174 point masses was derived from the average ice surface elevation rates for the accumulation and ablation zones of 87 glacier basins [2] (see Appendix A). However, these ice surface elevation rates were based on slightly different imagery acquisition time spans and processing algorithms for NPI [23] and SPI [2]. Therefore, a systematic bias between the icefields cannot be ruled out. We used the GRACE data to optimize the a-priori ice mass change pattern by deriving an individual scaling factor for each of the two subsets of point masses that represent SPI and NPI, respectively. Thus, two separate scaling factors were determined simultaneously, for which our adjustment yields a minimum residual gravity rate RMS. The monthly RMS values of the residual misfit between the gravity effect of the scaled optimum ice mass change pattern and the corrected gravity vector fields are included in Figure 4 (bottom).

A weighted fit of a linear plus periodic (annual and semi-annual sinusoid) model (red in Figure 4) to the time series yields the ice mass change rate for the SPI and NPI region. The GRACE Level-2 data quality degrades after March 2016, especially since the decommissioning of the accelerometer on the GRACE-B satellite in November 2016 [58] (see also Appendix A). We took this heterogeneity into account in the ice mass change rate estimation by an empirical downweightTing of the monthly ice mass change values after March 2016. For this purpose, the uncertainty of the monthly ice mass change estimates was calculated separately until and after that month. We obtained uncertainties (1σ) of ±22.7 Gt and ±53.5 Gt, respectively. This documents a noise increase by a factor of 2.4. The reciprocal of this value was introduced as weights for the monthly ice mass change values after March 2016 in the rate estimation. In addition, the Level-2 data for February 2015 was impaired by a two-day repeat orbital resonance [59]. Therefore, rate and accuracy estimates do not include this month.

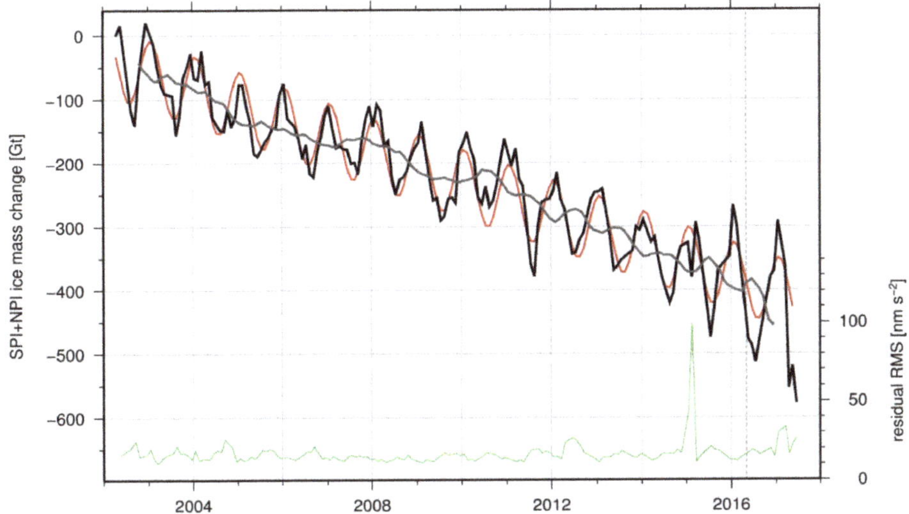

Figure 4. Time series of the ice mass change in the SPI and NPI region. Black: unfiltered time series of monthly relative ice mass change throughout the GRACE data period; red: linear plus periodic (annual and semi-annual) signal fitted to this time series; grey: low-pass filtered time series applying a running annual mean; green: residual root mean square (RMS) of the monthly gravity vectors after the adjustment of the ice mass scale factors. Dashed vertical line marks the month of March 2016, after which the GRACE data quality deteriorates and the monthly ice mass estimates were downweighted in the ice mass change rate determination.

3. Uncertainty Estimation

The uncertainty of the derived ice mass change rate is composed of the uncertainty of the individual monthly ice mass estimates and the uncertainties in the applied corrections. The residuals from the linear plus periodic model are used to estimate the precision of the ice mass change values [60]. The individual precision estimates for the two sub-periods, prior and after March 2016, lead to an uncertainty of the derived ice mass change rate of ±2.3 Gt/a.

The effect of all of the individual corrections applied on the ice mass change rate is summarized in Table 2. As a conservative estimate, we assume a 100% relative uncertainty for all of the corrections, that is, the corrections of the low-degree Stokes coefficients (lines 1–3 in Table 2) and the corrections for the surface mass change processes (lines 7–14 in Table 2). The total uncertainty of these corrections and the uncertainty of the monthly ice mass estimates (±2.3 Gt/a) is calculated as the root sum square of the individual components.

A crucial source of the uncertainty for the derived ice mass change rate is the applied GIA model. The GIA model La-A adopted in our correction is constrained by the GNSS observed uplift rates [6,7], and implies a (rock) mass gain equivalent to 9.1 Gt/a (Table 2). Employing the alternative GIA model La-B [6] (which also satisfies the uplift data; 10.5 Gt/a) in our GIA correction results in a decrease in the derived mass change rate by −1.4 Gt/a (Table 2). We adopted the standard deviation between both of the ice mass change rates as an estimate of the systematic impact of the uncertainty in the GIA correction. Adding this uncertainty contribution of ±1.0 Gt/a leads to a total uncertainty (1σ) of the ice mass change rate for the SPI and NPI region of ±4.7 Gt/a.

4. Results and Discussion

We obtained an ice mass change rate in the SPI and NPI region of −24.4 ± 4.7 Gt/a. The RMS of the residual gravity rate vectors over our grid domain (Figure 3, bottom) amounted to 0.38 nm s^{-2} a^{-1}

compared with the 2.52 nm s^{-2} a^{-1} RMS of the observed gravity vectors with all of the corrections applied (Figure 3, center). This corresponds to a signal variance reduction of 98% and demonstrates the efficiency of the scaled ice mass change pattern to explain the observed gravity changes. The determination of a common factor for the initial SPI and NPI a-priori pattern as described in Section 2 yields a residual RMS value of 0.44 nm s^{-2} a^{-1}. A significant reduction of the RMS (down to the above-mentioned value of 0.38 nm s^{-2} a^{-1}) is achieved when scaling the a-priori patterns of SPI and NPI separately. It is well-known that GRACE's ability to resolve masses close to each other is limited, and leakage effects may occur. In order to explore the sensitivity of the GRACE data concerning the separation of the SPI and NPI ice mass loss we carried out a forward computation of gravity rate vector components in our data grid according to the a-priori ice mass change pattern derived from Willis et al. [2]. This forward computation is repeated with systematically altered scaling factors for both SPI and NPI. For each pair of scaling factors (corresponding to a pair of mass change rates of the NPI and the SPI), the RMS of the residuals (gravity rate vector observed by GRACE after application of all of the corrections minus the computed gravity rate vector) is calculated. The results are shown in Figure 5. The RMS levels exhibit an elliptical shape with an optimum fit in the center. A movement from the RMS minimum along the direction of the semiminor axis of the ellipses would change the sum of both ice mass change rates and lead to a fast degradation of the fit. It is therefore an indication of the stability of our estimation of the sum of both ice mass change rates. A movement from the minimum along the semimajor axis of the ellipses would keep the sum of both of the mass rates constant, but would change the relation between the rates of the SPI and NPI. Here, a larger range of pairs of mass rates shows a reasonable fit. This demonstrates the degree of separability of the SPI and NPI ice mass change rates.

We interpret the tendency of our optimization, along with computational experiments applying different alternative a-priori ice mass change rate patterns, presented in Table A1 in the Appendix A, as an indication that the ice mass loss rates derived from the remote sensing results [2,26,28] (Figure 5) underestimate the ice loss contribution from NPI. Indeed, for the Benito glacier at the Western Flank of the NPI, a combination of field and satellite-derived elevation data suggests a dramatic surface lowering, attributed to a negative surface mass balance, with a mean rate of -3.0 ± 0.2 m/a between 1973 and 2017 [61].

The time series of the monthly ice mass change in the SPI and NPI region (Figure 4, top) is dominated by a steady decrease and an annual cycle. We estimated the annual and semi-annual amplitudes at 54.6 Gt and 0.8 Gt, respectively. The ice mass time series smoothed by the running annual mean shows small inter-annual variations and is dominated by a linear decrease. The long-term ice mass change is very well approximated by the linear model, in particular, we found no indication for an acceleration of the ice mass loss in the SPI and NPI region throughout the GRACE operation period (Figure 4). In fact, when including, in addition, a quadratic term in the model fitted to the ice mass change time series, this acceleration term was statistically not different from zero, according to a t-test with a 95% significance level. On the other hand, minimum observation periods of 10 years and 20 years are necessary for a statistically significant determination of ice mass loss acceleration of the AIS and Greenland ice sheet, respectively [62]. This makes a statistically sound inference of an acceleration of ice loss of the smaller Patagonian Icefields over the 15 year long GRACE observation period rather unlikely.

Prominent features in the time series of the residual gravity misfit (Figure 4, bottom) can be primarily explained by the orbital resonance appearing in the Level-2 solution for the month of February 2015 (RMS of almost 100 nm s^{-2}). A slight increase in misfit after March 2016 (on average 23.1 nm s^{-2}) compared to previous (on average 16.2 nm s^{-2}) months also becomes evident. This, together with the increase in the deviation from the linear plus periodic fit, gives a quantitative measure of the inevitable degradation of the Level-2 GRACE product emerging from the Level-1a and 1b data obtained during the final year of the GRACE mission. In our study, the data quality was interpreted to mean the ability to fit our mass-change pattern, and depends only on those combinations of spherical harmonic coefficients that dominate shaping the monthly gravity changes within our grid domain. Such mainly short-wavelength signals correspond to a great extent to coefficients of a higher degree and order. A degradation concentrated on low-degree coefficients leads to a long-wavelength bias, which is largely mitigated by our joint adjustment of the empirical ocean mass increment and the scaling factor for the ice mass change pattern with the full exploitation of the directional information of the 3D gravity vectors. A more detailed discussion of the GRACE Level-2 data degradation and its impact on the SPI and NPI ice mass estimation is given in Appendix A.

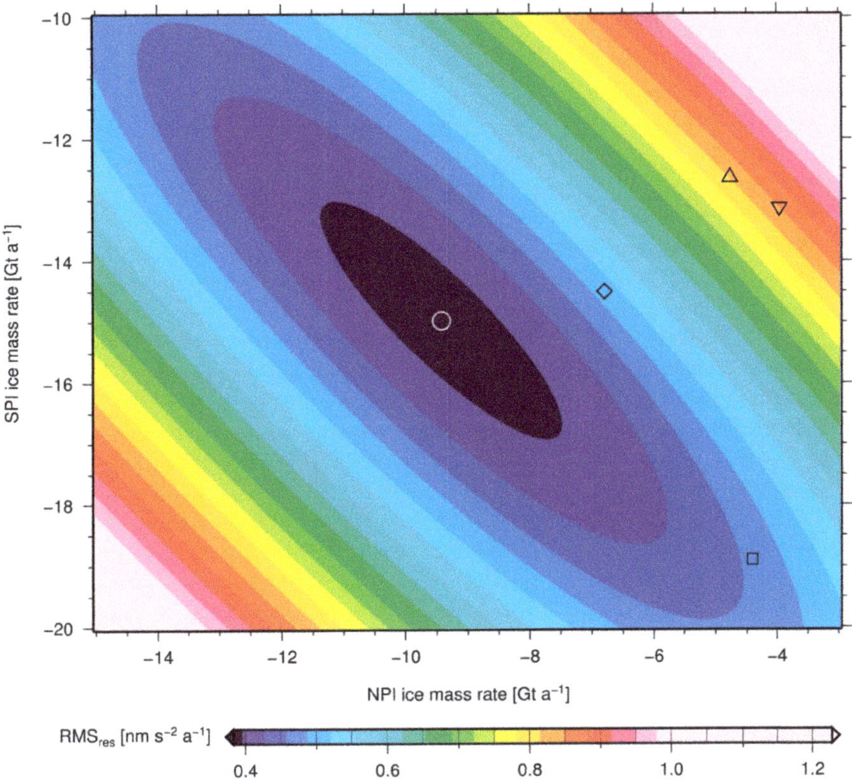

Figure 5. Separation of the ice mass loss at SPI and NPI. After the application of all corrections to the pseudo-observations derived from the GRACE Level-2 data, the residual gravity root mean square (RMS) is computed for pairs of ice mass change rates for SPI and NPI that are varied over plausible ranges. The residual RMS is shown in colors. White circle: ice mass change rates according to our pattern optimization; black symbols: ice mass change rates according to previous works; triangle: Braun et al. [26]; inverted triangle: Jaber [28]; diamond: Foresta et al. [4]; square: Willis et al. [2].

Table 2 shows the impact of the individual corrections applied in the course of our GRACE data analysis on the ice mass change rate for the SPI and NPI region. The GIA correction has by far the largest impact, contributing about one third to the derived ice mass change rate. Further corrections significant for the ice mass change rate are those for the mass changes in the ocean and the AIS. The ice mass loss and elastic solid Earth response at SPI and NPI cause a local lowering of the sea level adjacent to southern Patagonia through a loss of gravitational attraction coincident with the ice mass losses on land (Figure 1a). If this effect is not accounted for, the negative gravity pattern over the nearby ocean is absorbed into the ice mass estimate, while the long-wavelength sea level change is accommodated by the empirical regional ocean mass change rate correction. This effect has also been discussed in the context of the Antarctic Peninsula ice loss [63]. The far-field effect of the ice mass loss of the AIS manifests itself in a decrease in the N–S directed gravity component, with increasing intensity towards the south (Figure 2). Our correction model for the continental hydrology produces a signature of opposite orientation (intensity increasing towards N), yet smaller magnitude. Glaciers in the vicinity of the SPI and NPI are included in the derived ice mass change rate. The ice mass change of the extra-Patagonian glaciers in South America [26] results in a slight decrease in the SPI and NPI ice mass change rate, dominated by the glacial systems further south (Cordillera Darwin, Gran Campo Nevado). Mass variations in the adjacent lakes have a minor influence on the ice mass change rate; but are included because of their impact on the seasonal ice mass signal. The small magnitude of the empirical regional ocean mass change rate increment testifies to a high consistency of our explicit ocean mass correction with GRACE gravity. The last column in Table 2 demonstrates that the omission of any of the significant surface mass corrections degrades the fit of the adjusted model to the observations. The application of an identical data correction and analysis to alternative GRACE Level-2 CSR-RL06 harmonics (Center for Space Research, University of Texas [64]) over January 2003 to August 2016 (maximum degree and order 96) yields an SPI and NPI ice mass change rate within 3% of that derived from the ITSG-Grace2016 harmonics for the same period.

Other GRACE derived ice mass change rates as well as those based on Cryosat-2 swath altimetry, are essentially in agreement with our results (Table 1). A more substantive disagreement beyond the stated error bars is found with loss rates derived from remote sensing, especially for the SPI. The rates determined over the acquisition time spans not overlapping with the GRACE data period [3,25] shall not be discussed here. Furthermore, while remote sensing techniques are spatially limited, with separate results published for the two icefields, the GRACE derived rates represent integrals over a broader region. Therefore, the GRACE derived ice mass change rates can be compared only with those remote sensing results, which consistently sum the contributions of both icefields [2,26,28]. Synthetic-aperture radar interferometry (TanDEM-X) [26–28], in fact, seems to yield systematically smaller ice mass losses than optical imagery (ASTER, SPOT) [2,23,24]. Our results support the larger trend in ice loss as derived by Willis et al. [2] from the ASTER imagery.

5. Conclusions

Our analysis of the complete GRACE satellite gravimetry record between April 2002 and June 2017 yields an SPI and NPI ice mass change rate of -24.4 ± 4.7 Gt/a. The time series of the monthly ice mass change does not indicate an acceleration in ice mass loss throughout the GRACE period. This ice mass change rate corresponds to an average contribution to the eustatic sea level rise of 0.067 ± 0.013 mm/a, or about 3% of the mass contribution to the sea level implied by our ocean mass correction model. This confirms the disproportionally large impact of the Patagonian Icefields on sea level rise. The almost constant ice mass loss derived from GRACE over 15 years is also interesting in the context of the changes of ice flow velocities in the same time. In light of the very heterogeneous evolution in the glacier flow velocities [29], our result of a stable mass loss rate suggests that the differences among the individual glaciers basins in the non-linear mass balance variations cancel out to a great extent. We found strong indications that the NPI has a larger relative contribution to the total ice mass loss than previous remote sensing results have suggested [2,23,26,28].

We also show that the Level-2 data of the latest months of the GRACE mission provide valuable information on Patagonian ice mass changes. The derived ice mass change rate depends sensitively on the corrections for GIA and non-uniform ocean mass changes. Our new estimate of the SPI and NPI ice mass change rate is a step towards an improvement of the recent ice-unloading history, providing a strong link to the GIA and crustal deformation in this tectonically complex region, and, possibly, to a better determination of the effective viscoelasticity beneath southernmost South America.

Sustained and unabated ice mass loss over the GRACE observing period is remarkable, for there seem to be climate change processes (surface mass balance and ice discharge) that play out over a long time scale, especially when we consider the losses that can be traced into the middle of the 20th

Century [65]. This fact forebodes ill for the environment of southernmost South America in nearly every way. The loss of mountain glacier water resources will lead, inevitably, to irreversible changes in the habitability of the regional flora and fauna. This is the most important societal ramification that may be derived from the 15 year-long GRACE record in Patagonia.

Author Contributions: Conceptualization, A.R. and R.D.; data curation, A.R., A.G., E.M. and J.L.H.; formal analysis, A.R., A.G. and E.M.; investigation, A.R., A.G., M.H., E.M., J.L.H., R.P. and R.D.; methodology, A.G., M.H., E.I. and R.D.; resources, M.H. and R.P.; software, A.R., A.G. and E.M.; validation, A.G., E.M. and J.L.H.; writing (original draft), A.R.; writing (review and editing), A.G., M.H., E.I., R.P. and R.D.

Funding: Part of this research was funded by the German Research Foundation (DFG), grant number RI 2340/1-1. The APC was funded by the Open Access Publication Funds of the SLUB/TU Dresden. Part of the research was carried out at the Jet Propulsion Laboratory, California Institute of Technology, under a contract with the National Aeronautics and Space Administration, using ROSES awards 105526-967701.02.01.81 and 105526-281945.02.47.03.86 from the Earth Surface and Interior Program and the GRACE Science Team, respectively.

Acknowledgments: We thank Petra Döll for providing updated grids of the continental water storage of the WaterGAP 2.2c model. The lake-level time series of the Patagonian lakes was retrieved from the BDHI data base of the Argentine Subsecretaría de Recursos Hídricos. We thank the German Space Operations Center (GSOC) of the German Aerospace Center (DLR) for providing, continuously, nearly 100% of the raw telemetry data of the twin GRACE satellites.

Conflicts of Interest: The authors declare no conflict of interest. The funders had no role in the design of the study; in the collection, analyses, or interpretation of data; in the writing of the manuscript; or in the decision to publish the results.

Appendix A. Extended Description of GRACE Data Correction and Analysis Methods

GIA Correction

The monthly GRACE gravity fields are corrected for GIA according to the regional model La-A [6]. This model has been shown to fit best the surface deformation observed by GNSS [7]. The GIA uplift models are given as regional grids and include the elastic response to present-day ice mass changes [6]. This elastic contribution has to be removed before deriving the GIA correction models for GRACE. We applied surface densities according to the a-priori ice mass change rate pattern [2], as a load model and Green's functions of an elastic Earth model [66]. The elastic contribution to the vertical deformation is calculated for the grid nodes of the GIA model, and is subtracted from the model uplift rates. This reduced GIA model is expanded in spherical harmonics (degree/order 360). The deformation coefficients are converted to Stokes coefficients by an approximation [37] (Equation 9), enforcing mass conservation through a global layer of uniform mass outside Patagonia. These GIA Stokes rate coefficients are truncated to degree/order 2 to 120, multiplied by the time since the initial GRACE epoch (April 2002), and subtracted from the monthly GRACE Stokes coefficients sets.

Lange et al. [6] present, in addition, a second regional GIA model, termed model La-B, which fits, similarly well, the GNSS uplift observations, and has been deemed more plausible by those authors. An alternative GIA correction was derived from this GIA model La-B by the same procedure in order to quantify the impact of the uncertainty of the GIA models on the derived ice mass change (Table 2).

Surface Mass Variation Corrections

The monthly gravity vector grids are corrected for surface mass-variations of the AIS, the global ocean, continental water storage, and the great Patagonian lakes. In order to remove the far-field effect of ice mass change in Antarctica, the GRACE-derived time series of gridded Antarctic ice mass changes of the AIS_cci Gravimetric Mass Balance Product [52], generated in the scope of the European Space Agency's Climate Change Initiative project for the Antarctic Ice Sheet (AIS_cci), are applied. The latest months of the GRACE data period (November 2016 to June 2017) are not included in this product. These time series are therefore extrapolated for each AIS_cci grid cell, applying a linear plus periodic (periods: annual, semi-annual, and 161 days) model.

Our ocean mass correction consists in an updated solution of the gravitationally self-consistent distribution over the ocean of the water mass exchange with five major ice sheets/ice caps and the global continental hydrosphere [67] (following the pseudo-spectral approach [56]). The monthly ice mass change patterns of the AIS [52], Greenland ice sheet [52], and the Patagonian Icefields (this study, iterative solution), complemented by the ice mass rate patterns for Alaska and the Canadian Arctic [57], as well as the monthly patterns of the continental water storage according to the WaterGAP 2.2c model [53], are used as an input for the solution of the sea level equation. Because of the limited extent of the forcing models [52,53], the monthly ocean water mass distributions were computed through November 2016, and then extrapolated over the last GRACE mission months applying a linear plus periodic (annual, semi-annual) model. On a global average, this ocean mass model implies a relative sea level rise of 2.13 mm/a.

The effects of the changes in the continental water storage are removed by using total water storage time series on a global grid of the WaterGAP 2.2c model [53] with climate forcing WFDEI-CRU [54]. As these time series are available only until December 2015, they were extrapolated through June 2017, applying a linear plus periodic (annual and semi-annual) model. We took particular care of the effects of the hydrological mass variations associated with lake-level changes in Lago Buenos Aires/General Carrera (BAGC), Lago San Martín/O'Higgins, Lago Viedma, and Lago Argentino (Figure 1b). These large lakes receive the meltwater runoff from the adjacent NPI (BAGC) and SPI [55]. Brazo Rico and Brazo Sur of Lago Argentino were treated as an additional individual lake, because of the repeated damming by the Perito Moreno glacier during the GRACE data period. On short and seasonal time scales, we expect these lakes to produce a mass signal opposite to that of the icefields, but too close to be resolved separately by GRACE. We used tide gauge data in the lakes provided by the BDHI data base to derive the time series of the monthly mean lake levels. Before 2008, regular tide-gauge data are available only for Lago Argentino and Brazo Rico/Sur, the time series of the other three lakes were extrapolated backwards in time until April 2002, by stacking the annual lake-level signal. The mass signal of the lake-level changes was subtracted in the corresponding nodes of the global hydrological model grid prior to the application of the corrections to the GRACE data.

All of these surface mass-variation processes are represented by monthly time series of the surface density on global (ocean, hydrology) or regional (AIS, lakes) grids. For each of these processes and for each month, the corresponding 3D gravity vector components are derived for our regional grid and subtracted from the grids of the pseudo-observations.

Empirical Regional Ocean Mass Rate Correction

Our explicit ocean mass correction accounts for the mass exchange with five major glaciated regions and continental water storage. Additional sources may contribute to ocean mass variations in the region under investigation. Therefore, we apply an empirical, additional correction in order to accommodate the residual imperfections of our explicit ocean mass model. A set of point masses homogeneously distributed over the map extension in Figure 1a, spaced every 0.1° in longitude and in latitude increments that provide for an equal area between the point masses, is used for this purpose. Over all of the point masses falling within the ocean, the residual linear mass changes are modeled by three parameters describing a mean mass change rate and mass-rate gradients in a north and east direction, respectively. These three parameters are solved, simultaneously with the monthly scaling factors for the ice mass change pattern, in a joint least-squares adjustment to the monthly gravity vector grids.

In fact, this empiric correction would accommodate long-wavelength effects not only originating from regional deviations from our explicit ocean mass correction, but also due to residual imperfections in other correction models or low-degree GRACE data. The empirical regional ocean mass rate increment produces only a small change in the ice mass change rate (1%, see Table 2), indicating a high consistency between the applied correction models and GRACE data.

A-Priori Mass Change Pattern

An a-priori SPI and NPI ice mass change pattern (Wi12) is derived based on published remote sensing results [2]. We use ice surface elevation rates dh/dt, determined by differencing ASTER digital elevation models (DEM) from 2012 (SPI) and 2011 (NPI), relative to a DEM derived from the SRTM in February 2000 [2]. The dh/dt are given as the average values for accumulation and ablation zones of 87 (SPI: 49; NPI: 38) glacier basins. We applied an ice density of 0.9 g cm^{-3} to convert the dh/dt to mass change rates. Glacier boundary polygons from the GLIMS data base [68] were used together with equilibrium line altitudes [2,23] and a SRTM derived DEM [69] in order to compute the barycenter coordinates for the 174 accumulation/ablation zones. The ice mass change rates are treated as point masses located in the barycenter of the corresponding accumulation/ablation zone (Figure 1b). This preliminary ice mass change rate pattern is then optimized by searching separate scaling factors for SPI respectively NPI that minimize the residual gravity RMS in our GRACE data adjustment.

Our approach is largely insensitive to the choice of the a-priori mass pattern, because the upward continuation to orbit altitude acts as a low-pass filter in the space domain, damping the resolution and detail of the surface mass configuration. This demonstrated computations employing the alternative a-priori patterns Ja16, Br19, and Fo18 (Table A1). All three of these alternative a-priori patterns consist of only two point masses each, in which the ice mass change rates derived from the remote sensing results [26,28] or swath altimetry [4] are concentrated. Although these models represent only 73%, 75%, and 91%, respectively, of the Wi12 total ice mass change rate, its impact on the adjusted SPI and NPI ice-mass rate does not reach 1.5% (Table A1). However, the application of either of the alternative patterns reduces the residuals, despite their coarser resolution. The systematic decrease in the residual gravity RMS among the different a-priori patterns with a decreasing SPI/NPI ice mass rate ratio is interpreted as an indication that the Wi12 model underestimates the ice-mass loss at the NPI relative to that at the SPI, and motivates our optimization of the Wi12 pattern by two separate scaling factors for SPI and NPI.

Table A1. A-priori ice mass change rate patterns used to adjust GRACE gravity rate vectors.

A-Priori Pattern	Ice Mass Change Rate (Gt/a)				SPI/NPI Ratio	Reference	SPI and NPI$_{GRACE}$ (Gt/a)	RMSres (nm s^{-2} a$^{-1)}$
	SPI	NPI	Total	(%)				
Wi12	−18.934	−4.410	−23.344	100%	4.3	[2]	−23.8	0.44
Ja16	−13.135	−3.961	−17.096	73%	3.3	[28]	−23.5	0.43
Br19	−12.640	−4.770	−17.410	75%	2.6	[26]	−23.7	0.41
Fo18	−14.500	−6.790	−21.290	91%	2.1	[4]	−23.9	0.39

Impact of GRACE Level-2 Data Quality Degradation on SPI and NPI Ice Mass Determination

In order to assess the global temporal behavior of the GRACE data accuracy, we computed the residual equivalent water height anomalies for each spherical harmonic degree and order with respect to a model that includes linear, quadratic, and periodic terms. The standard deviation of these temporal residuals is interpreted as a measure of GRACE observation uncertainty. Figure A1 shows the degree amplitudes of these equivalent water height anomalies for the individual monthly Level-2 solutions. As expected, the maximum uncertainty is found at the highest degrees, while minimum standard deviations are found between degrees 10–30.

Among these degree amplitude curves, two outliers are observed, corresponding to the months of February 2015 (two-day repeat orbit; red) and March 2017. In addition, the deterioration of the data accuracy after March 2016 becomes evident, with degree amplitudes above the median (among all months of the entire GRACE data period), and substantially increased energy on the low degrees (<10).

In order to investigate the sensitivity of the SPI and NPI ice mass time series to this GRACE uncertainty evolution, the SPI and NPI region function (as applied in the classical regional integration approach) is expanded in the spherical harmonic coefficients. Unlike AIS and the Greenland ice sheet, which concentrate much of their signal on low degrees (<10), the small SPI and NPI region requiresa substantially broader band of degrees. Therefore, the uncertainty of those higher degrees is also more important. We derive degree amplitudes of the region functions for SPI and NPI, AIS, and the Greenland ice sheet, which are included in Figure A1. In general, only the portion of the error spectrum retained after filtering is integrated with the region function. In the present case, the filtering corresponds to the effect of using pseudo-observations at the orbit altitude. As the SPI and NPI region function distributes the energy over higher degrees, these are also less damped by the filtering (compared to AIS and Greenland). A comparison of the degree amplitudes of the observation uncertainty and the regional function reveals that the SPI and NPI region function has its strongest energy on the degrees for which the errors are smallest—degrees 10 to 30. Because of the small energy on the low degrees (<10), the accuracy degradation of the solutions after March 2016 has not had as efficient an effect on the SPI and NPI ice mass estimation, as, for example, for AIS or Greenland. Comparing the two outlier months reveals that February 2015 (repeat orbit) results in much larger residuals (green time series in Figure 4) than March 2017. This demonstrates that for SPI and NPI, the uncertainty on the higher degrees is much more important than the long-wavelength errors (unlike AIS or Greenland), but these higher degrees are not as severely affected by the accuracy degradation.

We conclude that the GRACE Level-2 data accuracy degradation affects mainly low degrees (<50), but the small SPI and NPI target is less sensitive to this low-degree degradation. Thus, also the latest months of the GRACE mission hold useful information about the Patagonian ice mass changes. In its exploitation, we took the data quality degradation into account by a cautious empirical downweighTing. An exclusion of the solutions affected by the data quality degradation has little impact on our final result—our time series over the period of April 2002 through March 2016 yields an ice mass change rate for the SPI and NPI region of −23.9 Gt/a.

A strategy is now in place at the Jet Propulsion Laboratory to model each of the thrusting events on the satellite lacking accelerometer data [70]. We therefore anticipate improvements in the Level-2 harmonics upon the next release that contains such modeling, and it will be interesting to further evaluate our scheme used here for the analysis of the late months of the GRACE mission.

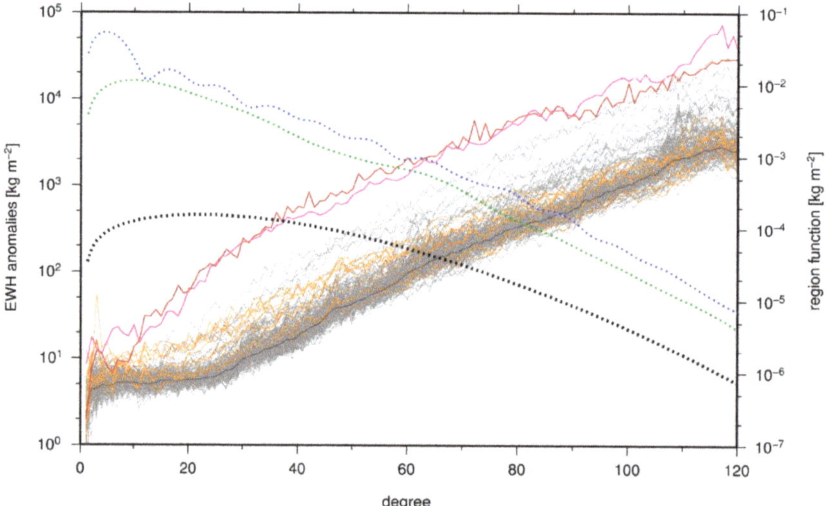

Figure A1. Degree amplitudes of equivalent water height anomalies for all of the solutions of the GRACE Level-2 data period (left axis). Light grey: monthly solutions from April 2002 through March 2016; orange: from April 2016 through June 2017; red: February 2015 (two-day repeat orbit); magenta: March 2017; dark grey: median among all monthly solutions. Dashed curves show degree amplitudes for filtered region functions (right axis) for SPI and NPI (black); AIS (blue) and Greenland ice sheet (green).

References

1. Casassa, G.; Rodríguez, J.L.; Loriaux, T. A New Glacier Inventory for the Southern Patagonia Icefield and Areal Changes 1986–2000. In *Global Land Ice Measurements from Space*; Kargel, J.S.S., Leonard, G.J.J., Bishop, M.P.P., Kääb, A., Raup, B.H.H., Eds.; Springer: Berlin/Heidelberg, Germany, 2014; pp. 639–660. ISBN 978-3-540-79818-7.
2. Willis, M.J.; Melkonian, A.K.; Pritchard, M.E.; Rivera, A. Ice loss from the Southern Patagonian Ice Field, South America, between 2000 and 2012. *Geophys. Res. Lett.* **2012**, *39*, 1–6. [CrossRef]
3. Rignot, E.; Rivera, A.; Casassa, G. Contribution of the Patagonia Icefields of South America to sea level rise. *Science* **2003**, *302*, 434–437. [CrossRef]
4. Foresta, L.; Gourmelen, N.; Weissgerber, F.; Nienow, P.; Williams, J.J.; Shepherd, A.; Drinkwater, M.R.; Plummer, S. Heterogeneous and rapid ice loss over the Patagonian Ice Fields revealed by CryoSat-2 swath radar altimetry. *Remote Sens. Environ.* **2018**, *211*, 441–455. [CrossRef]
5. Dietrich, R.; Ivins, E.R.; Casassa, G.; Lange, H.; Wendt, J.; Fritsche, M. Rapid crustal uplift in Patagonia due to enhanced ice loss. *Earth Planet. Sci. Lett.* **2010**, *289*, 22–29. [CrossRef]
6. Lange, H.; Casassa, G.; Ivins, E.R.; Schröder, L.; Fritsche, M.; Richter, A.; Groh, A.; Dietrich, R. Observed crustal uplift near the Southern Patagonian Icefield constrains improved viscoelastic Earth models. *Geophys. Res. Lett.* **2014**, *41*, 805–812. [CrossRef]
7. Richter, A.; Ivins, E.; Lange, H.; Mendoza, L.; Schröder, L.; Hormaechea, J.L.; Casassa, G.; Marderwald, E.; Fritsche, M.; Perdomo, R.; et al. Crustal deformation across the Southern Patagonian Icefield observed by GNSS. *Earth Planet. Sci. Lett.* **2016**, *452*, 206–215. [CrossRef]
8. Ivins, E.; James, T. Bedrock response to Llanquihue Holocene and present-day glaciation in southernmost South America. *Geophys. Res. Lett.* **2004**, *31*, L24613. [CrossRef]
9. Ivins, E.; James, T. Simple models for late Holocene and present-day Patagonian glacier fluctuations and predictions of a geodetically detectable isostatic response. *Geophys. J. Int.* **1999**, *138*, 601–624. [CrossRef]
10. Lliboutry, L. Glaciers of the wet Andes. In *Glaciers of South America*; Williams, R.S., Ferrigno, J.G., Eds.; U.S. Geological Survey Professional Paper 1386-I: Reston, VA, USA, 1998; pp. 148–206.
11. Glasser, N.F.; Hambry, S.; Winchester, V.; Aniya, M. Late Pleistocene and Holocene palaeoclimate and glacier fluctuations in Patagonia. *Glob. Planet. Chang.* **2004**, *43*, 79–101. [CrossRef]
12. Villalba, R.; Lara, A.; Boninsegna, J.A.; Masiokas, M.; Delgado, S.; Aravena, J.C.; Roig, F.A.; Schmelter, A.; Wolodarsky, A.; Ripalta, A. Large-scale temperature changes across the southern Andes: 20th-century variations in the context of the past 400 years. *Clim. Chang.* **2003**, *59*, 177–232. [CrossRef]
13. Kilian, R.; Lamy, F. A review of Glacial and Holocene paleoclimate records from southernmost Patagonia (49–55°S). *Quat. Sci. Rev.* **2012**, *53*, 1–23. [CrossRef]
14. Moy, C.M.; Dunbar, R.B.; Moreno, P.I.; Francois, J.-P.; Villa-Martínez, R.; Mucciarone, D.M.; Guildersond, T.P.; Garreaud, R.D. Isotopic evidence for hydrologic change related to the westerlies in SW Patagonia, Chile, during the last millennium. *Quat. Sci. Rev.* **2008**, *27*, 1335–1349. [CrossRef]
15. Li, X.; Holland, D.M.; Gerber, E.P.; Yoo, C. Impacts of the north and tropical Atlantic Ocean on the Antarctic Peninsula and sea ice. *Nature* **2014**, *505*, 538–542. [CrossRef] [PubMed]
16. Glasser, N.; Harrison, S.; Jansson, K.; Anderson, K.; Cowley, A. Global sea-level contribution from the Patagonian Icefields since the Little Ice Age maximum. *Nat. Geosci.* **2011**, *4*, 303–307. [CrossRef]
17. Lenaerts, J.T.M.; Van Den Broeke, M.R.; Van Wessem, J.M.; Van De Berg, W.J.; Van Meijgaard, E.; Van Ulft, L.H.; Schaefer, M. Extreme precipitation and climate gradients in Patagonia revealed by high resolution regional atmospheric climate modeling. *J. Clim.* **2014**, *27*, 4607–4621. [CrossRef]
18. Zemp, M.; Frey, H.; Gärtner-Roer, I.; Nussbaumer, S.U.; Hoelzle, M.; Paul, F.; Haeberli, W.; Denzinger, F.; Ahlstrøm, A.P.; Anderson, B.; et al. Historically unprecedented global glacier decline in the early 21st century. *J. Glaciol.* **2015**, *61*, 745–762. [CrossRef]
19. Yeh, S.-W.; Cai, W.; Min, S.-K.; McPhaden, M.J.; Dommenget, D.; Dewitte, B.; Collins, M.; Ashok, K.; An, S.-I.; Yim, B.-Y.; et al. ENSO atmospheric teleconnections and their response to greenhouse gas forcing. *Rev. Geophys.* **2015**, *56*, 185–206. [CrossRef]
20. Kennicutt, M.C.; Chown, S.L.; Cassano, J.J.; Liggett, D.; Massom, R.; Peck, L.S.; Rintoul, S.R.; Storey, J.W.V.; Vaughan, D.G.; Wilson, T.J.; et al. A roadmap for Antarctic and Southern Ocean science for the next two decades and beyond. *Antarct. Sci.* **2015**, *27*, 3–18. [CrossRef]

21. Schaefer, M.; Machguth, H.; Falvey, M.; Casassa, G. Modeling past and future surface mass balance of the Northern Patagonia Icefield. *J. Geophys. Res. Earth Surf.* **2013**, *118*, 571–588. [CrossRef]

22. Schaefer, M.; Machguth, H.; Falvey, M.; Casassa, G.; Rignot, E. Quantifying mass balance processes on the Southern Patagonia Icefield. *Cryosphere* **2015**, *9*, 25–35. [CrossRef]

23. Willis, M.J.; Melkonian, A.K.; Pritchard, M.E.; Ramage, J.M. Ice loss rates at the Northern Patagonian Icefield derived using a decade of satellite remote sensing. *Remote Sens. Environ.* **2012**, *117*, 184–198. [CrossRef]

24. Dussaillant, I.; Berthier, E.; Brun, F. Geodetic mass balance of the Northern Patagonian Icefield from 2000 to 2012 using two independent methods. *Front. Earth Sci.* **2018**, *6*, 8. [CrossRef]

25. Rivera, A.; Benham, T.; Casassa, G.; Bamber, J.; Dowdeswell, J.A. Ice elevation and areal changes of glaciers from the Northern Patagonia Icefield, Chile. *Glob. Planet. Chang.* **2007**, *59*, 126–137. [CrossRef]

26. Braun, M.H.; Malz, P.; Sommer, C.; Farías-Barahona, D.; Sauter, T.; Casassa, G.; Soruco, A.; Skvarca, P.; Seehaus, T.C. Constraining glacier elevation and mass changes in South America. *Nat. Clim. Chang.* **2019**, *9*, 130–136. [CrossRef]

27. Malz, P.; Meier, E.; Casassa, G.; Jana, R.; Svarca, P.; Braun, M. Elevation and mass changes of the Southern Patagonia Icefield derived from TanDEM-X and SRTM data. *Remote Sens.* **2018**, *10*, 188. [CrossRef]

28. Jaber, A.W. *Derivation of Mass Balance and Surface Velocity of Glaciers by Means of High Resolution Synthetic Aperture Radar: Application to the Patagonian Icefields and Antarctica*; DLR-Forschungsbericht 2016; ISBN 1434–8454. Available online: http://elib.dlr.de/109075/1/Thesis_AbdelJaber_final.pdf (accessed on 4 August 2018).

29. Mouginot, J.; Rignot, E. Ice motion of the Patagonian Icefields of South America: 1984–2014. *Geophys. Res. Lett.* **2015**, *42*, 1441–1449. [CrossRef]

30. Tapley, B.D.; Bettadpur, S.; Watkins, M.; Reigber, C. The gravity recovery and climate experiment: Mission overview and early results. *Geophys. Res. Lett.* **2004**, *31*, L09607. [CrossRef]

31. Ivins, E.R.; James, T.S.; Wahr, J.; Schrama, E.J.O.; Landerer, F.W.; Simon, K.M. Antarctic contribution to sea level rise observed by GRACE with improved GIA correction. *J. Geophys. Res.-Solid Earth* **2013**, *118*, 3126–3141. [CrossRef]

32. Gardner, A.S.; Moholdt, G.; Cogley, J.G.; Wouters, B.; Arendt, A.A.; Wahr, J.; Berthier, E.; Hock, R.; Pfeffer, W.T.; Kaser, G.; et al. A Reconciled Estimate of Glacier Contributions to Sea Level Rise: 2003 to 2009. *Science* **2013**, *340*, 852–857. [CrossRef]

33. Wouters, B.; Bonin, J.A.; Chambers, D.P.; Riva, R.E.M.; Sasgen, I.; Wahr, J. GRACE, time-varying gravity, Earth system dynamics and climate change. *Rep. Prog. Phys.* **2014**, *77*, 116801. [CrossRef]

34. Shepherd, A.; Ivins, E.; Rignot, E.; Smith, B.; van den Broeke, M.; Velicogna, I.; Whitehouse, P.; Briggs, K.; Joughin, I.; Krinner, G.; et al. Mass balance of the Antarctic Ice Sheet from 1992 to 2017. *Nature* **2018**, *558*, 219–222. [CrossRef]

35. Shepherd, A.; Ivins, E.R.; Geruo, A.; Barletta, V.R.; Bentley, M.J.; Bettadpur, S.; Briggs, K.H.; Bromwich, D.H.; Forsberg, R.; Natalia Galin, N.; et al. A reconciled estimate of ice sheet mass balance. *Science* **2012**, *338*, 1183–1189. [CrossRef]

36. Chen, J.L.; Wilson, C.R.; Tapley, B.D.; Blankenship, D.D.; Ivins, E.R. Patagonia icefield melting observed by gravity recovery and climate experiment (GRACE). *Geophys. Res. Lett.* **2007**, *34*, L22501. [CrossRef]

37. Ivins, E.R.; Watkins, M.M.; Yuan, D.N.; Dietrich, R.; Casassa, G.; Rülke, A. On-land ice loss and glacial isostatic adjustment at the Drake Passage: 2003–2009. *J. Geophys. Res.-Solid Earth* **2011**, *116*, B02403. [CrossRef]

38. Jacob, T.; Wahr, J.; Pfeffer, W.T.; Swenson, S. Recent contributions of glaciers and ice caps to sea level rise. *Nature* **2012**, *482*, 514–518. [CrossRef] [PubMed]

39. Schrama, E.J.O.; Wouters, B.; Rietbroek, R. A mascon approach to assess ice sheet and glacier mass balances and their uncertainties from GRACE data. *J. Geophys. Res. Solid Earth* **2014**, *119*, 6048–6066. [CrossRef]

40. Reager, J.T.; Gardner, A.S.; Famiglietti, J.S.; Wiese, D.N.; Eicker, A.; Lo, M.-H. A decade of sea level rise slowed by climate-driven hydrology. *Science* **2016**, *351*, 699–703. [CrossRef]

41. Mayer-Gürr, T.; Behzadpour, S.; Ellmer, M.; Kvas, A.; Klinger, B.; Zehentner, N. ITSG-Grace2016—Monthly and Daily Gravity Field Solutions from GRACE. GFZ Data Services 2016. Available online: ftp://ftp.tugraz.at/outgoing/ITSG/GRACE/ITSG-Grace2016/monthly/monthly_n120/ (accessed on 11 October 2018).

42. Forsberg, R.; Reeh, N. Mass change of the Greenland ice sheet from Grace. In *Proceedings of the 1st Meeting of the International Gravity Field Service*; Springer: Berlin, Germany, 2006; Volume 18, pp. 454–458.

43. Baur, O.; Sneeuw, N. Assessing Greenland ice mass loss by means of point-mass modeling: A viable methodology. *J. Geod.* **2011**, *85*, 607–615. [CrossRef]

44. Forsberg, R.; Sørensen, L.S.; Simonsen, S.B. Greenland and Antarctica Ice Sheet Mass Changes and Effects on Global Sea Level. *Surv. Geophys.* **2017**, *38*, 89–104. [CrossRef]

45. Ran, J.; Ditmar, P.; Klees, R.; Farahani, H.H. Statistically optimal estimation of Greenland Ice Sheet mass variations from GRACE monthly solutions using an improved mascon approach. *J. Geod.* **2018**, *92*, 299–319. [CrossRef]

46. Watkins, M.M.; Wiese, D.N.; Yuan, D.-N.; Boening, C.; Landerer, F.W. Improved methods for observing Earth's time variable mass distribution with GRACE using spherical cap mascons: Improved Gravity Observations from GRACE. *J. Geophys. Res. Solid Earth* **2015**, *120*, 2648–2671. [CrossRef]

47. Horwath, M.; Groh, A. Evaluation of recent GRACE monthly solution series with an ice sheet perspective. *Geophys. Res. Abstr.* **2016**, *18*. Available online: http://www.egsiem.eu/images/publication/EGU2016/horwath_egu2016_releases_b.pdf (accessed on 24 October 2018).

48. Kusche, J. Approximate decorrelation and non-isotropic smoothing of time-variable GRACE-type gravity field models. *J. Geod.* **2007**, *81*, 733–749. [CrossRef]

49. Bergmann-Wolf, I.; Zhang, L.; Dobslaw, H. Global Eustatic Sea-Level Variations for the Approximation of Geocenter Motion from Grace. *J. Geod. Sci.* **2014**, *4*. [CrossRef]

50. Cheng, M.; Tapley, B.D.; Ries, J.C. Deceleration in the Earth's oblateness. *J. Geophys. Res. Solid Earth* **2013**, *118*, 740–747. [CrossRef]

51. Wahr, J.; Nerem, R.S.; Bettadpur, S.V. The pole tide and its effect on GRACE time-variable gravity measurements: Implications for estimates of surface mass variations. *J. Geophys. Res. Solid Earth* **2015**, *120*, 4597–4615. [CrossRef]

52. Groh, A.; Horwath, M. The method of tailored sensitivity kernels for GRACE mass change estimates. *Geophys. Res. Abstr.* **2016**, *18*. Available online: https://data1.geo.tu-dresden.de/ais_gmb/ (accessed on 24 October 2018).

53. Döll, P.; Kaspar, F.; Lehner, B. A global hydrological model for deriving water availability indicators: Model tuning and validation. *J. Hydrol.* **2003**, *270*, 105–134. [CrossRef]

54. Weedon, G.P.; Balsamo, G.; Bellouin, N.; Gomes, S.; Best, M.J.; Viterbo, P. The WFDEI meteorological forcing data set: WATCH Forcing Data methodology applied to ERA-Interim reanalysis data. *Water Resour. Res.* **2014**, *50*, 7505–7514. [CrossRef]

55. Richter, A.; Marderwald, E.; Hormaechea, J.L.; Mendoza, L.; Perdomo, R.; Connon, G.; Scheinert, M.; Horwath, M.; Dietrich, R. Lake-level variations and tides in Lago Argentino, Patagonia: Insights from pressure tide gauge records. *J. Limnol.* **2016**, *75*, 62–77. [CrossRef]

56. Mitrovica, J.X.; Peltier, W.R. On postglacial geoid subsidence over the equatorial oceans. *J. Geophys. Res.* **1991**, *96*, 20053–20071. [CrossRef]

57. Harig, C.; Simons, F.J. Ice mass loss in Greenland, the Gulf of Alaska, and the Canadian Archipelago: Seasonal cycles and decadal trends. *Geophys. Res. Lett.* **2016**, *43*, 3150–3159. [CrossRef]

58. Flechtner, F. GRACE Science Data System Monthly Report, November 2016. Available online: ftp://rz-vm152.gfz-potsdam.de/grace/DOCUMENTS/NEWSLETTER/2016/GRACE_SDS_NL_1611.pdf (accessed on 4 October 2018).

59. Flechtner, F. GRACE Science Data System Monthly Report, February 2015. Available online: ftp://rz-vm152.gfz-potsdam.de/grace/DOCUMENTS/NEWSLETTER/2015/GRACE_SDS_NL_1502.pdf (accessed on 4 October 2018).

60. Groh, A.; Horwath, M.; Horvath, A.; Meister, R.; Sørensen, L.; Barletta, V.; Forsberg, R.; Wouters, B.; Ditmar, P.; Ran, J.; et al. Evaluating GRACE Mass Change Time Series for the Antarctic and Greenland Ice Sheet—ESA CCI Round Robin Methods and Results. *Geosciences*, under review.

61. Ryan, J.C.; Sessions, M.; Wilson, R.; Wündrich, O.; Hubbard, A. Rapid Surface Lowering of Benito Glacier, Northern Patagonian Icefield. *Front. Earth Sci.* **2018**, *6*, 47. [CrossRef]

62. Wouters, B.; Bamber, J.L.; van den Broeke, M.R.; Lenaerts, J.T.M.; Sasgen, I. Limits in detecting acceleration of ice sheet mass loss due to climate variability. *Nat. Geosci.* **2013**. [CrossRef]

63. Sterenborg, M.G.; Morrow, E.; Mitrovica, J.X. Bias in GRACE estimates of ice mass change due to accompanying sea-level change. *J. Geod.* **2012**, *87*, 387–392. [CrossRef]

64. Bettadpur, S. *GRACE 327-742: UTCSR Level-2 Processing Standards Document (For Level-2 Product Release 0006)*; Center for Space Research, The University of Texas at Austin: Austin, TX, USA, 2018. Available online: ftp://podaac.jpl.nasa.gov/allData/grace/docs/L2-CSR006_ProcStd_v5.0.pdf (accessed on 28 October 2018).

65. Aniya, M. Recent glacier variations of the Hielos Patagonicos, South America, and their contribution to sea-level rise. *Arct. Antarct. Alp. Res.* **1999**, *31*, 165–173. [CrossRef]
66. Farrel, W.E. Deformation of the Earth by surface loads. *Rev. Geophys. Space Phys.* **1972**, *10*, 762–795. [CrossRef]
67. Groh, A. Zur Bestimmung Eisinduzierter Massensignale aus der Kombination Geodätischer Daten. Ph.D. Thesis, Technische Universität Dresden, Dresden, Germany, 2014.
68. Raup, B.H.; Racoviteanu, A.; Khalsa, S.J.S.; Helm, C.; Armstrong, R.; Arnaud, Y. The GLIMS Geospatial Glacier Database: A New Tool for Studying Glacier Change. *Glob. Planet. Chang.* **2007**, *56*, 101–110. [CrossRef]
69. Farr, T.G.; Rosen, P.A.; Caro, E.; Crippen, R.; Duren, R.; Hensley, S.; Kobrick, M.; Paller, M.; Rodriguez, E.; Roth, L.; et al. The shuttle radar topography mission. *Rev. Geophys.* **2007**, *45*. [CrossRef]
70. Bandikova, T.; McCullough, C.M.; Kruizinga, G.L.H. GRACE Accelerometer Transplant. In Proceedings of the Transactions of the American Geophysical Union, Fall Meeting, New Orleans, LA, USA, 11–15 December 2017; abstract G31B-0405.

Article

Downscaling GRACE TWSA Data into High-Resolution Groundwater Level Anomaly Using Machine Learning-Based Models in a Glacial Aquifer System

Wondwosen M. Seyoum [1,*], Dongjae Kwon [1] and Adam M. Milewski [2]

[1] Department of Geography, Geology, and the Environment, Illinois State University, Normal, IL 61790, USA; dkwon@ilstu.edu

[2] Department of Geology, University of Georgia, Athens, GA 30602, USA; milewski@uga.edu

* Correspondence: wmseyou@ilstu.edu; Tel.: +1-309-438-2833

Received: 4 February 2019; Accepted: 2 April 2019; Published: 5 April 2019

Abstract: With continued threat from climate change and human impacts, high-resolution and continuous hydrologic data accessibility has a paramount importance for predicting trends and availability of water resources. This study presents a novel machine learning (ML)-based downscaling algorithm that produces a high spatial resolution groundwater level anomaly (GWLA) from the Gravity Recovery and Climate Experiment (GRACE) data by utilizing the relationship between Terrestrial Water Storage Anomaly (TWSA) from GRACE and other land surface and hydro-climatic variables (e.g., vegetation coverage, land surface temperature, precipitation, streamflow, and in-situ groundwater level data). The predicted downscaled GWLA data were tested using monthly in-situ groundwater level observations. Of the 32 groundwater monitoring wells available in the study site, 21 wells were used to develop the ML-based downscaling model, while the remaining 11 wells were used to assess the performance of the ML-based downscaling model. The test results showed that the model satisfactorily reproduces the spatial and temporal variation of the GWLA in the area, with acceptable correlation coefficient and Nash-Sutcliffe Efficiency values of ~0.76 and ~0.45, respectively. GRACE TWSA was the most influential predictor variable in the models, followed by stream discharge and soil moisture storage. Though model limitations and uncertainty could exist due to high spatial heterogeneity of the geologic materials and omission of human impact (e.g., abstraction), the significance of the result is undeniable, particularly in areas where in-situ well measurements are sparse.

Keywords: GRACE TWSA; groundwater level anomaly; downscaling; machine learning; boosted regression trees; glacial sediment

1. Introduction

With continued threat from climate change and human impacts, high-resolution and continuous hydrologic data availability is crucial for predicting trends and water resource availability. Given an increase in global population and water demand, measuring water storage and trends is valuable for water resources management, hazard analysis and mitigation, and food security [1–4]. These issues are exacerbated by the lack of hydrologic data, which leads to increased uncertainty in estimated water resource variables (e.g., river discharge, groundwater recharge) and hampers accurate predictions of water storage and trends [5–7]. Unfortunately, global in-situ data availability is poor due to malfunctioning of existing gauges, economic constraints, and the lack of data sharing among nations [8–11]. Conversely, the capability and demand of satellite applications in hydrology and water resources has grown in the past decades. Particularly, the availability of Terrestrial Water Storage

Anomaly (TWSA) data from the Gravity Recovery and Climate Experiment (GRACE) satellite since 2002 is revolutionizing the way we measure Earth's water storage variations.

GRACE has twin satellites that measure gravity. Through a series of processing steps, the gravity anomaly measured by GRACE satellite is converted into monthly TWSA, with a spatial resolution of several hundred kilometers. GRACE has provided unprecedented opportunities for the science community despite the limitations inhibited by its coarse spatial scale (e.g., 200,000 km^2). It has been used to understand regional and global trends in water storage changes [12–15], assess hydrologic extremes [16–18], and investigate water deficit, drought management [19–21], and surface reservoirs and lakes [22]. Combined with other satellite data (e.g., water level from satellite altimetry, vegetation from Moderate Resolution Imaging Spectroradiometer (MODIS)), GRACE has been used to understand the water dynamics of several river basins [23,24].

Given the spatial resolution challenges of GRACE, researchers have attempted to downscale the data to local or sub-regional scales using techniques such as data assimilation [25] and empirical methods [26–28]. Sun [26] used GRACE and Artificial Neural Networks (ANN) to predict monthly and seasonal water level changes for wells located in the US. This approach can be used to predict temporal changes in groundwater; however, it has limited skill in predicting groundwater change in space. Seyoum and Milewski [27] developed a robust approach using an ANN model that downscales TWSA from the GRACE scale (~200,000 km^2) to a watershed scale (watershed areas range between 5000 and 20,000 km^2) in the Northern High Plains Aquifer. They predicted spatial TWSA from GRACE and other hydro-climate variables (e.g., land surface temperature, precipitation, soil moisture, etc.); however, extracting high-resolution groundwater storage information requires further disaggregation of the downscaled product. As a result, the uncertainty of the derivate product (e.g., groundwater storage anomaly) is amplified. In addition, the ability of the ANN downscaling model to capture the spatial variability within the watershed is limited. In both the Seyoum and Milewski [27] and Miro and Famiglietti [28] methods, because the models are empirical, replicating the results to other regions may be limited. Unlike physical models, the models in both the above approaches lack capturing information about the spatial variation of groundwater, which results in lower prediction values of water storage anomaly.

The objective of this study was to test the efficacy of downscaling groundwater level anomaly from GRACE TWSA anomaly and hydro-climate and land surface variables in a glacial system, in order to further improve the approach by Seyoum and Milewski [27], as well as to minimize existing limitations of downscaling methods (e.g., method replicability). This work presents a novel downscaling approach by using multiple variables and a two-stage Machine Learning (ML) method. A two-stage Boosted Regression Tree (BRT) Machine Learning (ML) method was used to develop the downscaling model. BRT is less sensitive to large magnitude differences between input variables, is flexible in working with datasets with missing data (e.g., GRACE), and is suitable for developing multi-stage models and ensembles to create one representative model.

In order to incorporate seasonal variability in both the training and testing sets, the entire dataset, which ranges from 2002 to 2016, was split into two subsets: Training (ranges from 2002 to 2014) and testing (ranges from 2015 to 2016) subsets. This represents an improvement to the previous ANN approach by [27], where the data for training and testing were randomly selected from the entire dataset. Lastly, variable importance and comparison analysis were conducted to understand groundwater variation in the glacial system. Furthermore, a spatial randomization of the GRACE pixels was tested to check the effect of GRACE grid size on the overall accuracy. The results will improve the use of coarse resolution GRACE data in small-scale studies (e.g., groundwater trend analysis in localized aquifer systems) and facilitate integration of GRACE data into local water resources management applications. This is especially important in data-scarce regions where in-situ monitoring of the groundwater system is problematic [11]. Further, promising results have been reported involving assessment of groundwater resources and depletion using global scale hydrologic models and data assimilation [29] such as using the PCR–GLOBWB model [30] and the WaterGAP model [31], and regional scale groundwater models using MODFLOW [32] in data-scarce regions. Integration of the proposed study in such large-scale models, for example, to improve model performances is vital. In addition, it can be used to improve model performances in data assimilation techniques. GRACE TWSA have proven to be good supplements and calibration data for the models but their use has been restricted due to coarse spatial resolution [33].

2. Methods and Data

This study presents a machine learning approach to map high-resolution groundwater level anomaly (GWLA) from GRACE TWSA and other land surface and hydro-climate data (e.g., vegetation coverage, land surface temperature, precipitation, streamflow). A two-stage ML model: (1) Individual ML models for each groundwater level measurement station and (2) an ensemble model, which combines multiple ML models representing each well, was designed to predict high-resolution GWLA at a finer resolution than GRACE's current resolution. Various publicly available satellite and model-based datasets were used to set up the downscaling model and were tested in time (2002–2016) and space (at 11 different locations) using in-situ groundwater level measurement data, which were excluded from model training. Predictor variables importance, interactions, and the influence of predictor variables on the predictand were analyzed in this study. In addition, GRACE TWSA averaged over the entire study area, TWSA from each $1° \times 1°$ grid, and TWSA average over a $3° \times 3°$ moving window were used and spatial randomization of the grids were implemented to test the effect of scale of GRACE data.

Previous research has demonstrated that storage in the vadose zone contributes to the total GRACE signal in some areas [34]; however, due to the poor hydraulic characteristics of the glacial till, we assumed that the TWSA signal from GRACE is primarily dominated by groundwater. Therefore, the GRACE TWSA was directly mapped to groundwater storage anomaly. This minimizes further disaggregation of GRACE TWSA and reduces accumulation of errors associated with processing during the disaggregation of GRACE signal. Further, the downscaling approach was developed only considering the glacial aquifer system found covering the bedrocks in the study area, thus excluding the bedrock aquifers. This is because the bedrock aquifers are deep confined aquifer systems in most parts of the study area with little or no modern recharge, specifically in the central and southern parts [35]. Therefore, the deep bedrock aquifers have less influence on the total water storage change from GRACE compared to the surficial glacial sediments.

2.1. Study Area

The study area is located in the state of Illinois, a part of the Midwestern U.S., and covers an area of ~150,000 km^2 (Figure 1). The surficial geology is mostly comprised of glacial drift sediments, while the land use is primarily rainfed agricultural row crop, corn, and soybeans. The major soil type in Illinois has a silty loam texture with a porosity of around 50%. Topographically, Illinois is characterized by mainly flat land with alternating ridges and plains (end moraines and till plains) in some areas that formed as a result of the retreating Wisconsin Episode glaciers [36]. A continental climate with cold winters and warm summers characterizes the region. The long-term average annual precipitation of Illinois is ~960 mm, whereas the long-term mean, minimum, and maximum annual temperature of Illinois are 11 °C, 5 °C, and 17 °C, respectively [37].

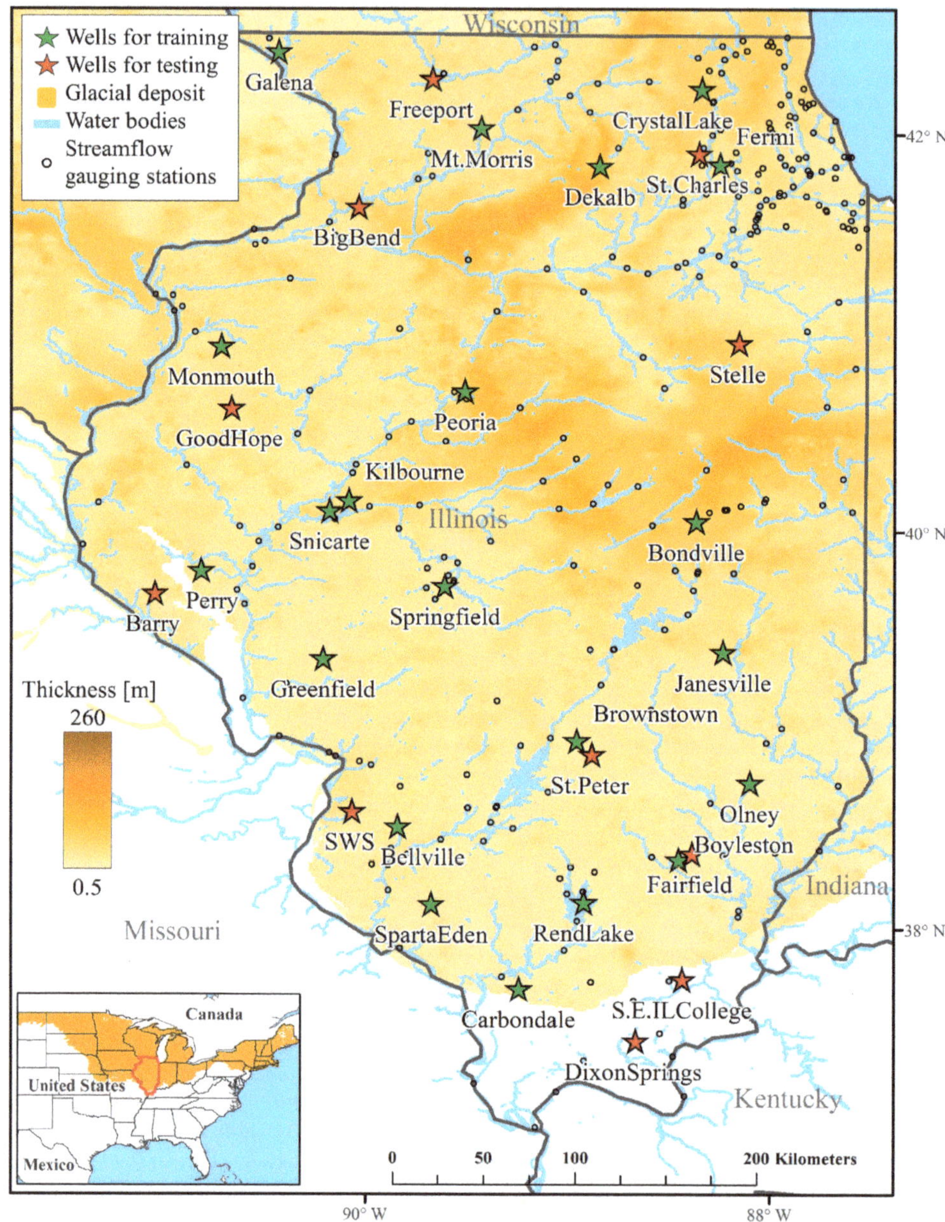

Figure 1. Location map of the study area and location of stream gauging stations and shallow groundwater wells (green: training wells, and red: testing wells) used in the study.

2.2. Data Source and Processing

Ten predictor variables (e.g., TWSA, precipitation, vegetation cover, etc.) and one predictand variable (well-based groundwater level anomaly) were used to train and test the multi-stage ML models. The variables, data sources, and processing steps are summarized in Figure 2, with details provided in the subsections below. All the data sources, with variable spatial and temporal resolutions were converted into uniform temporal (monthly) and spatial (0.25° × 0.25°) resolutions. Following the uniform resolution process, a distance weighted average scheme, based on the location of the groundwater level measurement stations, was implemented to extract timeseries input data from gridded raster datasets (Figure 3). If a given groundwater level measurement station was within the threshold distance of 0.1° from the grid cell node (Figure 3a), the value of that grid cell node was used as input data in to the model. However, if a groundwater measurement station was located outside of the threshold distance from the grid cell node, a distance weighted average scheme was used to get the average of the nearest four grid cell nodes surrounding the station (Figure 3b).

Type	Variables	Sources and processing		Unit	Type
Grid	GRACE TWSA	Averaging (GFZ, CSR, and JPL)		TWSA $[mm]$	Grid
Grid	TRMM 0.25°×0.25° (27.8 km)	From $[mm\ hour^{-1}]$ to $[mm\ month^{-1}]$	Resampling $[0.25° \times 0.25°]$	Precipitation $[mm]$	Grid
Point	Stream flow (Q)	From $[ft^{-3}sec^{-1}]$ to $[mm\ month^{-1}]$	Interpolation & Resampling $[0.25° \times 0.25°]$	Normalized Q $[mm]$	Grid
Grid	MODIS NDVI (MOD13A3)1km	Mosaic 4 scenes	N. of pixels above 0.7 in a 0.25° grid cell	Vegetation Cover [%]	Grid
Grid	MODIS LST (MOD11C3) 5.6 km	Day & Night average & Convert from [K] to [°C]	Resampling $[0.25° \times 0.25°]$	Temperature [°C]	Grid
Grid	NLDAS accumulated snow depth	Resampling $[0.25° \times 0.25°]$		Snow water equivalent $[kg\ m^{-2}]$	Grid
Grid	NLDAS soil moisture	Resampling $[0.25° \times 0.25°]$		Soil moisture $[kg\ m^{-2}]$	Grid
Grid	NLDAS plant canopy surface water	Resampling $[0.25° \times 0.25°]$		Plant canopy water $[kg\ m^{-2}]$	Grid
Grid	Glacial deposit Hydraulic conductivity	Resampling $[0.25° \times 0.25°]$		K $[ft/day]$	Grid
Grid	Glacial deposit Total thickness	Resampling $[0.25° \times 0.25°]$		Thickness $[ft]$	Grid
Point	Depth to groundwater level	From depth $[ft]$ to GWL anomaly $[mm]$		GWLA $[mm]$	Point

Figure 2. Summary of variables, data type and sources, and processing applied on the dataset used in the first and second stages of the downscaling model.

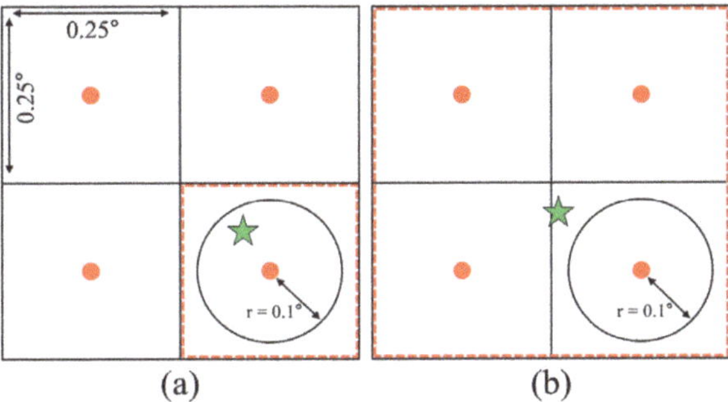

Figure 3. Distance-weighted average scheme showing how input data were extracted from grid cell nodes (symbol: red dot) based on the location of groundwater measurement station (symbol: green star) and a distance threshold value of 0.1° (**a**) when a groundwater measurement station is within the threshold distance from grid cell node and (**b**) a groundwater measurement station is located outside of the threshold distance from the grid cell node.

2.2.1. Terrestrial Water Storage Anomaly (TWSA)

The GRACE mission, launched in 2002, is designed to track changes in the Earth's gravity field using two identical satellites separated 220 km from each other at a 500 km orbital altitude [38,39]. GRACE detects change in gravity anomaly by using the high precision aboard instrument with the K-band ranging system capable of measuring the disturbance in orbital distances due to gravity variations (or mass changes) with an accuracy of 1 micron [39]. After atmospheric and oceanic effects are accounted for, the remaining signal at a monthly timescale on land is mostly related to variations of terrestrial water storage [40]. An ensemble mean of the latest release version (RL-05 gridded; $1° \times 1°$) level-3 GRACE product from three processing centers: The Center for Space Research at the University of Texas, Austin (CSR), the Jet Propulsion Laboratory (JPL), and the GeoforschungsZentrum Potsdam (GFZ) were used to ensure the highest level of accuracy [27,40]. To restore the signal attenuated during filtering and truncation of GRACE TWSA observations during processing [40], the GRACE TWSA was multiplied by the scaling factor supplied with the GRACE data. The study time span is restricted by GRACE data availability that ranges from April 2004 to July 2016. GRACE land data are available at http://grace.jpl.nasa.gov, supported by the NASA MEaSUREs Program [40,41].

2.2.2. Precipitation (P)

Precipitation data used in the model were obtained from the Tropical Rainfall Measurement Mission (TRMM) satellite. The TRMM Multisatellite Precipitation Analysis (TMPA) 3B43 product that provides monthly global (between 50 °N to 50 °S) precipitation intensity at $0.25° \times 0.25°$ (~27.8 km) spatial resolution [42] was used in the study. The TRMM 3B43 product is available on the web from NASA Giovanni (Geospatial Interactive Online Visualization and Analysis Infrastructure) service (URL: https://giovanni.gsfc.nasa.gov/giovanni/; access date: 26 September 2017).

2.2.3. Stream Flow (Q)

Stream flow data used in the model was collected from the U.S. Geological Survey (USGS) streamflow database, the National Water Information System (NWIS) [43]. The USGS dataRetrieval R package was used for retrieving daily stream flow data from the study area [44,45]. The daily flow rate (cms) for each station was converted into monthly discharges per unit area, with simplified units in mm per month, by considering the contributing drainage area of each gaging station. Further, this calculated point discharge per unit area data was linearly interpolated and resampled to obtain gridded raster discharge data at the appropriate scale. Location of gages used to get stream flow data is shown in Figure 1.

2.2.4. Vegetation Coverage (VEC)

As vegetation plays a role in the terrestrial water storage, the Moderate Resolution Imaging Spectroradiometer (MODIS) satellite MOD13A3 (MODIS/Terra Vegetation Indices Monthly L3 Global 1 km SIN Grid V006) product that provides a monthly normalized vegetation index (NDVI) at a 1 km spatial resolution was used in the study [46]. Four scenes (h10v04, h10v05, h11v04, and h11v05) of the MOD13A3 product covered the study area and were downloaded from the NASA Earth Data Search website (URL: https://search.earthdata.nasa.gov/; access date: 28 August 2017). The mosaicked NDVI images were converted into percent vegetation coverage [%] using a threshold greenness index value of 0.7 [27], percent greenness (NDVI $\geq$ 0.7) was calculated for each grid cell size of $0.25° \times 0.25°$.

2.2.5. Land Surface Temperature (LST)

Similarly, the MODIS MOD11C3 (MODIS/Terra Land Surface Temperature and Emissivity Monthly L3 Global 0.05Deg CMG) product was used to obtain monthly land surface temperatures. It has a 5.6 km spatial resolution (0.05°) [47]. The earlier product versions (V4 and V41) of this data were used over the latest version (V5), because the algorithm used in V5 shows underestimation of

LST under heavy aerosol condition [48,49]. Night and day LST layers of each monthly MOD11C3 were averaged into a single image to represent the mean temperature of a month. The MOD11C3 product is available on the NASA Earth Data Search website (URL: https://search.earthdata.nasa.gov/; access date: 31 August 2017).

2.2.6. Snow Water Equivalent (SWE), Soil Moisture (SM), and Plant Canopy Water (CW)

Data for Snow Water Equivalent, Soil Moisture, and Plant Canopy Water were obtained from the Noah land surface model (LSM) based on the North American Land Data Assimilation System (NLDAS) [50]. The NLDAS-2 monthly climatology Noah model provides information at a monthly time scale with a spatial resolution of $0.125° \times 0.125°$ for the conterminous US. The products were downloaded from the NASA Giovanni (Geospatial Interactive Online Visualization and Analysis Infrastructure) service (URL: https://giovanni.gsfc.nasa.gov/giovanni/; access date: 6 October 2017).

2.2.7. Drift Thickness (H) and Aquifer Characteristics (K)

Since aquifer heterogeneity is significant in the glacial deposit, integrating subsurface information about the aquifer material in the ML model is crucial. Gridded hydrogeologic properties for the glaciated United States were developed by Bayless, et al. [51] from water-well drillers' records. From this product, hydrogeologic information, such as hydraulic conductivity and thickness of the glacial deposits, was extracted and used in the model. The data can be obtained from the USGS's Science Base website, https://www.sciencebase.gov/catalog/item/58756c7ee4b0a829a3276352; access date: 8 August 2017). A sample gridded map showing all the input variables used in the methodology is shown in Figure 4.

2.2.8. Groundwater Level Data (GWL)

In-situ groundwater level data were used as a target variable (predictand) in the ML model. Continuous groundwater level measurements are available from 32 stations in the study area, and were collected from the Illinois State Water Survey [52]. Daily data were averaged into monthly groundwater level data. Furthermore, to mimic GRACE's long-term anomaly data, the Groundwater Level Anomaly (GWLA) was calculated by subtracting the long-term average (according to GRACE's long-term mean, 2004–2009) from individual GWL observations. The Shallow Groundwater Wells Network data is available on the Illinois State Water Survey website (URL: https://www.isws.illinois.edu/warm/groundwater/: access date: 21 June 2017).

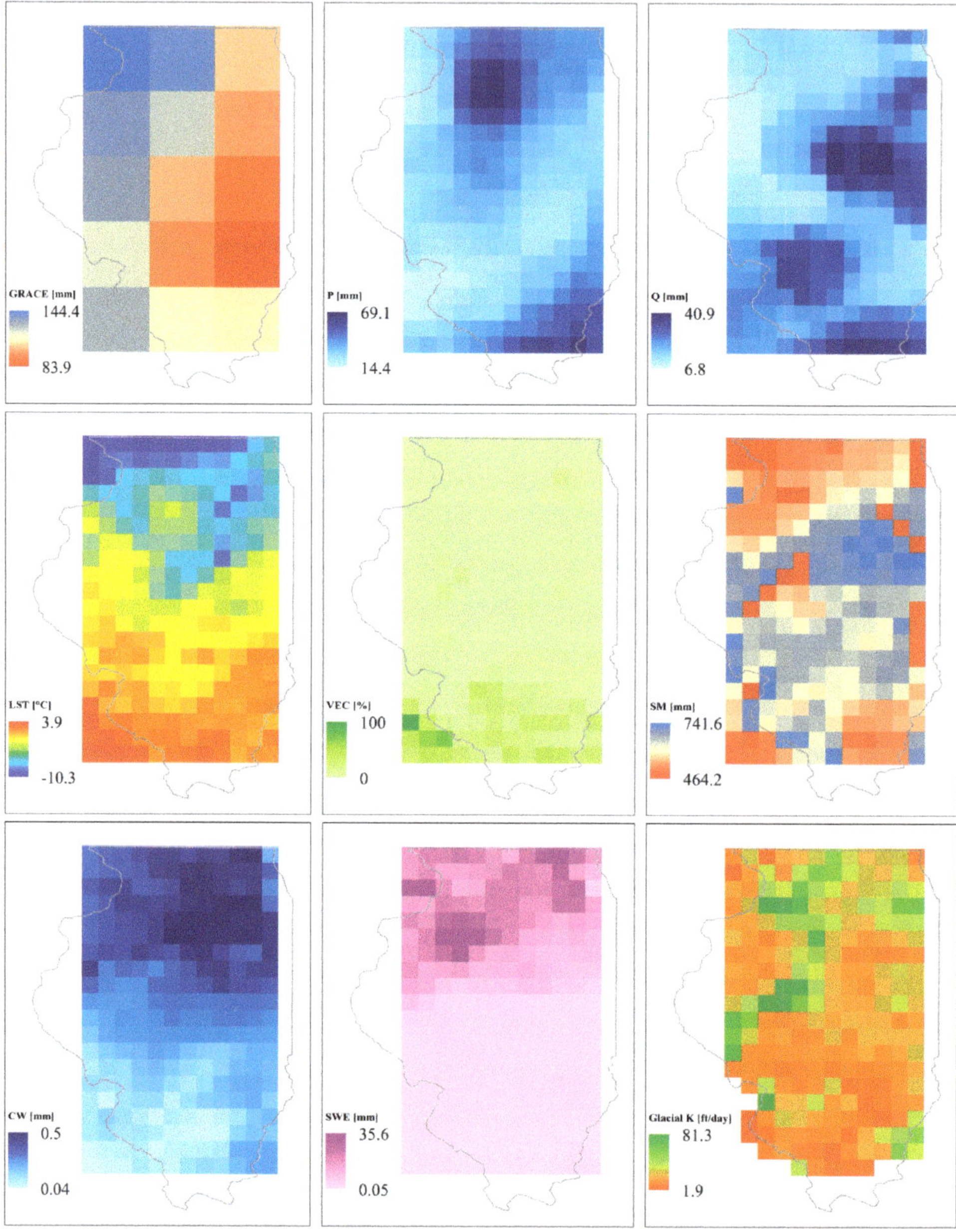

Figure 4. Maps showing sample input variables used in the methodology for the month of January 2016. All the data are resampled to 0.25° × 0.25° grid resolution, except Gravity Recovery and Climate Experiment (GRACE) and Terrestrial Water Storage Anomaly (TWSA) data, which are a 1° × 1° grid.

2.3. Modeling Algorithms

Downscaling GRACE to predict a high-resolution groundwater level anomaly and determination of variable importance was accomplished using a Boosted Regression Tree (BRT) also known as a gradient boosting technique. BRT, which is a combination of statistical and machine learning techniques, aims to improve the performance and prediction accuracy of a single tree-based model produced by fitting and combining a large number of tree-based models [53,54]. First, a sequence of several simple trees (weak models) is fitted to the data, and an ensemble (boosting) technique is applied to produce a robust model from several models [53]. A tree-based method is conceptually simple yet powerful [55] and has advantages over other predictive learning techniques in that the tree-based method is easy to interpret, handles missing values, is not affected by outliers and/or does not need prior data transformation, and it has the ability to deal with irrelevant inputs. For further details, refer to Friedman, Hastie and Tibshirani [55] and Elith, Leathwick and Hastie [53]. In addition, BRT works best with continuous and categorical data. The simplified theory of boosted regression method is briefly explained below.

Figure 5 shows how a tree-based regression divides a covariant space by partitioning the predictor variables [56] and creates a simple tree-based model. A dataset space indicates a relationship between predictors (X_1 and X_2) and predictand (Y) so that the predictand value can be approximated by many rectangular subspaces (Figure 5a) according to values of X_1 and X_2 and the splitting that forms the tree (Figure 5b). The method builds binary trees that partition the data into two samples at each split node. The variable for splitting and its split value are selected based on the conditions that best fit the data. In the example in Figure 5b, at first, the covariant space can be divided at $X_1 = 8$. This is based on the splitting ($X_1 \leq 8$ and $X_1 > 8$) that can represent the covariant space most effectively in terms of residual error. Consequently, the second split points will be selected in both the $X_1 \leq 8$ and $X_1 > 8$ regions. The number of split nodes determines the tree size; the increase in tree size is controlled either by a predetermined tree size or by residual error.

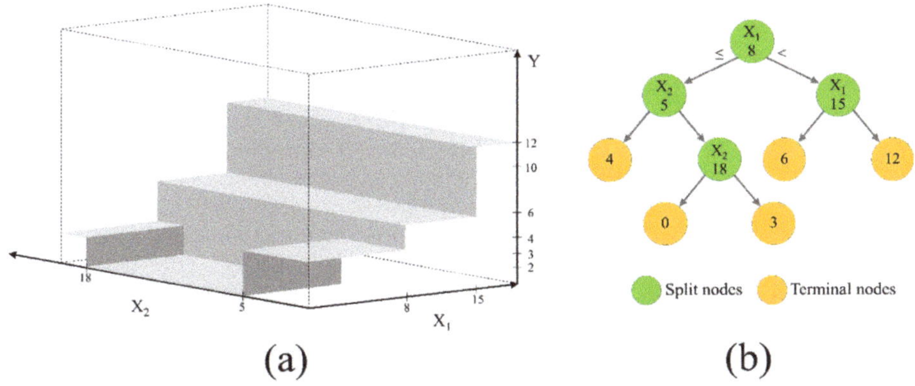

Figure 5. A tree-based regression showing (**a**) a rectangular covariant space created based on predictors X_1 and X_2 and the predictand (target) Y variable, and (**b**) split nodes of the decision tree partitioned based on predictor variables and residual error.

During boosting, a series of several simple trees is built based on the results of the prediction residuals of the preceding tree. Figure 6 shows a schematic of successive regression trees based on residuals. As shown in Figure 6, Decision Tree (2) is built based on a tree (partition) fitted to the residuals from Decision Tree (1), while Decision Tree (3) is a tree (partition) fitted to the residuals from Decision Tree (2). This helps to find a partition (Decision Tree (M)) that will further reduce the residual (error) variance for the data (as shown in the bottom graph of Figure 6). As a result, boosting decision trees improves their accuracy, often dramatically [55]. In boosting, decision trees are fitted iteratively to the training data and the algorithms vary in how they quantify lack of fit and select settings for the next iteration [53]. In this study, LS Boost (least-squares regression) algorithm was used for BRT modeling (see Friedman [57]). Optimization of hyperparameters (e.g., number of learners, maximum number of splits, and learning rate) (see Elith, Leathwick and Hastie [53] for details) and overall model design and simulation was conducted using MATLAB® 2017b. The model was constructed and tested using 10-fold cross-validation to determine the optimal hyperparameters, number of learners, maximum number of splits, and learning rate.

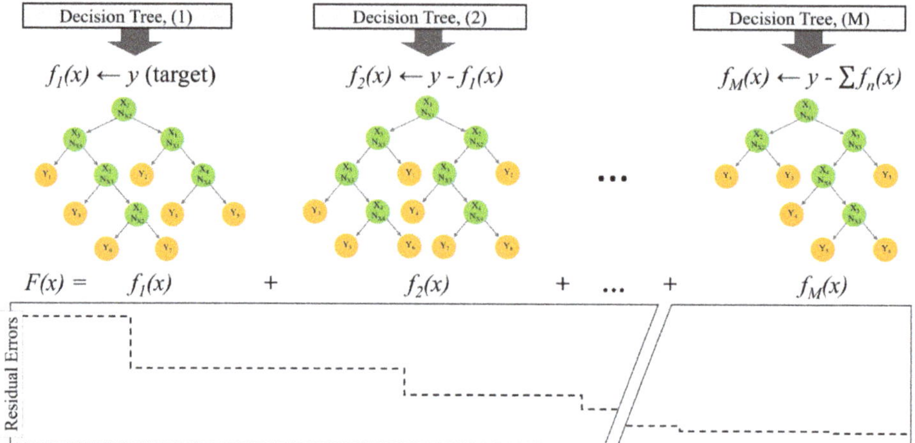

Figure 6. A schematic diagram showing the boosting processes in the Boosted Regression Tree (BRT) method.

2.4. Downscaling Model Design

A two-stage ML approach was used in this study to generate a downscaled GWLA grid from GRACE TWSA. The first stage involved building multiple temporal GWLA prediction ML models (M_1, M_2, ..., M_{21}) for individual groundwater wells in the study area (top left section of Figure 7). The second stage comprised of integrating these multiple models (from the previous stage), along with additional spatial information about the glacial deposit (e.g., hydraulic conductivity), into a single spatial downscaling model. In the second stage, a single spatial GWLA prediction model (M') was produced for the entire study area (bottom left of Figure 7) and used to generate a downscaled GWLA at 0.25° x 0.25° resolution (the right section of Figure 7). As shown in the figure, the GWLA at a given month can be predicted (1) using GWLA predicted based on individual temporal models of training wells and water-budget variables (X) and (2) using two dimensional GWLA pattern generation based on intermediate variables (Y') from the first models and spatial characteristics (Z) for each grid cell.

The GWLA prediction models in the first stage, in addition to GRACE TWSA, use terrestrial water cycle variables (e.g., vegetation coverage, LST, P, Q, SWE, plant canopy water, and soil moisture) (the specific variables used in each stage are shown in Figure 2) as predictors, and in-situ-based GWLA as a predictand. While in the second stage, predictor variables are restricted to GRACE TWSA and aquifer characteristics (e.g., glacial thickness, hydraulic conductivity). Consequently, the spatial pattern of GWLA can be explained by glacial deposit characteristics in the second stage model.

The model was tested using data independent of training (calibration). Eleven groundwater well stations, excluded from the training sets, were used to test the performance of the spatial downscaling model (M'). The strength of the relationship between individual predictor variables and predictand was analyzed using variable importance (VI). The measure of variable importance is based on how many times the variable is selected for splitting and how much the model is improved as a result of the splitting (see Friedman and Meulman [58] for details). Relative VI is calculated for each input variable and converted to a 0 to 100 scale. In addition, the sensitivity of each predictor variable was assessed using partial dependence plots that show the dependence between the predictand and a given predictor variable, while averaging the influence of all other predictors [58].

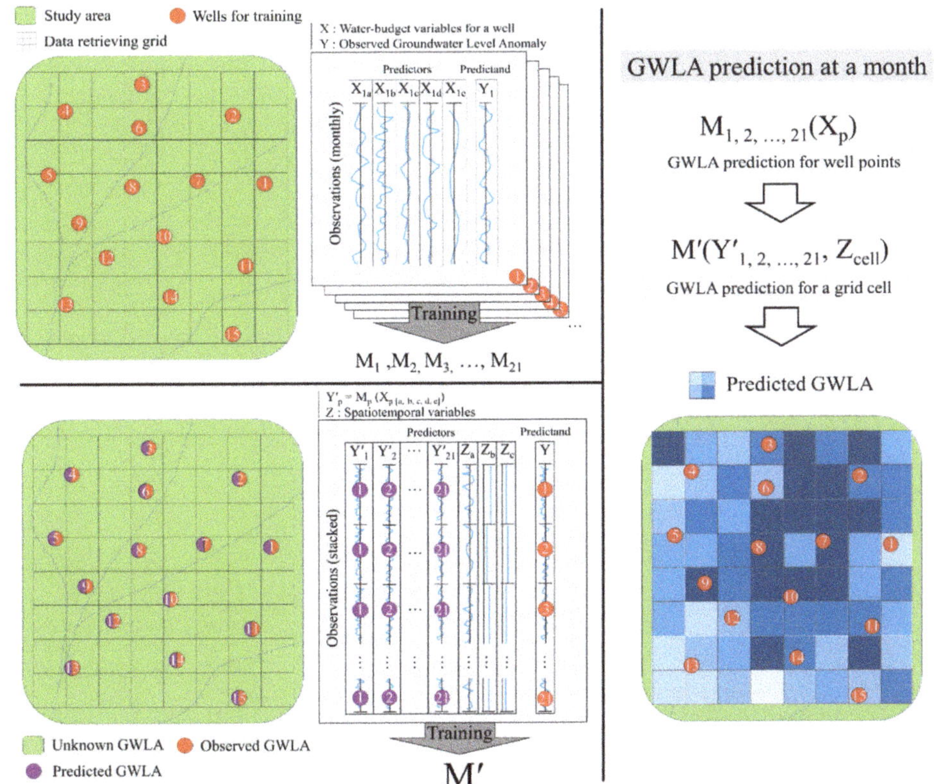

Figure 7. The conceptual design of the downscaling model. The process begins at the top left side of the figure (represents models in the first stage), followed by the bottom left (the spatial downscaling ensemble model in the second stage), and the right side explains how the prediction of the downscaled and gridded GWLA product is produced.

During training and testing error metrics such as Mean (residual) Error (ME), Mean Absolute Error (MAE), Root-Mean-Square Error (RMSE), correlation coefficient (R), and Nash-Sutcliffe Efficiency (NSE) were used to evaluate the performance of the models by comparing model-simulated results with the observed GWLA.

3. Results

The results are presented according to the approaches applied in the downscaling procedure (2.4). Various GRACE grid sizes and spatial randomization of the GRACE grids were tested to standardize input from GRACE data and to test its effect on model accuracy. The results showed there is no clear improvement in the downscaling model. This could be due to the data where adjacent GRACE time series grids are highly correlated. As a result, the grid size and location have no impact in the model accuracy of the downscaling model. Lastly, the limitations of the method were discussed with respect to the actual condition in the study area; implications for future studies are also suggested.

3.1. Model Simulation Results

From the total of 32 groundwater level measurement stations in the study area, individual ML models were constructed for 21 of them, which were selected based on their spatial distribution. As the sample size, which is limited by GRACE sample size, was small to train the models, a ten-fold cross validation method was used. This method divided the entire data into ten subsets. Each subset was used nine times as a training set and one time as a test set. A model was fitted to the training set and evaluated based on the test set. As a result, the error estimation for model training was provided averaged over all trials to get the total effectiveness of the model. Table 1 shows the overall statistics for model training of individual models (in the first stage ML models) and the ensemble downscaling model in the second stage. Generally, the results showed that the models exhibited excellent performances. The average ME, RMSE, R, and NSE values are 4.5 mm, 306 mm, 0.91, and 0.82, respectively. Similarly, the ensemble downscaling model shows an excellent performance with ME, RMSE, R, and NSE values of 0.0 mm, 192.0 mm, 0.98, and 0.96, respectively. Wells with good performance metrics (e.g., M7 and M10) imply the predictor variables explained well the observed GWLAs. On the other hand, wells with low performance metrics (e.g., M6, M12, M15, and M19) indicate that the predictor variables were short of fully explaining the observed GWLA. Note that the units are in mm for GWLA.

Table 1. The overall statistics showing model performance by comparing model-predicted GWLA with the in-situ-derived GWLA data (data ranges from 2002 to 2016) for the training (calibration), M1-M21 include models in the first stage, and M' is the ensemble downscaling model.

Models	Stations Name	ME (mm)	MAE (mm)	RMSE (mm)	R	NSE
M1	Bellville	0.27	317.71	450.12	0.88	0.76
M2	Bondville	−0.20	146.29	216.70	0.96	0.91
M3	Boyleston	−0.21	279.70	405.13	0.81	0.64
M4	Brownstown	13.39	131.48	163.73	0.98	0.92
M5	Carbondale	0.00	209.15	303.28	0.95	0.90
M6	CrystalLake	0.00	206.99	322.47	0.74	0.54
M7	Dekalb	0.00	0.74	2.78	1.00	1.00
M8	Fermi	0.22	277.96	420.64	0.85	0.71
M9	Galena	2.13	115.41	212.85	0.93	0.86
M10	Greenfield	0.00	0.39	1.07	1.00	1.00
M11	Janesville	0.01	96.84	150.91	0.91	0.82
M12	Kilbourne	0.00	410.54	557.37	0.83	0.65
M13	Monmouth	0.00	118.51	149.77	0.98	0.96
M14	Mt.Morris	19.31	671.85	975.44	0.87	0.72
M15	Olney	20.11	274.65	553.39	0.75	0.54
M16	Peoria	0.05	1.69	3.20	1.00	1.00
M17	Perry	0.16	130.43	161.14	1.00	0.99
M18	RendLake	9.15	125.30	204.90	0.94	0.86
M19	Snicarte	0.18	241.00	346.57	0.83	0.66
M20	SpartaEden	35.18	194.03	295.29	0.95	0.85
M21	Springfield	−0.24	368.14	643.00	0.85	0.71
M'	Downscaling	0.00	104.52	192.06	0.98	0.96
Average for individual models		4.52	201.06	305.99	0.91	0.82

Figure 8 shows selected time series model-simulated GWLA vs in-situ-based GWLA. The models predicted the variation in GWLA very well for most of the groundwater measurement locations in the study area (e.g., Figure 8a). However, a few wells showed a slight underprediction (or overprediction) of the GWLA; specifically, underprediction in the picks of positive anomalies and overprediction in the picks of negative anomalies (Figure 8b). Other exceptions include GWLA models for Galena (Figure 8c) and Kilbourne (Figure 8d) groundwater wells. Where the model for Galena station overpredicted the peak GWLA, whereas the model for Kilbourne station fairly captured the long-term GWLA variability; however, the model produced time-varying seasonal signal compared to the observed GWLA data. This is attributed to the fact that the Kilbourne well is located in an irrigated agricultural region of Illinois. Due to the lack of data, human impacts (e.g., pumping) were not considered as an input in this study. As shown in the figure, the hydrograph from the in-situ groundwater level data exhibited typical characteristics of pumping water level where little or no seasonal fluctuations were observed.

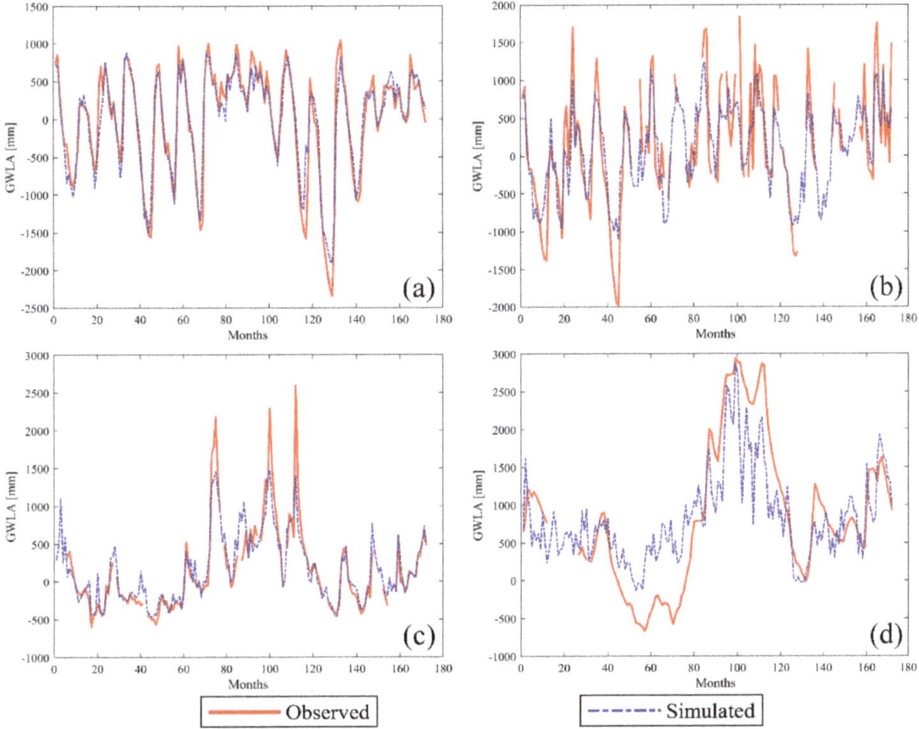

Figure 8. Time series graphs showing model-simulated and in-situ-derived GWLA for selected groundwater measurements stations: (**a**) Bondville, (**b**) Fermi, (**c**) Galena, and (**d**) Kilbourne. The blue line shows model-simulated while red is observed GWLA.

3.2. Variable Contributions and Sensitivity

Several studies indicated the GRACE signal is dominated by sub-surface storage (e.g., groundwater storage) [59,60]. The variable importance corroborated that TWSA from GRACE is an influential predictor variable of the predicted GWLA. Discharge is the second most important predictor variable. All the models (in both the first and second stages) indicated TWSA as a primary predictor variable (Figure 9a,b) for the downscaled GWLA. Figure 9a shows the percentage contribution of each predictor variable across all models in the first stage where individual models have been built for each groundwater level measurement station. Figure 9b shows the percentage contribution of predictors in the second stage of the downscaling approach. In the first stage, stream discharge and soil moisture storage are the second and third most influential predictor variables, respectively. This can be explained by the glacial aquifer system in Illinois, which is mainly characterized by a shallow groundwater level, which influences baseflow contributions to the streams. The remaining predictor variables (e.g., precipitation (P), land surface temperature (LST), snow water equivalent (SWE)) are less and equally influential predictor variables. The relationship of these variables, though they directly or indirectly affect the amount of recharge to the groundwater, may not be simple and direct. In the second stage (downscaling model), the GRACE TWSA is relatively the most influential predictor variable, followed by a few predicted GWLA from the models in the first stage (Figure 9b).

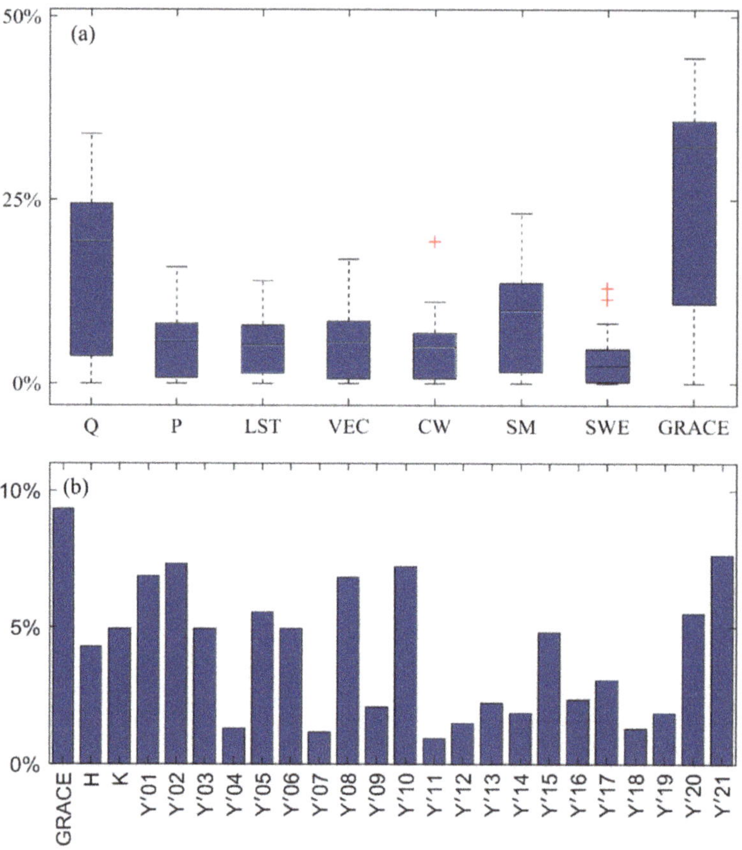

Figure 9. Percentage contribution of each predictor variable. (**a**) Boxplots for each predictor across all the models in the first stage and (**b**) bars for each predictor in the second stage, downscaling model.

Figure 10 displays the Partial Dependence (PD) plots for selected models, which show the effect of a given variable on the response (predictand) after accounting for the average effect of all other predictor variables. Generally, there is a non-linear relationship between the predictor variables and the predictand (GWLA). The partial responses show a direct relationship between GWLA and GRACE TWSA, stream discharge, precipitation, and soil moisture. However, the magnitude effect of these predictor variables on the GWLA is different (see the y-axis of the graphs in Figure 10). The GRACE TWSA and soil moisture control a wider range of the GWLA. The GWLA has an inverse relationship with land surface temperature and plant canopy water, as these predictor variables increase as the GWLA decreases. LST and CW favors evapotranspiration while reducing recharge. No clear relationship exists with vegetation coverage (greenness index) or snow water equivalent. The relationship between vegetation coverage and GWLA is not direct (Figure 10f), where GWLA responds to little to no vegetation and high plant periods.

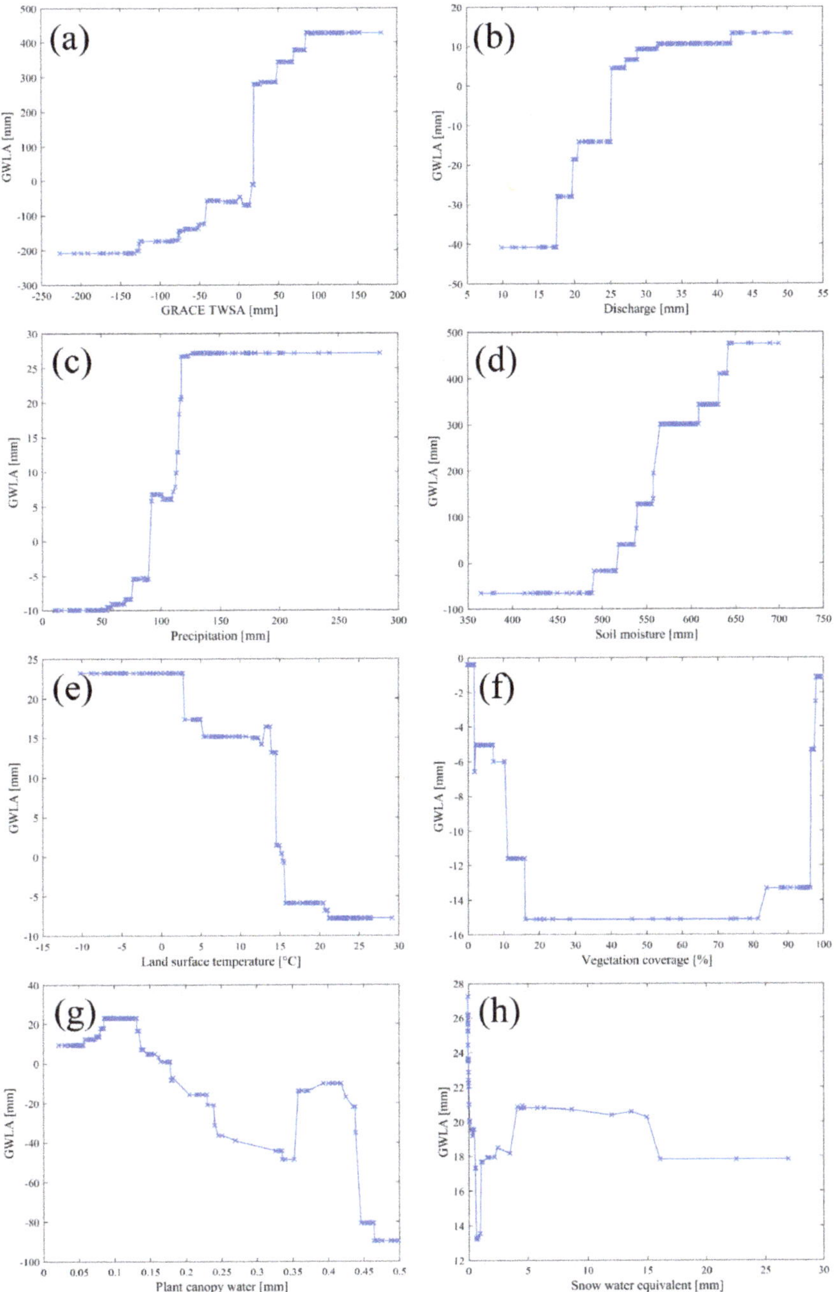

Figure 10. Partial dependence plots for predictor variables in selected GWLA prediction models (**a–h**). The cross marks indicate distribution of the data points.

3.3. Testing and Spatial Accuracy

This study presents a novel approach that allows simulating higher resolution GWLA at different times and locations in space. The method was verified by comparing groundwater level data that was independent of model construction. Figure 11 displays the distribution of test wells (red stars) and their statistical values (labeled next to the stations, the circles are also drawn to scale). Generally, the testing results showed that the statistical values are within the acceptable ranges. Half of the test wells have NSE values from 0.5 to 0.6, the remaining have values between 0.23 and 0.48. One well, Big Bend station, has shown a very low NSE value of -1.37; however, it has a high correlation coefficient value of 0.75. This well is located within the floodplain of the Rock River where the groundwater interacts with the river via baseflow. The direction of seasonal fluctuation (indicated by high correlation coefficient) can be simulated well by the downscaling model mainly influenced by stream discharge. However, the model falls short of predicting the magnitude of the GWLA compared to the observed GWLA data, which is different from the magnitude of discharge. Generally, the correlation coefficients for testing are within the range of 0.66 to 0.82. The test results show that the downscaling approach presented here satisfactorily simulates the GWLA data at $0.25° \times 0.25°$ spatial and monthly temporal scale from GRACE. The output scale is defined based on the scale of input data. TRMM has the lowest resolution of the input data at $0.25° \times 0.25°$. Though not tested in this study, it is possible to downscale further to a higher resolution using high-resolution input data.

The background image in Figure 10 shows a sample-predicted GWLA (for the month of December 2007) for the study site at $0.25° \times 0.25°$ resolution. We can see that there is significant GWLA spatial variation during that month. A negative groundwater level anomaly of more than 500 mm (areas in yellow and red) was observed in the southern and northwestern parts of the study site, whereas the north and northeastern parts gained a positive groundwater level anomaly up to 500 mm (areas in blue). Overall, the similar degree of heterogeneity in groundwater behavior is observed in other months from model output compared to that of GRACE data.

Figure 12 shows the timeseries model-predicted versus in-situ-derived GWLA for each test well in the study area. Differences in model performances are observed between different test wells; good predictive performances are observed in, e.g., test wells at Fairfield and Freeport stations (Figure 12c,d) and poor performance in the test well at Big Bend station (Figure 12b). As seen in Figure 12, Big Bend station data availability is low, resulting in poor NSE and R values for this station due to sample size. In some of the test wells, the model underestimates the GWLA (e.g., Figure 12a,d,e). The reverse, an overestimate of the GWLA, is observed in the St. Peter test well data (Figure 12g). Generally, the model predicted the timing and seasonal variability (direction) of the GWLA. However, in some instances (as described above), the model less accurately predicted the magnitude of the high and low groundwater anomalies. This is due to the skill of the models mainly controlled by input data, specifically, data for the predictor variables. The authors believe the variances in the predictor variables may not be capable of fully explaining the magnitudes of variability in the predictand variable (GWLA). Despite some limitations, as indicated by the statistical measures above (Figure 11), the model predicts the monthly variations in the groundwater level moderately well and better predicts the long-term (seasonal to interannual) variability in the groundwater level. For example, in stations such as Barry, Big Bend, Freeport, Good Hope, and SWS (Figure 12a,b,d,e,i), the model predicted the dry anomalies (e.g., periods from 40 to 60 and 120 to 140) and wet anomalies (e.g., periods from 80 to 120).

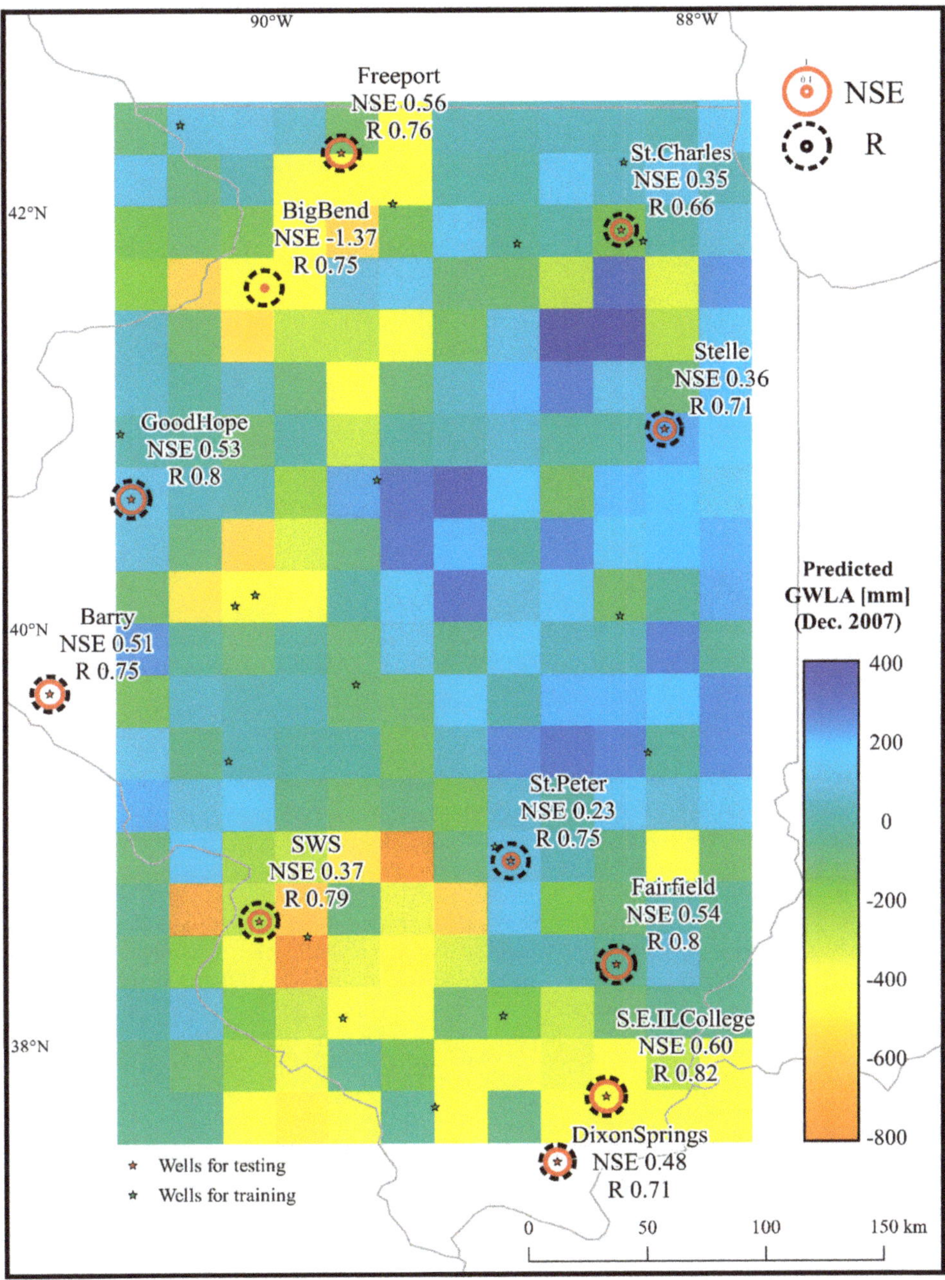

Figure 11. Map showing predicted GWLA at 0.25° × 0.25° scale, overlain circles (drawn to scale) with performance metrics (NSE: Nash-Sutcliffe Efficiency and R: Correlation coefficients) are the locations of in-situ groundwater level measurement stations. Performance metrics indicate the comparison between model-predicted and in-situ GWLA.

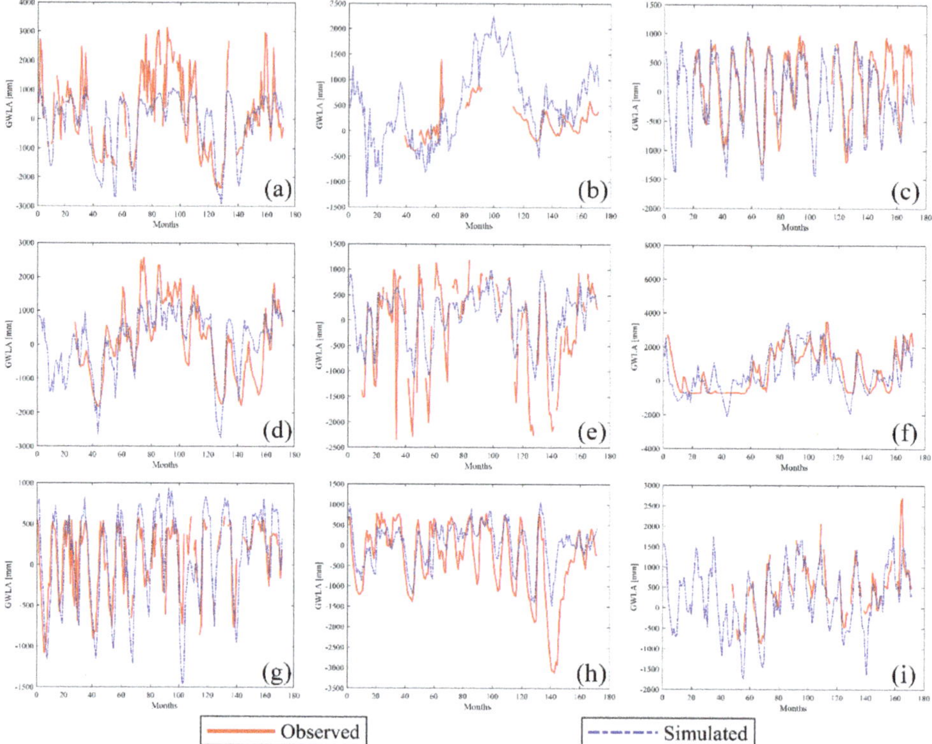

Figure 12. Monthly timeseries model-predicted (blue line) vs. in-situ-derived GWLA (red) for test wells in the study area: (**a**) Barry, (**b**) Big Bend, (**c**) Fairfield, (**d**) Freeport, (**e**) Good Hope, (**f**) St. Charles (**g**) St. Peter, (**h**) Stelle, and (**i**) SWS stations.

4. Discussion

Overall, the approach developed in this study demonstrates the integration of satellite- and model-based hydrological variables along with GRACE data to predict high-resolution groundwater level variation in a glacial aquifer system. The results clearly demonstrate that downscaled GWLA could be predicted at a resolution of $0.25° \times 0.25°$ in new periods and/or space in the study area, a limiting factor in previous studies. Additional tests have been conducted to increase replicability of this study, such as testing the spatial scale of input variables (e.g., GRACE grid). Using the methodology developed in this study (a generic code is available from the authors upon request), end-users can implement it elsewhere to predict GWLA, given all the assumptions are valid. The two main assumptions are (1) GRACE TWSA is mainly controlled by the groundwater storage variation in the shallow aquifer system and (2) human influence (e.g., groundwater abstraction for irrigation from the aquifer) is not considered as an input variable in the models.

The purpose of using publicly available data (e.g., satellite data) is to establish the applicability of this study in data-scarce regions where hydrologic variables are limited, such as lack of groundwater recharge rate and poorly distributed well data in space and time. However, significant uncertainty is expected from these data sources. For example, GRACE data has its own uncertainty that arises from GRACE processing. Other input data, some model-based and satellite-based, have their own contribution to the total input data uncertainty and potential error propagation. These errors are mitigated by using ML methods, as they do a good job of handling errors associated with input data.

Only the shallow groundwater system of the glacial sediment was considered in this study by assuming the GRACE signal is strongly related to shallow systems. As a result, shallow bedrock aquifers overlain by no or thin layers of glacial sediment, specifically in the northern part of the study area, were not included. The authors believe that the poor performance of the ML model in this part of the study region (e.g., at Crystal Lake and Fermi stations (see Table 1 and Figure 1)) could be due to exclusion of the shallow bedrock. Further, exclusion of human impact as a predictor variable has also affected the performance of wells located in the irrigated regions of Illinois (e.g., Kilbourne and Sincarte wells) where the main source of irrigation water is groundwater. As it has been seen in the hydrograph of Kilbourne stations (Figure 8d), the in-situ groundwater data shows more of a flat declining or rising groundwater level anomaly; however, the model that simulated the natural seasonal fluctuations was probably influenced by seasonal variability of precipitation and other predictor variables. Further, the spatial pattern of groundwater level anomalies may not be fully captured due to heterogeneity of the glacial system. The predictor variables used in this study may not be used to sufficiently describe this heterogeneity in order to predict the GWLA accurately in these instances. Adding more variables (e.g., storage characteristics of the aquifer, including the bedrock), as well as improved resolution of the input variables, potentially minimizes the bias and improves the prediction capacity of the ML model.

The methodology developed in this study can be directly applied in shallow sedimentary aquifer systems with similar settings where (1) GWLA response to climate variables is relatively quick and (2) anthropogenic influences (e.g., groundwater abstraction) are not extensive. For example, the downscaling model can be applied in the Chad Basin, the Congo Basin, and the Rift Valley Basin in Africa. The central part of the Chad Basin has a Quaternary unconfined sediment aquifer system made up of fluvio-lacustrine deposits and Aeolian sands and covers an area of 500,000 km^2 [32]. This basin is one of the most poorly gauged basin in the world but provides fresh water resources for the people of several nations in the region. Similarly, it can also applied in the fluvio-lactustrine sediment aquifers covering the Rift Valley in East Africa and the Cenozoic sediment aquifers in the Congo Basin, among others.

To apply the methodology in a different setting, it is possible to omit or include predictor variables. For example, snow water equivalent and vegetation coverage are insignificant in arid regions. As a result, researchers can omit these predictor variables from the downscaling model. Likewise, if additional variables are necessary to explain the predictand variable, they can be easily included as predictor variables. For example, if the geological setting is a fractured (Karstic) bedrock aquifer system, it is possible to include additional predictor variables, such as fracture density and fracture spacing, to explain the predictand variable, the GWLA. In addition, as BRT works well with both categorical and numerical variables, it is possible to include categorical data (e.g., yes or no type data or qualitative data) by assigning numeric values such as 0 (no) and 1 (yes).

5. Conclusions

With continued threat from climate change and human impacts, high-resolution and continuous hydrologic data availability is crucial for predicting trends and water resource availability. GRACE has provided unprecedented opportunities to assess regional and global water resources, despite the limitations due to its coarse spatial scale. This study presented a two-stage machine learning approach to map high-resolution groundwater level anomaly (GWLA) from GRACE TWSA and other publicly available land surface and hydro-climate data (e.g., vegetation coverage, land surface temperature, precipitation, streamflow, etc.) in a glacial aquifer system in Illinois. In both stages, the models were validated and tested. The study finds GRACE to be the most influential predictor variable in the ML-based models, and the scale of GRACE or spatial location of GRACE grids have little to no impact on the performance of the ML-based downscaling model. This is due to the characteristics of GRACE grids, where they have a coarse resolution and where adjacent GRACE grids are spatially highly correlated. Stream flow and soil moisture storage are the second and third most influential predictor variables in the model.

Generally, the model training and testing results, in the first stage, showed excellent performances. The average ME, RMSE, R, and NSE values are ~4.5 mm, 306 mm, 0.91, and 0.82, respectively. Similarly, the ensemble downscaling model (in the second stage) shows an excellent performance with ME, RMSE, R, and NSE values of 0.0 mm, 192.0 mm, 0.98, and 0.96, respectively. Further, the downscaling model was tested satisfactorily using the monthly in-situ-based GWLA data excluded during model construction. A few test wells exhibit relatively poor statistical performances, which is attributed to the lack of anthropogenic influences (e.g., pumping) in groundwater-irrigated regions and insufficient representation of aquifer heterogeneity in the predictor variables. Various sources of uncertainties in the model and downscaled product are expected, such as uncertainty from the input satellite and model-based data and uncertainty from the model assumption (e.g., exclusion of the shallow bedrock aquifer system located in the northern part of the study area). However, noises due to input data were often reduced (or minimized) during the machine learning process.

Author Contributions: W.M.S. was responsible for conceptualization and design, overseeing the project, analysis and interpretation of the results, and manuscript preparation. D.K. involved in data processing, analysis, and interpretation of the result, and manuscript preparation. A.M.M. aided conceptualization of the project, interpretation of the results, and reviewing the manuscript.

Funding: This research was funded by the College of Arts and Sciences, Illinois State University. The APC was funded by the Powell fund, Department of Geography, Geology, and the Environment and the Office of Research and Graduate Studies, Illinois State University.

Conflicts of Interest: The authors declare no conflict of interest.

References

1. Bouman, B.A.M.; Tuong, T.P. Field water management to save water and increase its productivity in irrigated lowland rice. *Agric. Water Manag.* **2001**, *49*, 11–30.
2. Barua, S.; Muttil, N.; Ng, A.W.M.; Perera, B.J.C. Rainfall trend and its implications for water resource management within the yarra river catchment, australia. *Hydrol. Process.* **2012**, *27*, 1727–1738. [CrossRef]
3. Chen, J.L.; Wilson, C.R.; Tapley, B.D. The 2009 exceptional amazon flood and interannual terrestrial water storage change observed by grace. *Water Resour. Res.* **2010**, *46*. [CrossRef]
4. Brutsaert, W. Long-term groundwater storage trends estimated from streamflow records: Climatic perspective. *Water Resour. Res.* **2008**, *44*. [CrossRef]
5. Bjerklie, D.M.; Lawrence Dingman, S.; Vorosmarty, C.J.; Bolster, C.H.; Congalton, R.G. Evaluating the potential for measuring river discharge from space. *J. Hydrol.* **2003**, *278*, 17–38.
6. Milewski, A.; Sultan, M.; Jayaprakash, S.M.; Balekai, R.; Becker, R. Resdem, a tool for integrating temporal remote sensing data for use in hydrogeologic investigations. *Comput. Geosci.* **2009**, *35*, 2001–2010. [CrossRef]
7. Seyoum, W.M.; Milewski, A.M.; Durham, M.C. Understanding the relative impacts of natural processes and human activities on the hydrology of the central rift valley lakes, east africa. *Hydrol. Process.* **2015**, *29*, 4312–4324. [CrossRef]
8. Nijssen, B.; O'Donnell, G.M.; Lettenmaier, D.P.; Lohmann, D.; Wood, E.F. Predicting the discharge of global rivers. *J. Clim.* **2001**, *14*, 3307–3323. [CrossRef]
9. Wells, B.; Toniolo, H. Hydrologic modeling of three sub-basins in the Kenai river watershed, Alaska, USA. *Water* **2018**, *10*, 691.
10. Her, Y.; Jeong, J. SWAT+ versus SWAT2012: Comparison of sub-daily urban runoff simulations. *Am. Soc. Agric. Biol. Eng.* **2018**, *61*, 1287–1295. [CrossRef]
11. MacDonald, A.M.; Bonsor, H.C.; Dochartaigh, B.É.Ó.; Taylor, R.G. Quantitative maps of groundwater resources in Africa. *Environ. Res. Lett.* **2012**, *7*, 024009. [CrossRef]
12. Shamsudduha, M.; Taylor, R.G.; Jones, D.; Longuevergne, L.; Owor, M.; Tindimugaya, C. Recent changes in terrestrial water storage in the upper nile basin: An evaluation of commonly used gridded GRACE products. *Hydrol. Earth Syst. Sci.* **2017**, *21*, 4533–4549. [CrossRef]
13. Voss, K.A.; Famiglietti, J.S.; Lo, M.; Linage, C.; Rodell, M.; Swenson, S.C. Groundwater depletion in the middle east from GRACE with implications for transboundary water management in the Tigris-Euphrates-Western Iran region. *Water Resour. Res.* **2013**, *49*, 904–914. [CrossRef] [PubMed]

14. Seyoum, W.M.; Milewski, A.M. Monitoring and comparison of terrestrial water storage changes in the northern high plains using grace and in-situ based integrated hydrologic model estimates. *Adv. Water Resour.* **2016**, *94*, 31–44. [CrossRef]

15. Lezzaik, K.; Milewski, A. A quantitative assessment of groundwater resources in the middle east and north Africa region. *Hydrogeol. J.* **2018**, *26*, 251–266. [CrossRef]

16. Sun, Z.; Zhu, X.; Pan, Y.; Zhang, J. Assessing terrestrial water storage and flood potential using GRACE data in the Yangtze river basin, China. *Remote Sens.* **2017**, *9*, 1011. [CrossRef]

17. Sun, Y.A.; Scanlon, R.B.; AghaKouchak, A.; Zhang, Z. Using GRACE satellite gravimetry for assessing large-scale hydrologic extremes. *Remote Sens.* **2017**, *9*, 1287. [CrossRef]

18. Seyoum, W.M. Characterizing water storage trends and regional climate influence using grace observation and satellite altimetry data in the Upper Blue Nile River Basin. *J. Hydrol.* **2018**, *566*, 274–284. [CrossRef]

19. Ma, S.; Wu, Q.; Wang, J.; Zhang, S. Temporal evolution of regional drought detected from GRACE TWSA and CCI SM in Yunnan Province, China. *Remote Sens.* **2017**, *9*, 1124. [CrossRef]

20. Thomas, A.C.; Reager, J.T.; Famiglietti, J.S.; Rodell, M. A GRACE-based water storage deficit approach for hydrological drought characterization. *Geophys. Res. Lett.* **2014**, *41*, 1537–1545. [CrossRef]

21. Lezzaik, K.; Milewski, A.; Mullen, J. The groundwater risk index: Development and application in the middle east and north Africa region. *Sci. Total Environ.* **2018**, *628–629*, 1149–1164. [CrossRef]

22. Hassan, A.A.; Jin, S. Lake level change and total water discharge in East Africa rift valley from satellite-based observations. *Glob. Planet. Chang.* **2014**, *117*, 79–90. [CrossRef]

23. Lee, H.; Beighley, R.E.; Alsdorf, D.; Jung, H.C.; Shum, C.K.; Duan, J.; Guo, J.; Yamazaki, D.; Andreadis, K. Characterization of terrestrial water dynamics in the Congo basin using GRACE and satellite radar altimetry. *Remote Sens. Environ.* **2011**, *115*, 3530–3538. [CrossRef]

24. Papa, F.; Frappart, F.; Güntner, A.; Prigent, C.; Aires, F.; Getirana, A.C.V.; Maurer, R. Surface freshwater storage and variability in the Amazon basin from multi-satellite observations, 1993–2007. *J. Geophys. Res. Atmos.* **2013**, *118*, 11951–11965. [CrossRef]

25. Zaitchik, B.F.; Rodell, M.; Reichle, R.H. Assimilation of GRACE terrestrial water storage data into a land surface model: Results for the Mississippi River Basin. *J. Hydrometeorol.* **2008**, *9*, 535–548. [CrossRef]

26. Sun, A.Y. Predicting groundwater level changes using GRACE data. *Water Resour. Res.* **2013**, *49*, 5900–5912. [CrossRef]

27. Seyoum, W.M.; Milewski, A.M. Improved methods for estimating local terrestrial water dynamics from grace in the northern high plains. *Adv. Water Resour.* **2017**, *110*, 279–290.

28. Miro, E.M.; Famiglietti, S.J. Downscaling GRACE remote sensing datasets to high-resolution groundwater storage change maps of California's Central Valley. *Remote Sens.* **2018**, *10*, 143. [CrossRef]

29. Eicker, A.; Schumacher, M.; Kusche, J.; Döll, P.; Schmied, H. Calibration/data assimilation approach for integrating GRACE data into the watergap global hydrology model (WGHM) using an ensemble Kalman filter: First results. *Surv. Geophys.* **2014**, *35*, 1285–1309. [CrossRef]

30. Wada, Y.; van Beek, L.P.H.; van Kempen, C.M.; Reckman, J.W.T.M.; Vasak, S.; Bierkens, M.F.P. Global depletion of groundwater resources. *Geophys. Res. Lett.* **2010**, *37*. [CrossRef]

31. Döll, P.; Müller Schmied, H.; Schuh, C.; Portmann, F.T.; Eicker, A. Global-scale assessment of groundwater depletion and related groundwater abstractions: Combining hydrological modeling with information from well observations and GRACE satellites. *Water Resour. Res.* **2014**, *50*, 5698–5720. [CrossRef]

32. Leblanc, M.; Favreau, G.; Tweed, S.; Leduc, C.; Razack, M.; Mofor, L. Remote sensing for groundwater modelling in large semiarid areas: Lake Chad Basin, Africa. *Hydrogeol. J.* **2006**, *15*, 97–100. [CrossRef]

33. Lo, M.-H.; Famiglietti, J.S.; Yeh, P.J.F.; Syed, T.H. Improving parameter estimation and water table depth simulation in a land surface model using GRACE water storage and estimated base flow data. *Water Resour. Res.* **2010**, *46*, W05517. [CrossRef]

34. Brookfield, A.E.; Hill, M.C.; Rodell, M.; Loomis, B.D.; Stotler, R.L.; Porter, M.E.; Bohling, G.C. In Situ and grace-based groundwater observations: Similarities, discrepancies, and evaluation in the High Plains Aquifer in Kansas. *Water Resour. Res.* **2018**, *54*, 8034–8044.

35. Panno, S.V.; Askari, Z.; Kelly, W.R.; Parris, T.M.; Hackley, K.C. Recharge and groundwater flow within an Intracratonic basin, Midwestern United States. *Ground Water* **2018**, *56*, 32–45. [CrossRef]

36. Piskin, K.; Bergstrom, R.E. *Glacial Drift in Illinois: Thickness and Character*; Illinois State Geological Survey, Circular 490: Urbana, IL, USA, 1975; p. 37.

37. NOAA. National Centers for Environmental Information, Climate at A Glance: Statewide Time Series, Published April 2018. Available online: http://www.Ncdc.Noaa.Gov/cag/ (accessed on 10 May 2018).

38. Tapley, B.D.; Bettadpur, S.; Watkins, M.; Reigber, C. The gravity recovery and climate experiment: Mission overview and early results. *Geophys. Res. Lett.* **2004**, *31*. [CrossRef]

39. Steitz, D.; O'Donnell, F.; Buis, A.; Chandler, L.; Baguio, M.; Weber, V. *Grace Launch*; Press Kit; 2002. Available online: https://www.jpl.nasa.gov/news/press_kits/gracelaunch.pdf (accessed on 30 May 2018).

40. Landerer, F.W.; Swenson, S.C. Accuracy of scaled GRACE terrestrial water storage estimates. *Water Resour. Res.* **2012**, *48*, W04531.

41. Swenson, S.; Wahr, J. Post-processing removal of correlated errors in GRACE data. *Geophys. Res. Lett.* **2006**, *33*. [CrossRef]

42. Huffman, G.J.; Bolvin, D.T.; Nelkin, E.J.; Wolff, D.B.; Adler, R.F.; Gu, G.; Hong, Y.; Bowman, K.P.; Stocker, E.F. The TRMM multisatellite precipitation analysis (TMPA): Quasi-global, multiyear, combined-sensor precipitation estimates at fine scales. *J. Hydrometeorol.* **2007**, *8*, 38–55. [CrossRef]

43. Hirsch, R.M.; Fisher, G.T. Past, present, and future of water data delivery from the US Geological Survey. *J. Contemp. Water Res. Educ.* **2014**, *153*, 4–15. [CrossRef]

44. Ryberg, K.R.; Vecchia, A.V. *Waterdata—An R Package for Retrieval, Analysis, and Anomaly Calculation of Daily Hydrologic Time Series Data*; Version 1.0; US Geological Survey: Reston, VA, USA, 2012; pp. 1258–2331.

45. Hirsch, R.M.; De Cicco, L.A. *User Guide to Exploration and Graphics for River Trends (Egret) and Dataretrieval: R Packages for Hydrologic Data*; US Geological Survey: Reston, VA, USA, 2015; pp. 2328–7055.

46. Didan, K. *MOD13A3: MODIS/Terra vegetation indices monthly l3 Global 1 km SIN grid V006.* NASA EOSDIS Land Processes DAAC. 2015. Available online: http://doi.org/10.5067/MODIS/MOD13A3.006 (accessed on 28 August 2017).

47. Wan, Z.; Hook, S.; Hulley, G. *MOD11C3 MODIS/Terra Land Surface Temperature/Emissivity Monthly L3 Global 0.05 Deg CMG V006.* 2015. Available online: http://doi.org/10.5067/MODIS/MYD11C3.006 (accessed on 31 August 2017).

48. Hulley, G.C.; Hook, S.J. Intercomparison of versions 4, 4.1 and 5 of the MODIS land surface temperature and emissivity products and validation with laboratory measurements of sand samples from the Namib desert, Namibia. *Remote Sens. Environ.* **2009**, *113*, 1313–1318. [CrossRef]

49. Wan, Z. New refinements and validation of the MODIS land-surface temperature/emissivity products. *Remote Sens. Environ.* **2008**, *112*, 59–74.

50. Rui, H.; Mocko, D. *Readme Document for North America Land Data Assimilation System Phase 2 (NLDAS-2) Products*; The National Aeronautics and Space Administration: Greenbelt, MD, USA, 2013.

51. Bayless, E.R.; Arihood, L.D.; Reeves, H.W.; Sperl, B.J.S.; Qi, S.L.; Stipe, V.E.; Bunch, A.R. *Maps and Grids of Hydrogeologic Information Created from Standardized Water-Well Drillers' Records of the Glaciated United States*; U.S. Geological Survey Scientific Investigations Report 2015-5105; US Geological Survey: Reston, VA, USA, 2017; p. 34.

52. Water and Atmospheric Resources Monitoring Program. *Shallow Groundwater Network*; Illinois State Water Survey: Champaign, IL, USA, 2015.

53. Elith, J.; Leathwick, J.R.; Hastie, T. A working guide to boosted regression trees. *J. Anim. Ecol.* **2008**, *77*, 802–813. [CrossRef] [PubMed]

54. Schapire, R.E. The boosting approach to machine learning: An overview. In *Nonlinear Estimation and Classification*; Springer: New York, NY, USA, 2003; pp. 149–171.

55. Friedman, J.; Hastie, T.; Tibshirani, R. *The Elements of Statistical Learning*; Springer Series in Statistics New York; Springer: New York, NY, USA, 2001; Volume 1.

56. Breiman, L.; Friedman, J.; Olshen, R.; Stone, C.J. *Classification and Regression Trees*; Taylor & Francis: Boca Raton, Fl, USA, 1984.

57. Friedman, J.H. Greedy function approximation: A gradient boosting machine. *Ann. Stat.* **2001**, *29*, 1189–1232. [CrossRef]

58. Friedman, J.H.; Meulman, J.J. Multiple additive regression trees with application in epidemiology. *Stat. Med.* **2003**, *22*, 1365–1381. [CrossRef] [PubMed]

59. Rodell, M.; Chen, J.; Kato, H.; Famiglietti, J.S.; Nigro, J.; Wilson, C.R. Estimating groundwater storage changes in the Mississippi River Basin (USA) using GRACE. *Hydrogeol. J.* **2006**, *15*, 159–166. [CrossRef]
60. Rodell, M.; Famiglietti, J.S. The potential for stellite-based monitoring of groundwater storage changes using GRACE: The High Plains Aquifer, Central US. *J. Hydrol.* **2002**, *263*, 245–256. [CrossRef]

 *remote sensing* 

Article

Hydrologic Mass Changes and Their Implications in Mediterranean-Climate Turkey from GRACE Measurements

Gonca Okay Ahi [1,*] **and Shuanggen Jin** [2,3,*]

1   Department of Geomatics Engineering, Hacettepe University, Beytepe Campus, 06800 Ankara, Turkey
2   Shanghai Astronomical Observatory, Chinese Academy of Sciences, Shanghai 200030, China
3   School of Remote Sensing and Geomatics Engineering, Nanjing University of Information Science and Technology, Nanjing 210044, China
*   Correspondence: goncaokayahi@hacettepe.edu.tr (G.O.A.); sgjin@shao.ac.cn (S.J.); Tel.: +90-312-297-6990 (G.O.A.)

Received: 14 November 2018; Accepted: 7 January 2019; Published: 10 January 2019

Abstract: Water is arguably our most precious resource, which is related to the hydrological cycle, climate change, regional drought events, and water resource management. In Turkey, besides traditional hydrological studies, Terrestrial Water Storage (TWS) is poorly investigated at a continental scale, with limited and sparse observations. Moreover, TWS is a key parameter for studying drought events through the analysis of its variation. In this paper, TWS variation, and thus drought analysis, spatial mass distribution, long-term mass change, and impact on TWS variation from the parameter scale (e.g., precipitation, rainfall rate, evapotranspiration, soil moisture) to the climatic change perspective are investigated. GRACE (Gravity Recovery and Climate Experiment) Level 3 (Release05-RL05) monthly land mass data of the Centre for Space Research (CSR) processing center covering the period from April 2002 to January 2016, Global Land Data Assimilation System (GLDAS: Mosaic (MOS), NOAH, Variable Infiltration Capacity (VIC)), and Tropical Rainfall Measuring Mission (TRMM-3B43) models and drought indices such as self-calibrating Palmer Drought Severity (SCPDSI), El Niño–Southern Oscillation (ENSO), and North Atlantic Oscillation (NAO) are used for this purpose. Turkey experienced serious drought events interpreted with a significant decrease in the TWS signal during the studied time period. GRACE can help to better predict the possible drought nine months before in terms of a decreasing trend compared to previous studies, which do not take satellite gravity data into account. Moreover, the GRACE signal is more sensitive to agricultural and hydrological drought compared to meteorological drought. Precipitation is an important parameter affecting the spatial pattern of the mass distribution and also the spatial change by inducing an acceleration signal from the eastern side to the western side. In Turkey, the La Nina effect probably has an important role in the meteorological drought turning into agricultural and hydrological drought.

Keywords: terrestrial water storage (TWS); GRACE; GLDAS; TRMM; drought; ENSO; NAO; Turkey

1. Introduction

Traditional methods of monitoring hydrological processes (e.g., in situ measurements of precipitation and soil moisture content) have generally been inadequate to characterize extreme hydrologic events [1,2]. Their temporal and spatial resolutions are not good enough to characterise water mass variations at a regional or global scale. In order to improve our knowledge to predict and monitor these water mass changes in the scope of drought analysis, flood potential assessment, groundwater changes, soil moisture analysis, etc., there are an increasing number of available datasets being produced, especially from remote sensing techniques. These techniques offer information on vegetation, precipitation, surface water storage, evapotranspiration, soil moisture, groundwater, and snow components. Tropical Rainfall Measuring Mission (TRMM) [3], TRMM Multisatellite Precipitation Analysis (TMPA), Precipitation Estimation from Remotely Sensed Information using Artificial Neural Networks (PERSIANN) [4], the CPC MORPHing technique (CMORPH) [5], Climate Hazards Group InfraRed Precipitation with Station data (CHIRPS) [6], Global Satellite Mapping of Precipitation (GSMaP) [7–9], and Global Precipitation Measurement (GPM) [3] are some of the available satellite precipitation missions offering multiple products in real-time. While missions like Soil Moisture Active Passive (SMAP) provide satellite soil moisture values, the U.S. Department of Agriculture's (USDA) global reservoir and lake monitoring service provides information on satellite surface water levels by using near-real-time radar altimeter data. Finally, to study satellite surface/subsurface water, since 2002, the Gravity Recovery and Climate Experiment (GRACE) has provided entirely new observations suitable for quantifying and monitoring continental or regional TWS changes [10–13], groundwater changes [14,15], drought monitoring [16–19], and flood potential assessments [20] at a spatial resolution of a few hundred kilometers with uniform data coverage. The approaches used for estimating groundwater storage variations with the main applications of GRACE data for groundwater monitoring can be found in [21]. Moreover, since 2018, GRACE Follow-On (GRACE-FO) satellite gravity mission has been established to continue tracking Earth's water movements at different spatial scales. The results retrieved from the satellite gravity measurements are independent of the in situ data and might be interpreted independently. However, to increase the accuracy and make a better interpretation, the trend needs to be assessed based on a combination of additional data sets provided by other remote sensing technics. Moreover, as observed in many studies, the comparison of global hydrological models' results with GRACE data supports the analyses. The Global Land Data Assimilation System (GLDAS) [22], which is developed jointly by NASA and NOAA, simulates Terrestrial Water Storage through four main land surface models: VIC [23], NOAH [24], Mosaic [25], and CLM [26]. The Climate Prediction Center (CPC) model [27], Land Dynamics model (LadWorld) [28], WaterGAP Global Hydrology Model (WGHM) [29–31], and Organizing Carbon and Hydrology in Dynamic Ecosystems (ORCHIDEE) Land surface model [32] are examples of such global hydrological models [33], which provide a general overview on the use of global hydrological models' results (e.g., water storage change) as a reference to calibrate/validate GRACE data. Additionally, [33] found inconsistencies in the previous studies between hydrological model simulation results and GRACE-based observations and provided possible explanations for these inconsistencies.

In Turkey, besides traditional hydrological studies, TWS is poorly investigated at a continental scale with the new satellite techniques (e.g., GRACE). Previous studies in Turkey revealed about 0.7 cm/year the TWS variation [34]. Among them, the study of TWS variations in Turkey [35] with GRACE and GLDAS (NOAH) data sets [22] for the period from 2003 to 2009, showed a significant decrease of up to a rate of 4 cm/year for both data sets in the southern part of the central Anatolian region. This decrease present in both datasets is explained by decreasing groundwater variations confirmed by the existing well in the above mentioned regions. More recently, the effect of drought and water extraction on groundwater storage in central Turkey are also described [36]. They also showed how the groundwater storage can affect the TWS. In addition, long-term TWS changes in Turkey during the 2004–2014 period by associating GLDAS/NOAH data are studied and accounted for TWS variation between -17 and 16 cm in amplitude, with an important decrease in 2008 [37]. The requirement of further studies to isolate model errors and anthropogenic effects for Turkey, in order to explain the GRACE signal, which points out a robust acceleration in TWSA is emphasized also by other studies [38]. These prior studies point out the need for a summarizing and extended up-to-date study which accounts for a longer time period, associating different auxiliary data and parameters to understand and interpret the TWS change mechanism and possible drought events at a national scale.

In this paper, the water storage variation is studied over Turkey at a seasonal time scale for the period from April 2002 to January 2016 using monthly GRACE land mass grids (Level 3-RL05) from CSR. To compare the results, monthly grids of GLDAS data with a $1° \times 1°$ resolution and three GLDAS hydrological models: MOS, NOAH, VIC, are also analyzed. To estimate precipitation over Turkey within the studied period, the TRMM-3B43 model is also considered, with a 0.25° × 0.25° resolution. In addition, ENSO, the self-calibrating Palmer Drought Severity Index, and NAO are compared with the satellite-derived GRACE TWS data.

2. Data and Methods

2.1. Studied Area and Its Hydrological Characteristics

Turkey is geographically located at approximately 36–42° N and 26–45° E, with an approximate area of 783.562 km^2. A glance at a digital elevation model of Turkey reveals that mountains encircle the peninsula of Anatolia in four directions (see Figure 1).

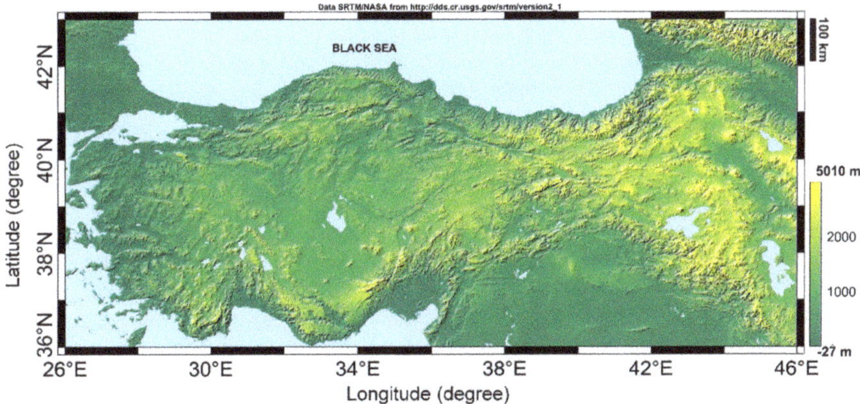

Figure 1. Digital Elevation Model of Turkey.

In Turkey, the yearly mean precipitation rate is approximately 643 mm (501 billion m^3 water). A total of 274 billion m^3 of that water is transferred to the atmosphere by evaporation from soil, water, and vegetation surfaces. Additionally, 158 billion m^3 flow to the sea and lakes in closed basins through streams. Furthermore, 28 billion m^3 out of the remaining 69 billion m^3 water feeding the groundwater contributes to the surface water. Besides, an additional water input of about 7 billion m^3 comes from neighboring countries. Thus, the surface water potential of the country is 193 billion m^3, or 234 billion m^3 by adding the 41 billion m^3 of water contributing to the groundwater [37,39].

Important and severe drought events experienced in Turkey are recorded between 1971–1974, 1983–1984, 1989–1990, 1996, 2001, and 2007–2008 [40,41]. There are also other studies investigating droughts in Turkey revealing an agricultural and hydrological drought starting from November/December 2006 to December 2008, so the drought period is recorded as 2007–2008 [42]. Especially in 2008, there was no variation in snow and precipitation for nine to 10 months [43].

2.2. Terrestrial Water Storage (TWS) from GRACE

Three processing centers, including CSR (Center for Space Research, Texas), JPL (Jet Propulsion Laboratory, California), and GFZ (GeoForschungsZentrum, Potsdam), provide official releases of GRACE gravity data at three different levels (Level 1, 2, 3), depending on the expertise and needs of the users for both the time-averaged and time-variable fields. In this study, Level 3 (RL05) land data of the CSR processing center have been used, which were ready to use as many necessary preprocessing steps had already been applied (removal of atmospheric pressure/mass changes, replacement of the C20 (degree 2 order 0) coefficients with the solutions from Satellite Laser Ranging [44], the estimation of the degree-1 coefficients (geocenter) from [45], the correction of glacial isostatic adjustment (GIA), destriping [46], Gaussian filtering). Level 3 data are in the form of GRACE-derived mass grids expressed as the TWS function of the gravity fields with 1 degree in both latitude and longitude (approx. 111 km at the equator) spatial sampling and estimates over land from the gravity coefficient anomalies for each month (ΔC_{lm}, ΔS_{lm}) [47], as below:

$$\Delta \eta_{land}(\theta, \phi, t) = \frac{a\rho_{ave}}{3\rho_w} \sum_{l=0}^{\infty} \sum_{m=0}^{l} \widetilde{P}_{lm}(\cos\theta) \frac{2l+1}{1+k_l}(\Delta C_{lm}\cos(m\phi) + \Delta S_{lm}\sin(m\phi)) \tag{1}$$

where ρ_{ave} is the average density of the Earth, ρ_w is the density of fresh water, a is the equatorial radius of the Earth, $\widetilde{P}_{lm}$ is the fully-normalized Legendre associated function of degree l and order m, k_l is the Love number of degree l [48], θ is the spherical co-latitude (polar distance), and ϕ is the longitude. All grids are obtained from the following link: https://grace.jpl.nasa.gov/data/get-data/monthly-mass-grids-land/.

The first, additionally applied data processing, also recommended by the processing center for the Level 3 land data grid, in order to prevent possible attenuation of the surface mass variations due to the sampling and post-processing of GRACE observations (destriping, gaussian) and to regain part of the information loss in prior data processing, is the multiplication of one for each 1-degree land grid by a set of provided scaling coefficients, as shown below in Equation (2):

$$g'(x, y, t) = g(x, y, t) \times s(x, y) \tag{2}$$

where x is the longitude index, y is the latitude index, t is time (month) index, $g(x, y, t)$ is the grid node, $s(x, y)$ is the scaling grid, and $g'(x, y, t)$ is the gain-corrected time series. Moreover, additionally applied data processing, leakage error correction (residual errors after filtering and rescaling), has been performed (as below in Equation (3)) with the provided file obtained from the following link: ftp://podaac-ftp.jpl.nasa.gov/allData/tellus/L3/land_mass/RL05/netcdf/, containing scaling coefficients (as mentioned previously) and leakage error estimates.

$$g'_{leak\_corr}(x, y, t) = g'(x, y, t) + leakage\_err(x, y) \tag{3}$$

where x is the longitude index, y is the latitude index, t is the time (month) index, $g'(x, y, t)$ is the gain-corrected time series, $leakage\_err(x, y)$ is the leakage error estimates, and $g'_{leak\_corr}(x, y, t)$ is the scaled and leakage error corrected time series. In this study, monthly mass grids of GRACE land data (Level-3 RL05) from the CSR processing center concerning the period from April 2002 to January 2016 with a $1° \times 1°$ spatial resolution have been used after applying these additional processing steps.

2.3. Global Land Data Assimilation System (GLDAS) Models Data

GLDAS has been developed jointly by scientists at the National Aeronautics and Space Administration (NASA)-Goddard Space Flight Center (GSFC) and the National Oceanic and Atmospheric Administration (NOAA)-National Centers for Environmental Prediction (NCEP). GLDAS is a global, high-resolution, offline (uncoupled to the atmosphere) terrestrial modeling system that incorporates satellite- and ground-based observations in order to produce optimal fields of land surface states (e.g., soil moisture, snow water equivalent, and canopy water storage...) and fluxes (e.g., rainfall, snowmelt, evapotranspiration...) in near–real time [22]. Currently, GLDAS drives four land surface models: MOS, NOAH, the Community Land Model (CLM), and the VIC. In this study, GLDAS version1 (GLDAS-1) monthly data of the four land surface models are downloaded from https://disc.gsfc.nasa.gov/datasets?keywords=gldas&page=1, with a $1° \times 1°$ spatial resolution, concerning the period from January 2002 to January 2016. In these data sets, the GLDAS provides a time series of land surface states and fluxes (25 variables), which can be used to study water storage. The anomalies corresponding to the major part of the signal to TWS can be assumed to arise from the change in soil moisture (kg/m^2), snow water equivalent (kg/m^2), and canopy water storage (kg/m^2).

Hence, firstly, these land surface state variables are derived from the file covering Turkey and then, the TWS from GLDAS models is calculated, as shown by Equation (4):

$$TWS_{GLDAS} = \Delta SM + \Delta SWE + \Delta CWS \tag{4}$$

where, TWS_{GLDAS} is the change in terrestrial water storage from GLDAS, ΔSM is the change in soil moisture, ΔSWE is the change in the snow moisture equivalent, and ΔCWS is the change in canopy water storage. Soil moisture values are averaged before integrating them into the TWS calculation, according to the three-layer model for VIC and MOS, and the four-layer model for NOAH. In this study, CLM models have not been used.

2.4. Tropical Rainfall Measuring Mission (TRMM) Data

TRMM is a joint mission between NASA and the Japan Aerospace Exploration (JAXA) Agency in order to study rainfall for weather and climate research (1997–2015). With the help of several space-borne instruments, TRMM satellite data allow precipitation from diurnal to interannual time scales to be measured, which led to improving our understanding of tropical cyclone structure and evolution, including important variability associated with the Madden-Julian Oscillation and with El Nino Southern Oscillation (ENSO), convective system properties, lightning-storm relationships, climate and weather modeling, and human impacts on rainfall. The data also supported operational applications such as flood and drought monitoring and weather forecasting (https://trmm.gsfc.nasa.gov/).

In our study, we used the TRMM-3B43 Level 3 gridded monthly satellite-gauge (SG) combination data set with a $0.25° \times 0.25°$ degree spatial resolution downloaded from http://mirador.gsfc.nasa.gov/# to estimate the precipitation variations over Turkey.

2.5. In-Situ Precipitation Data

The Turkish state meteorological service provides an annual cumulative rainfall distribution (1981–2010) map produced on a GIS platform by kriging the in-situ rainfall data of 255 meteorological stations. This map has been downloaded from the following website: https://mgm.gov.tr/eng/forecast-cities.aspx, and used further to compare/validate TRMM data and rainfall from ground meteorological stations.

2.6. Self-Calibrating Palmer Drought Severity Index (SCPDSI) Data

The SCPDSI [49] can estimate the departure relative to normal conditions in the surface water balance by using a hydrological accounting system [50,51]. The PDSI is primarily considered a meteorological drought indicator, and sometimes, an agricultural drought indicator [52]. The needed drought index data was downloaded from the following website: https://crudata.uea.ac.uk/cru/data/drought/, as global land data covering the time period from 1901 to 2016 with a $0.5°$ latitude-longitude spatial resolution. Then, grids corresponding to Turkey and the time period from 2002 to 2016 were extracted from global data.

2.7. El Niño–Southern Oscillation (ENSO) Index Data

ENSO is described as warming on the ocean surface, or above-average sea surface temperatures (SST), in the central and eastern tropical Pacific Ocean. This is one of the most important climate phenomena on Earth due to its ability to change the global atmospheric circulation, which in turn, influences temperature and precipitation across the globe. The magnitude of the ENSO is often expressed by the Niño SST3.4 index, derived from the normalized Sea Surface Temperature (SST). El Nino (warm phase) and La Niña (cold phase) are two contrary phases of ENSO [53]. In order to understand the magnitude of ENSO which influences precipitation and to improve our understanding of the occurrence of drought events at a national scale, SST data were downloaded from the following website: http://www.cpc.ncep.noaa.gov/data/indices/, where monthly ERSSTv5 (1981–2010 base period) Niño 1 + 2 (0–10° South) (90° W–80° W) Niño 3 (5° N–5° S) (150° W–90° W) Niño 4 (5° N–5° S) (160° E–150° W) Niño 3.4 (5° N–5° S) (170 W–120° W) is available. The time period (from 2002 to 2016) was extracted from global data.

2.8. North Atlantic Oscillation (NAO) Index Data

NAO is the variability in atmospheric mass circulation especially observed in the cold season months (November–April) over the middle and high latitudes of the Northern Hemisphere (from central North America to Europe and much into Northern Asia). The understanding of its mechanism on the surface temperature, storms, precipitation, ocean, and ecosystem results in understanding global climate change [54]. Strong positive phases (+) of the NAO tend to be associated with below-average precipitation over southern and central Europe. Conversely, above average temperature and precipitation anomalies are typically observed during strong negative phases (-) of the NAO (https://www.cpc.ncep.noaa.gov/data/teledoc/nao.shtml). The monthly mean NAO index data from 2002 to 2017 were downloaded from the following website: https://www.cpc.ncep.noaa.gov/products/precip/CWlink/pna/nao.shtml.

To conclude the data and methods section and proceed with the results and analysis section, Table 1 below summarizes the studied research topics, used input data, methodology, and additional processing applied in this paper.

Table 1. Studied topics and general structure of the paper.

Research Topic	3.1. Drought Analysis from Time Series	3.2. Spatial Mass Distribution and Its Causes	3.3. Long-Term Mass Change	3.4. Impact on TWS Variations	3.5. Understanding Drought and Its Relation with Climatic Change
Input data (global)	GRACE and GLDAS TWS	GRACE, GLDAS TRMM, In-situ	GRACE and GLDAS TWS	GRACE(TWS), GLDAS **, TRMM (precipitation)	GRACE(TWS), SCPDSI, ENSO, NAO
Methodology	from the analyses of TWS time series	Harmonic analyses and least squares fitting according to [31] for both GRACE and GLDAS data Annual amplitude (cm), phase (degree) and trend (cm/yr) values are derived	from the analyses of TWS time series	Evapotranspiration(t) = P(t) − R(t) − TWS (GRACE)	from the analyses of TWS time series
Additional processing	-GRACE: Scaling and leakage error correction -GLDAS: soil moisture values are summed with respect to the number of models layers as following: three-layer model for MOS and VIC and four-layer model for NOAH TWS(GLDAS) = ΔSM + ΔSWE + ΔCWS *	*	*	GLDAS: soil moisture values are summed with respect to the number of models layers as following: three-layer model for MOS and VIC and four-layer model for NOAH *	*

* Interested time (~2002–2016) and land grids (Turkey) are extracted for all global data sets. ** Rainfall rate, Evapotranspiration, Soil moisture. P(t) = Precipitation from TRMM. R(t) = Runoff derived from MOS, NOAH, and VIC models.

3. Results and Analysis

3.1. Drought Analysis from Time Series

In order to understand important drought periods, TWS time series derived from GRACE data and from GLDAS models were analyzed in the studied time period from 2002 to 2016. Figure 2 indicates residual (GRACE TWS-mean GRACE TWS) monthly TWS variations in Turkey (cm) according to GRACE and GLDAS models (MOS, NOAH, VIC).

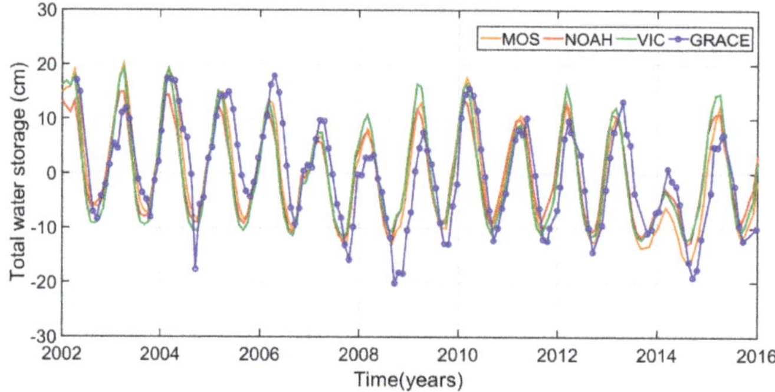

Figure 2. Residual monthly TWS variations in Turkey (cm) according to GRACE (GRACE TWS-mean GRACE TWS) (in blue with circle), GLDAS-MOS TWS—mean (GLDAS MOS TWS) (in orange), GLDAS NOAH TWS—mean (GLDAS NOAH TWS) (in red), and GLDAS VIC TWS—mean (GLDAS VIC TWS) (in green) models.

Figure 2 indicates that GRACE TWS time series and GLDAS models are consistent. According to the GRACE TWS signal, the studied time period shows some sudden TWS decreases in September 2004 and 2008, which falls after dry summer periods, and in October 2014. However, to be prudent, the decrease in September 2004 may be an over estimation of GRACE solutions. Additionally, there are some significant increasing and decreasing trends in TWS during some time intervals. Firstly, GRACE TWS time series indicate an important decreasing trend period from February 2006 to November 2008 [42]. This underlines the beginning of an agricultural and hydrological drought, which occurred due to the below-average precipitation [55] from November/December 2006 to December 2008. In this context, GRACE time series provide an earlier warning (~9 months before) for the beginning of a decreasing trend in TWS. Briefly, GRACE indicates that 2008 is a remarkable year in terms of drought records. This finding is also supported with no snow and precipitation variation in 2008 [43]. After this decreasing time interval, Turkey is exposed to an increasing trend in TWS from November 2008 to March 2010. According to Figure 2, the second decreasing period begins in March 2010 and lasts until October 2014. During this time interval, historical records indicate the beginning of a meteorological drought in 2012, intensified with dry summers, as is usual for the Mediterranean climate. Even in 2013, the observed above-average precipitation levels (32 cm in 2013 where average is 27 cm) could not stop the decreasing trend in TWS because of the successively observed below-average precipitation that occurred in 2014 (17 cm) [56]. To resume, the analysis of GRACE TWS time series and the results of previous works revealed that the GRACE signal is mostly sensitive to agricultural drought (insufficient soil moisture content) and hydrological drought (significant reduction in winter precipitation) (2006–2008) compared to meteorological drought (little rain combined with increased temperature and lower humidity) (2012–2014) for the studied region. Following this prior understanding about the important increasing and decreasing trends in TWS, leading to important drought events in the time period from 2002 to 2016, we now focus (Section 3.2) on the spatial mass distribution of the TWS and its causes at the national scale.

3.2. Spatial Mass Distribution and Its Causes

The TWS variations have significant seasonal signals. As we are interested in seasonal variations with significant annual and semiannual periods, a mathematical function/model (Equation (5)), which includes the annual and semi-annual variations with linear trend terms, has to be used to fit the TWS time series [57]:

$$M(t) = a + bt + \sum_{k=1}^{2} c_k \sin(\omega_k(t - t_0) + \phi_k) + \varepsilon(t) \tag{5}$$

where $M(t)$ is the time series; t is the time; t_0 is a reference time; a is the constant; b is the trend; c_k, ϕ_k, and ω_k are annual amplitude, phase, and frequency, respectively; k = 1 is for the annual variation and k = 2 is for the semi-annual variation; and $\varepsilon(t)$ is the un-modeled residual term. During the analysis of time series, strong annual signals are found, and k = 1 year is used here. Using the least-squares method to fit the time series of GRACE data at each point, the annual amplitude, annual phase, and trend terms of TWS are estimated. Figures 3 and 4 show the annual amplitude (cm) and annual phase (degree) map of TWS variation of Turkey from GRACE data and GLDAS models (MOS, NOAH, VIC) concerning the 2002–2016 period, respectively. The TWS variation estimates of both methods show a good agreement in annual amplitudes and phases. The mean annual amplitude is 11.06 cm and 11.19 cm from GRACE and GLDAS models (MOS), respectively. Additionally, the mean annual phase is 21.90 and 22.98° from GRACE and GLDAS models (MOS), respectively. The larger annual amplitude values >20 cm (in red, Figure 3a,d) are observed in the eastern parts of Turkey. However, significant amplitude values also appear at shorelines (also Black Sea, Aegean, Mediterranean). The smaller annual amplitude of TWS variations of nearly 2.89 cm is seen in the middle of Turkey (central Anatolia, Figure 3a–d). According to phase plots, there is lateral variation (especially concerning GRACE phase plot, Figure 4a) from eastern part to the western part of Turkey, with a small increase through to the west side contrary to the amplitude plots. This means that lower phase values are observed in the eastern parts of Turkey, where larger amplitude values have been previously estimated. Table 2 indicates the mean annual amplitude, phase, and trend variations in Turkey according to GRACE and GLDAS models.

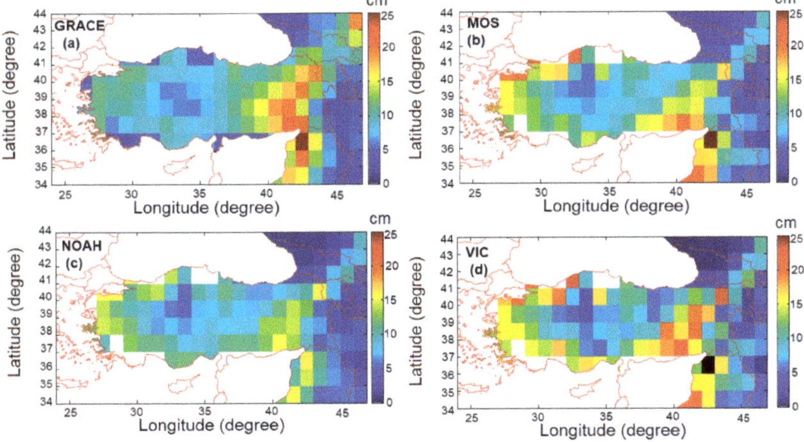

Figure 3. Annual amplitude (cm) of TWS variation of Turkey from (**a**) GRACE data and GLDAS models, (**b**) MOS, (**c**) NOAH, and (**d**) VIC concerning the 2002–2016 period.

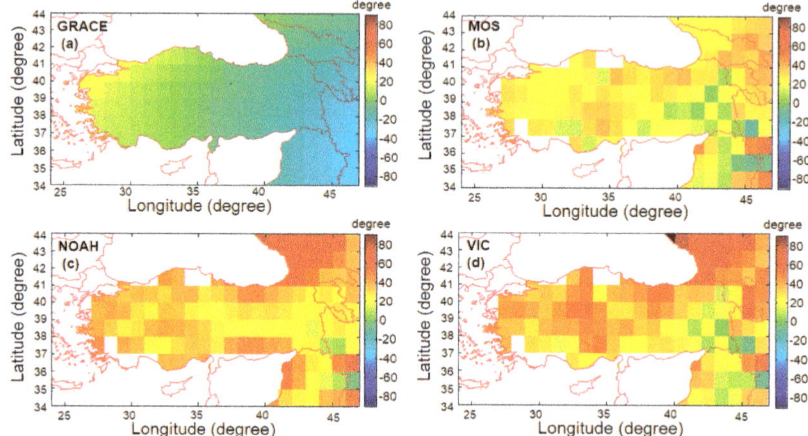

Figure 4. Annual phase (degree) of TWS estimated from (**a**) GRACE data and GLDAS models, (**b**) MOS, (**c**) NOAH, and (**d**) VIC concerning the 2002–2016 period. In order to see small variation, the color bar is limited.

Table 2. Mean annual amplitude, phase, and trend variations in Turkey according to GRACE and GLDAS models.

Groundwater Storage	Annual Amplitude (cm)	Annual Phase (degree)	Trend (cm/yr)
GRACE	11.06	21.90	−0.77
GLDAS-MOS	11.19	22.98	−0.74
GLDAS-NOAH	10.00	32.25	−0.31
GLDAS-VIC	11.81	32.22	−0.39

Finally, Figure 5 shows the trend (cm/yr) of TWS of Turkey according to GRACE and GLDAS models (MOS, NOAH, VIC) concerning the 2002–2016 period.

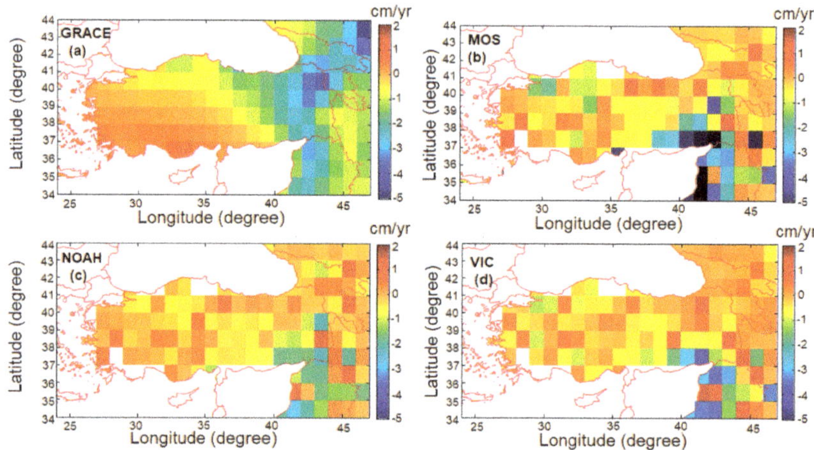

Figure 5. Linear trend of TWS variation of Turkey according to (**a**) GRACE data and GLDAS (cm/yr) models, (**b**) MOS, (**c**) NOAH, and (**d**) VIC concerning the 2002–2016 period.

Trend plots of Turkey according to GRACE data (Figure 5a) also show lateral increasing variation from the eastern part to western part, as observed in phase plots (Figure 4a), while GLDAS models plots also show lateral variation, but in a mostly heterogeneous way (Figure 5b). Distinctive trend values (≥1 cm/yr) appear in the eastern part of Turkey (Figure 5a,b), in accordance with amplitude values (Figure 3a,d).

Even GRACE and GLDAS 2D plots agree with each other and show the spatial distribution of TWS; however, they do not explain the reason for the above-presented spatial distribution. To study the possible reasons for this, GRACE 2D plots are compared with the TRMM-derived precipitation data shown here as 2D sections. Figure 6 shows the (a1) annual amplitude (cm) of TWS variation from GRACE data and (b1) annual amplitude of precipitation estimated from TRMM model, (a2) annual phase of TWS (degree) estimated from GRACE data and (b2) annual phase of precipitation (degree) estimated from TRMM model, (a3) annual trend of TWS variation from GRACE data (cm/yr), (b3) annual trend of precipitation (cm/yr) estimated from the TRMM model between 2002–2016 in Turkey, and (c1) in-situ meteorological rainfall data.

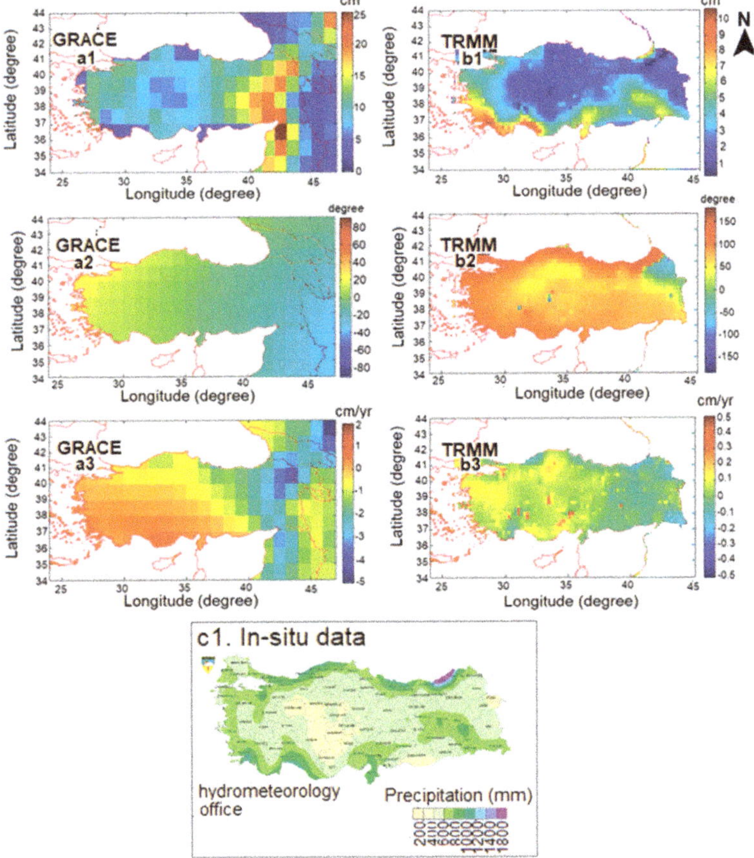

Figure 6. (**a1**) Annual amplitude (cm) of TWS variation from GRACE data and (**b1**) annual amplitude of precipitation estimated from TRMM model (scales are different), (**a2**) annual phase of TWS (degree) estimated from GRACE data, (**b2**) annual phase of precipitation (degree) estimated from TRMM model, (**a3**) annual trend of TWS variation from GRACE data (cm/yr), and (**b3**) annual trend of precipitation (cm/yr) estimated from TRMM model between 2002–2016 in Turkey. (**c1**) Turkish state meteorological service's annual cumulative rainfall distribution (1981–2010) map produced by kriging the in-situ rainfall data of 255 meteorological stations on a GIS platform. Scales are different.

Figure 6a1,b1 show larger GRACE-TWS (in red, ≥20 cm) and TRMM anomalies (~7 cm) in the Eastern part of Turkey (see also (Figure 3b–d). The smallest TRMM-precipitation anomalies are observed in central Anatolia (in blue, ≤3 cm), in agreement with GRACE TWS (Figure 6a1). In addition, the TRMM 2D amplitude (Figure 6b1) and phase plot (Figure 6b2) reveal the input of the water because of the precipitation observed on the shorelines of Turkey (Black Sea, Aegean, Mediterranean). This result shows agreement with the GLDAS models' TWS annual amplitude plots (Figure 3b–d) and also with the in-situ annual cumulative rainfall distribution map of the Turkish state meteorological service (Figure 6c1). Trend plots in Figure 6a3,b3 indicate a positive trend through the western part, especially to the Aegean and Mediterranean shorelines of Turkey. Previous studies focused on groundwater loss and reservoir/lake storage change [58,59] also support an acceleration signal in GRACE analyses over the western part of Turkey [38], which has been spatially mapped in this paper. This acceleration seems to be related to the precipitation patterns or flow (from the precipitation-rich eastern part to precipitation-lacking Central Anatolia). However, this might not be the only reason for this and has to be studied in more depth (e.g., human-induced groundwater withdrawal, as mentioned in [36]). After studying the spatial mass distribution and its causes, we will now have a general look and try to focus on the analysis of the nature of the long-term mass change in TWS in Turkey (Section 3.3).

3.3. Long-Term Mass Change

The long-term variations of TWS are estimated and investigated according to GRACE and GLDAS models (MOS, NOAH, VIC) in Turkey. Figure 7 indicates residual mean monthly TWS variations in Turkey (cm) from GRACE and GLDAS models between 2002 and 2016. The mean TWS values of a specific month are first calculated by taking the average of all grids (77 grids) with different geographic coordinates in Turkey for each month of a specific year (e.g., 2002/04, 2002/05), and then all corresponding months of a specific year are averaged between them (January 2002, January 2003...). According to GRACE data, Figure 7 indicates an increasing behavior of the TWS variation from January to April. The monthly maximum of mean TWS at about 10 cm is reached in April. The TWS variation is negative from April to September, with a minimum of −12.88 cm in September. This corresponds to possible dry summer periods, as is normal for the Mediterranean climate. This decrease is followed by a gradual increase from September to December, when we expect more precipitation or snow for the central and eastern parts of the country. The seasonal variations of TWS estimated from GLDAS models are in good agreement with the results from GRACE.

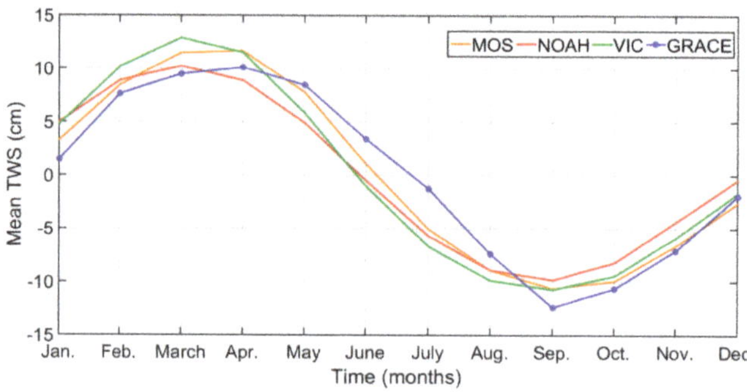

Figure 7. Residuals mean monthly TWS changes for Turkey land calculated from GLDAS-MOS (in orange), GLDAS-NOAH (in red), and GLDAS-VIC (in green) models, and from GRACE data (blue circled) for 2002–2016.

In Figure 8, the long-term variation of mean monthly TWS in the spring, summer, autumn, and winter periods from 2002 to 2016 is studied from GRACE-derived TWS data (cm/month). The larger amplitudes of the long-term seasonal TWS signal are observed in spring (max: 20.64 cm, mean:

12 cm), corresponding to April (Figure 7). These orders of amplitudes are followed by the winter period, which is larger than the summer period. Surprisingly, the weaker TWS values (max: 3 cm, mean: −7 cm) are observed in autumn, corresponding to September (Figure 7), instead of summer, as one might expect. This can be explained by the emphasized drought conditions, arising after dry summers (see Section 3.4.1), which are systematically observed in September almost every four years during the studied time period (September/2004, September/2008, October/2014 see also Figure 2). For all seasons, an important decrease is observed in August 2008. This corresponds to the severe drought events experienced in Turkey, especially in 2008 (also Figure 2). To conclude, there is a major decreasing trend close to −1 cm/yr in Turkey for the studied period which has to be taken into account. The exhibition of the negative trend signal of GRACE time series can be attributed to drought conditions and groundwater withdrawal [35] for the years 2003–2008. The proper study of the long-term change of signal can reveal a drought prediction with the amplitude of a decreasing trend (cm/yr), as shown in our results, and can lead to a drought preparedness. In the next section, we will study the parameters (e.g., precipitation, rainfall rate, evapotranspiration, soil moisture) which might have influenced/contributed to the amplitude of the above-presented spatial distribution (Section 3.2) and long-term mass change (Section 3.3).

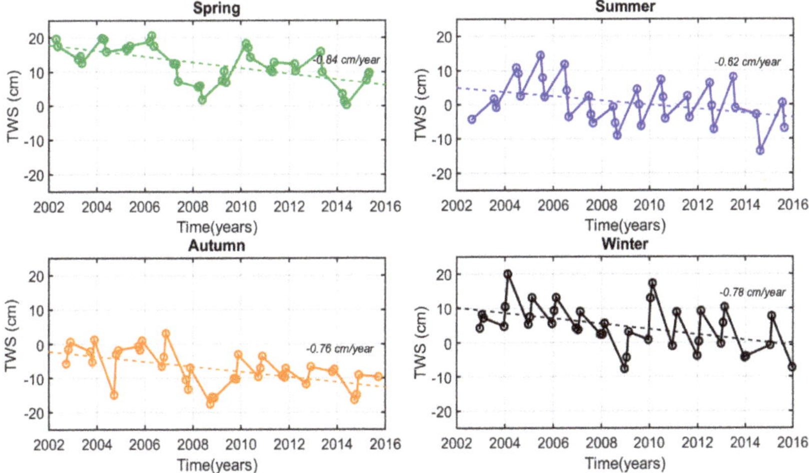

Figure 8. Long-term variation of mean monthly TWS of Turkey concerning (**a**) spring, (**b**) summer, (**c**) autumn, and (**d**) winter period from 2002 to 2016.

3.4. Impacts on TWS Variations

In this section, we performed statistical analysis with IBM SPSS25 software in order to understand descriptive measures and correlations between studied variables and GRACE TWS.

3.4.1. Impact of Precipitation

In order to understand the impact on TWS variations, we used estimated precipitation (cm/month) from TRMM, which is an average rate over a month, and compared the result to the GRACE data. The comparison is shown in Figure 9, which is concerned with the mean monthly precipitation from the TRMM model and the mean monthly TWS derived from GRACE. The TRMM-derived precipitation ranges from 0.71 (July 2015) to 15.36 cm/month (December 2012), while the GRACE-derived TWS ranges from −17.48 (August 2008) to 20.64 cm (April 2006). The amplitude of the TRMM is smaller compared to GRACE data. This can be explained by the fact that GRACE data, displaying more of an increasing and decreasing trend, is not only affected by the precipitation, but also by the other parameters, as we further investigate in this paper.

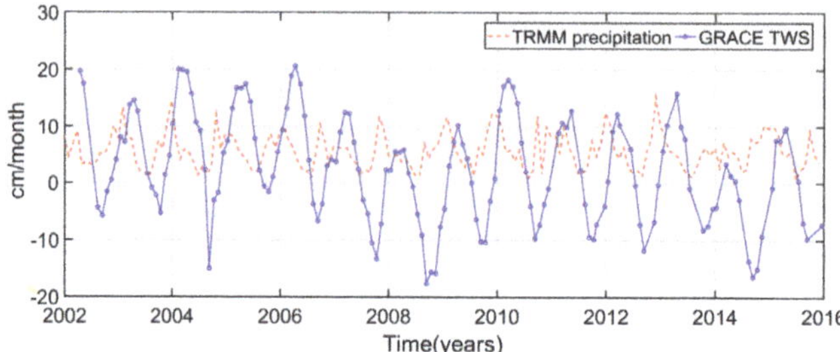

Figure 9. Mean monthly precipitation (cm/month) from TRMM model (in red) and mean monthly GRACE TWS (cm, in blue with circle) of Turkey during 2002–2016.

To give numerical orders of magnitude concerning the correlation between TRMM and GRACE TWS time series, firstly, with the Kolmogorov-Smirnov method, the data is checked to assess whether the distribution is normal or not. The results show that the TRMM distribution is not normal (p = 0.036 < 0.05). Hence, Spearman's rho method is preferred to more appropriately study the correlation between variables. As seen Table 3, there is a significantly positive correlation between TRMM and GRACE TWS with 0.34.

Table 3. Correlation between TRMM precipitation and GRACE TWS according to Spearman's rho method.

Correlations			TRMM	GRACE
Spearman's rho	TRMM precipitation	Correlation Coefficient	1.000	0.337 **
		Sig. (2-tailed)		0.000
	GRACE TWS	Correlation Coefficient	0.337 **	1.000
		Sig. (2-tailed)	0.000	

** Correlation is significant at the 0.01 level (2-tailed).

3.4.2. Impact of Rainfall Rate

Rainfall rate values (kg/m^2) are available in the GLDAS models data files. We extracted corresponding values in the studied region, and converted (cm/month), averaged, and calculated the residual. Figure 10 shows the residual mean monthly rainfall rate (cm/month) extracted from GLDAS models (MOS, NOAH, VIC), from TRMM and residual mean monthly TWS variation from GRACE data from 2002 to 2016 for Turkey. The rainfall data changes between −3.5 and 6.5 cm/month (MOS) and −5 and 9.5 cm/month (TRMM). In Figure 10, a good agreement between all models of GLDAS (MOS, NOAH, VIC) and TRMM data is observed. The correlation between GRACE TWS, TRMM precipitation, and GLDAS models' rainfall rate is given in the Table 4.

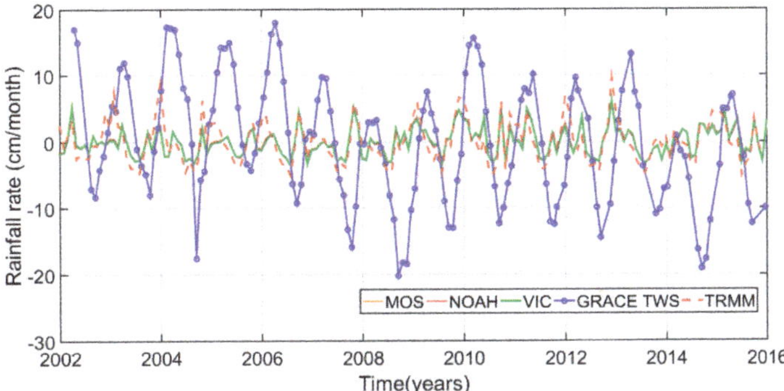

Figure 10. Residual mean monthly rainfall rate (cm/month) from GLDAS-MOS (in orange), GLDAS-NOAH (in red), and GLDAS-VIC (in green) models, and from TRMM (in red striped) and residual mean monthly TWS variation from GRACE data (in blue with circle) from 2002 to 2016 for Turkey.

Table 4. Correlation between GRACE TWS, TRMM precipitation, and GLDAS models (MOS, NOAH, VIC) rainfall rate according to Spearman's rho method.

Correlations				
			GRACE	**TRMM**
	GRACE TWS	Correlation Coefficient	1.000	0.338 **
		Sig. (2-tailed)		0.000
	MOS rainfall rate	Correlation Coefficient	0.243 **	0.826 **
		Sig. (2-tailed)	0.003	0.000
Spearman's rho	NOAH rainfall rate	Correlation Coefficient	0.244 **	0.826 **
		Sig. (2-tailed)	0.003	0.000
	VIC rainfall rate	Correlation Coefficient	0.227 **	0.788 **
		Sig. (2-tailed)	0.005	0.000
	TRMM precipitation	Correlation Coefficient	0.338 **	1.000
		Sig. (2-tailed)	0.000	

** Correlation is significant at the 0.01 level (2-tailed).

There is a positive correlation between GRACE TWS and the rainfall rate derived from GLDAS models (r = ~0.24) and a strong positive correlation between TRMM precipitation and the rainfall rate derived from GLDAS models (r = ~0.82).

3.4.3. Impact of Evapotranspiration

The estimation of evapotranspiration (ET) has been performed by using the water balance equation [60,61] as is traditional for a closed basin area, in the case of available observed streamflow data (see Equation (6)):

$$ET(t) = P(t) - R(t) - TWS_{GRACE} \tag{6}$$

where t is the time; TWS_{GRACE} is the water storage derived from GRACE data; and $P(t)$, $R(t)$, and $ET(t)$ are the precipitation provided from TRMM, runoff extracted from available GLDAS files of different models, and evapotranspiration, respectively. Figure 11 compares the residual mean monthly evapotranspiration values calculated from Equation (6) and GRACE TWS data in order to understand the role of the evapotranspiration effect on the TWS parameter. Evapotranspiration values range from −3.8 to 8.6 cm/month (VIC). This finding shows that both data (GLDAS evapotranspiration and GRACE TWS) are in-phase. The amplitudes of evapotranspiration seem to impact the rainfall rate equally to the TWS amplitudes.

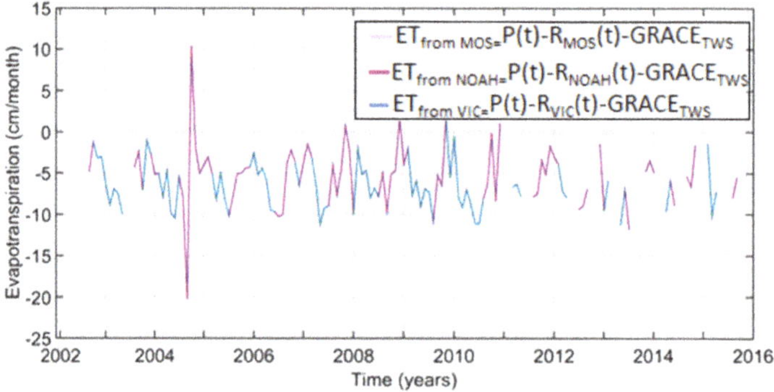

Figure 11. Mean monthly evapotranspiration (cm/month) calculated from precipitation (TRMM) minus runoff (from model (MOS, NOAH, VIC)) minus GRACE TWS variations between 2002 and 2016 for Turkey.

3.4.4. Impact of Soil Moisture

First, for each GLDAS land model, soil moisture values (kg/m^2) available in the GLDAS data files are extracted from the global grids, and then reduced to the studied region scale. In addition, according to the models, soil moisture values are summed with respect to the number of model layers, as follows: three-layer model for MOS and VIC and four-layer model for NOAH are converted to cm and averaged, and the residual mean monthly variations are finally calculated. Figure 12 shows the comparison between residual mean monthly soil moisture variation (cm) obtained from GLDAS models and residual mean monthly GRACE TWS of Turkey from 2002 to 2016. Soil moisture values ranges from −15.4 to 18.9 cm (MOS), while GRACE TWS data ranges from −16 to 20 cm. According to Figure 12, it can be concluded that for Turkey, the most efficient parameter producing the important part of the GRACE TWS signal is the soil moisture. This statement is supported not only by the MOS model, but also by other GLDAS models (NOAH, VIC). Table 5 shows the correlation coefficients between GRACE TWS and soil moisture obtained from GLDAS models (MOS, NOAH, VIC).

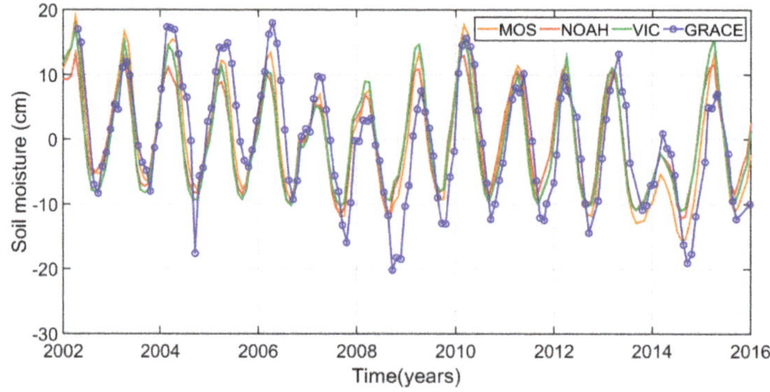

Figure 12. Residual mean monthly soil moisture variation (cm) of Turkey from GLDAS MOS (in orange), GLDAS-NOAH (in red), and GLDAS-VIC (in green) models, and residual mean monthly TWS variation from GRACE data (in blue with circle) from 2002 to 2016.

Table 5. Correlation between GRACE TWS and soil moisture derived from GLDAS models (MOS, NOAH, VIC) according to Spearman's rho method.

Correlations			
			GRACE
Spearman's rho	GRACE TWS	Correlation Coefficient Sig. (2-tailed)	1.000
	MOS soil moisture	Correlation Coefficient Sig. (2-tailed)	0.840 ** 0.000
	NOAH soil moisture	Correlation Coefficient Sig. (2-tailed)	0.797 ** 0.000
	VIC soil moisture	Correlation Coefficient Sig. (2-tailed)	0.790 ** 0.000

** Correlation is significant at the 0.01 level (2-tailed).

The correlation coefficient between GRACE TWS and soil moisture derived from the MOS model with r = 0.84 indicates a strong positive relation between these two times series and supports that the GRACE TWS signal is mostly dependent on soil moisture content in the studied region. To sum up, soil moisture is a key parameter in terms of drought monitoring because, as mentioned previously in Section 3.1, the insufficiency of soil moisture is a turning point indicating the change from classical drought, observed because of the lack of precipitation, to the agricultural drought. GRACE TWS time series are very sensible to agricultural drought (2006–2008).

3.5. Understanding the Drought and Its Relation with Climatic Change

In Section 3.4, we investigated the impact and correlation of different parameters (e.g., precipitation, soil moisture, etc.) on the signal amplitude, specifically on the TWS time series. We found that the precipitation is an important parameter which governs the pattern of spatial mass distribution (Section 3.2) and soil moisture produces the most important part of the GRACE TWS signal (Section 3.4.4). In this part, we decided to combine the available data and our findings with the self-calibrating Palmer Drought Severity Index, ENSO, and NOA index, which use also precipitation, temperature, soil moisture, and so on [51,52] to understand the type, the variability, and the severity of a drought event.

3.5.1. Self-Calibrating Drought Severity Index (SCPDSI)

In Figure 13, non-seasonal GRACE TWS data are compared with the SCPDSI drought severity index.

The analysis of the SCPDSI time series shows a good agreement in terms of the increasing and decreasing trend of the signal with GRACE-derived residual TWS. From a general point of view, in Figure 13, except for some small variations, SCPDSI time series show that progressive decreasing periods (2002–2008, 2010–2014) and increasing periods are (2008–2010, 2014–2016) followed up with the GRACE TWS. Nevertheless, the SCPDSI index is not as sensitive as GRACE data to small variation, especially during 2002–2006 (miss the increase 02/2003–02/2004). September 2004 (may be also an overestimation of GRACE solution), 2008, and October 2014 appear as dramatic drought cases for both time series.

Statistically speaking, according to the Kolmogorov-Smirnov method, p is equal to 0.2 and the Pearson Correlation can be used here to study the correlation coefficients between GRACE TWS and the self-calibrating Palmer Drought Severity Index (SCPDSI), as shown in Table 6.

A positive correlation is found to be r = 0.25. To sum up, according to GRACE data, the SCPDSI index, and historical data, the reason for the drought can be primarily categorized as meteorological. In this step, we decided to extend our results and to more deeply study the underlying processes of the drought event, i.e., the causes resulting in a lack of precipitation from the point of view of climatic change. For this reason, in the next section (Section 3.5.2), GRACE TWS anomalies are compared with the ENSO and NAO index, improving our understanding of the occurrence of drought events.

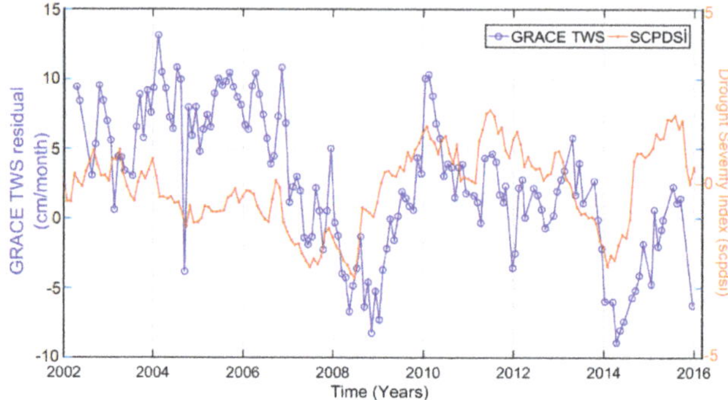

Figure 13. Mean monthly residual TWS variation from GRACE data (blue circled line, seasonal signal removed) and mean monthly self-calibrating Palmer Drought Severity Index (orange dotted line) from ~2002 to 2016 of Turkey.

Table 6. Correlation between residual GRACE TWS and self-calibrating Palmer Drought Severity Index (SCPDSI).

Correlations		SCPDSI	GRACE TWS
SCPDSI	Pearson Correlation Sig. (2-tailed)	1	0.246 ** 0.002
GRACE TWS	Pearson Correlation Sig. (2-tailed)	0.246 ** 0.002	1

** Correlation is significant at the 0.01 level (2-tailed).

3.5.2. El Niño Southern Oscillation (ENSO) and North Atlantic Oscillation (NAO) Indices

Figure 14 compares the SST3.4 and NAO time series with the non-seasonal GRACE TWS anomaly time series.

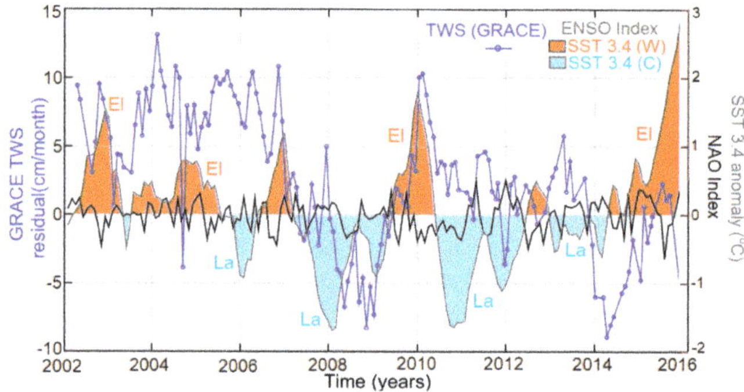

Figure 14. Mean monthly residual TWS variation from GRACE data (blue circled line, seasonal signal removed) compared with the ENSO index Niño SST3.4 (El: El Niño and La: La Niña; Warm: orange shading and Cool: cyan shading) Niño SST3.4. (prepared as in [62]) and with monthly mean NAO index (in black).

According to [63], ENSO also affects the Mediterranean winter climate. During El Niño events, the Mediterranean cyclone track is shifted northward, which affects precipitation. Moreover, less precipitation in southwestern Europe, as well as the Black Sea area, during cold events, but more precipitation in the same regions during warm events are founded [64]. Figure 14 reveals that cold events (La Nina: cyan shading) corresponding to La Nina effect occur between 2006–2009 and 2010–2014. This can be interpreted as a climatic impact creating a lack of precipitation in Turkey. This statement is also validated by the decreasing trend of GRACE TWS and SCPDSI (Figure 13) time series. This lack of precipitation due to the possible effect of the La Nina phase first results in a meteorological drought (Figure 13), which then turns into agricultural and hydrological drought, as mentioned in [55]. Spatially speaking, as a possible proof of the impact of ENSO, as mentioned in [64], Black sea coasts in Figure 6b1 show a low ($\leq$3 cm) annual amplitude of precipitation estimated from the TRMM model, unlike the other coasts.

According to Figure 14, the warm phase El Nino (orange shading) is observed between 2002 and the beginning of 2007 (with some interaction of the cold phase creating instantaneous drops) and 2014–2016. The climatic impact of the warm phase, resulting in an increase in precipitation, seems to appear as the increase of GRACE TWS and ENSO, while SCPDSI is more sensitive to the decrease of signal amplitude in this period due to the interaction of the cold phase (Figure 13). GRACE TWS anomalies (Figure 14) and rainfall rate (Figure 10) are important for these time intervals cited above. Spatially speaking, the bigger annual amplitudes of precipitation ($\geq$7 cm) estimated from the TRMM model (Figure 6b1) are observed in Southeastern, Mediterranean, and Aegean parts of Turkey.

Concerning the NAO index, the time series do not show any strong positive or negative anomalies or trends. As a reminder, strong positive phases (+) tend to be associated with below-average precipitation, while strong negative phases (-) are related with the above average temperature and precipitation anomalies. In this case, the amplitude of the NAO time series is small. It can be concluded that NAO does not have any significative impact on the studied area. Table 7 shows the correlation between GRACE TWS with ENSO and the NAO index. There is a positive correlation in the order of r = 0.3 between GRACE TWS and the ENSO index. Additionally, GRACE TWS and the NAO index show a small negative correlation (r = -0.05).

Table 7. Correlation between residual GRACE TWS and ENSO Index.

Correlation		GRACE TWS
GRACE TWS	Pearson Correlation Sig. (2-tailed)	1.000
ENSO index	Pearson Correlation Sig. (2-tailed)	0.295 ** 0.000
NAO index	Pearson Correlation Sig. (2-tailed)	-0.049 0.552

** Correlation is significant at the 0.01 level (2-tailed).

4. Discussion

Our results provide a broad context for the current hydrological status in Turkey by combining various external data sets (e.g., hydrological models, remote sensing techniques, drought indices) and reveal new drought events, spatial extension of the mass change, long-term variation, impacts on TWS, and the effect of climatic change, in addition to the previous studies within the GRACE mission operation time. The limitations of the work are related to the spatial resolution of the GRACE mission, which does not allow monitoring of the very high resolution surface mass change (e.g., dams, reservoirs); the lack or inaccessibility of the in-situ rainfall data. Which could provide information about groundwater withdrawal; and additionally and more specifically, the limited satellite mission lifetime that ended in 2017. Even though there has been a new following mission, "GRACE-FO", which started operating on 22 May, there is a data gap between the two missions. For this reason, as future research to generate a new perspective to drought analysis, there is the aim to develop statistical modelling from GRACE time series. This connection between GRACE and GRACE-FO revealed by statistical modelling will be valuable in terms of the continuity of drought monitoring and prediction, especially in the case of missing data within the GRACE-FO operation period. We also demonstrated that GRACE is more sensitive to agricultural and hydrological drought and less sensitive to meteorological drought, which occurs in the case of a lack of precipitation, increase of temperature, and decrease in humidity. GRACE data might be combined with the datasets derived from remote sensing techniques that measure above-cited external data sets to conduct a more sensitive analysis and to predict meteorological droughts [65]. The combined solutions can be assessed in different regions to test the sensibility of the GRACE data to differentiate different types/states of drought.

5. Conclusions

According to the drought analysis studied in this paper from GRACE-derived TWS time series, Turkey experienced dramatic drought events in 09/2004 (may also be an overestimation of GRACE solution), 09/2008, and 10/2014. Moreover, TWS decreasing periods are recorded as follows: 04/2002–09/2004; 02/2006–09/2008; 03/2010–10/2014. In terms of assessment of the drought, GRACE can help to better predict the possible drought (starting from 02/2006) nine months before, with a decreasing trend observed in GRACE TWS times series compared to previous studies which do not take satellite gravity data (see only since 11/2006) into account. Moreover, the GRACE signal is more sensitive to agricultural and hydrological drought compared to meteorological drought. Spatially, mass amplitudes are larger (20 cm) in the eastern part compared to central (2 cm, smaller) or Aegean parts, related to received precipitation levels. Shorelines also show distinctive values compared to the central part. There is an acceleration signal from the eastern side to the western side, which is related to the precipitation. Concerning long-term mass change, Turkey experiences a decreasing trend in the order of 1 cm/yr. Rainfall rate, evapotranspiration, and precipitation constitute a small part of the signal, while soil moisture is the parameter most affecting the GRACE signal in the studied region according to soil moisture values derived from GLDAS models and GRACE TWS results having a strong correlation (r = 0.84). Precipitation has a specific impact on the pattern of the spatial mass distribution. In Turkey, we observed a meteorological drought turning into agricultural and hydrological drought due to the climatic impact of the La Nina effect (cold phase) resulting in a lack of precipitation in Turkey. The GRACE signal is very sensitive to this climatic change. It is worth mentioning that the NAO index does not show any meaningful anomalies and correlation with GRACE TWS (r = −0.05). Finally, in order to real-time monitor and estimate possible drought conditions in the future, either in Turkey or in another region, we propose the combination of the new and up-to-date satellite gravity mission data (GRACE-FO), offering more accurate measurements and providing information about mass decreasing and increasing trends; the precipitation to understand the spatial mass distribution patterns; soil moisture data (models and also in-situ) to monitor the occurrence of a possible agricultural drought; the ENSO index to predict possible excess or deficiency in precipitation; and the drought indices, which provide information about the type and the variability of the drought.

Author Contributions: Data curation, G.O.A.; Funding acquisition, S.J.; Methodology, G.O.A. and S.J.; Resources, G.O.A.; Software, G.O.A.; Supervision, S.J.; Writing – original draft, G.O.A.; Writing – review & editing, G.O.A. and S.J.

Funding: The work is supported by the Strategic Priority Research Program of Chinese Academy of Sciences (Grant No. XDA23040102) and Startup Foundation for Introducing Talent of NUIST (Grant No. 2243141801036).

Acknowledgments: We thank the following organizations for providing the data used in this work: the GRACE Project and CSR (Center for Space Research, Univ. Texas), the TRMM projects, and a Global Land Data Assimilation System (GLDAS) developed jointly by scientists at the National Aeronautics and Space Administration (NASA) Goddard Space Flight Center (GSFC) and the National Oceanic and Atmospheric Administration (NOAA) National Centers for Environmental Prediction (NCEP). We are very thankful to three anonymous reviewers who helped us in improving the quality of our manuscript and also to Assist. Kamil Teke, Assoc. Prof. Semra Türkan from Hacettepe University, and Mustafa Serkan Işık from Istanbul Technical University for their help.

Conflicts of Interest: The authors declare no conflict of interest.

References

1. Dai, A. Drought under global warming: A review. *Wiley Interdiscip. Rev. Clim. Chang.* **2011**, *2*, 45–65. [CrossRef]
2. Famiglietti, J.S.; Rodell, M. Water in the balance. *Science* **2013**, *340*, 1300–1301. [CrossRef] [PubMed]
3. Huffman, G.J.; Bolvin, D.T.; Braithwaite, D.; Hsu, K.; Joyce, R.; Kidd, C.; Nelkin, E.J.; Sorooshian, S.; Tan, T.; Xie, P. Algorithm Theoretical Basis Document (ATBD) Version 5.2 NASA—NASA Global Precipitation Measurement (GPM) Integrated Multi-satellitE Retrievals for GPM (IMERG). Available online: https://go.nasa.gov/2M1webR (accessed on 10 January 2019).
4. Ashouri, H.; Hsu, K.L.; Sorooshian, S.; Braithwaite, D.K.; Knapp, K.R.; Cecil, L.D.; Nelson, B.R.; Prat, O.P. PERSIANN-CDR: Daily precipitation climate data record from multisatellite observations for hydrological and climate studies. *Bull. Am. Meteorol. Soc.* **2015**, *96*, 69–83. [CrossRef]
5. Joyce, R.J.; Janowiak, J.E.; Arkin, P.A.; Xie, P. CMORPH: A Method that Produces Global Precipitation Estimates from Passive Microwave and Infrared Data at High Spatial and Temporal Resolution. *J. Hydrometeorol.* **2004**, *5*, 487–503. [CrossRef]
6. Funk, C.; Peterson, P.; Landsfeld, M.; Pedreros, D.; Verdin, J.; Shukla, S.; Husak, G.; Rowland, J.; Harrison, L.; Hoell, A.; et al. The climate hazards infrared precipitation with stations—A new environmental record for monitoring extremes. *Sci. Data* **2015**, *2*, 150066. [CrossRef] [PubMed]
7. Okamoto, K.; Ushio, T.; Iguchi, T.; Takahashi, N.; Iwanami, K. The Global Satellite Mapping of Precipitation (GSMaP) project. In *International Geoscience and Remote Sensing Symposium (IGARSS)*; Institute of Electrical and Electronics Engineers (IEEE): Seoul, South Korea, 2005.
8. Kubota, T.; Hashizume, H.; Shige, S.; Okamoto, K.; Aonashi, K.; Takahashi, N.; Ushio, T.; Kachi, M. Global precipitation map using satelliteborne microwave radiometers by the GSMaP project: Production and validation. In *International Geoscience and Remote Sensing Symposium (IGARSS)*; Institute of Electrical and Electronics Engineers (IEEE): Denver, CO, USA, 2006.
9. Aonashi, K.; Awaka, J.; Hirose, M.; Kozu, T.; Kubota, T.; Liu, G.; Shige, S.; Kida, S.; Seto, S.; Takahashi, N.; et al. GSMaP Passive Microwave Precipitation Retrieval Algorithm: Algorithm Description and Validation. *J. Meteorol. Soc. Jpn.* **2009**, *87*, 119–136. [CrossRef]
10. Long, D.; Pan, Y.; Zhou, J.; Chen, Y.; Hou, X.; Hong, Y.; Scanlon, B.R.; Longuevergne, L. Global analysis of spatiotemporal variability in merged total water storage changes using multiple GRACE products and global hydrological models. *Remote Sens. Environ.* **2017**, *192*, 198–216. [CrossRef]
11. Luo, Z.; Yao, C.; Li, Q.; Huang, Z. Terrestrial water storage changes over the Pearl River Basin from GRACE and connections with Pacific climate variability. *Geod. Geodyn.* **2016**, *7*, 171–179. [CrossRef]
12. Long, D.; Longuevergne, L.; Scanlon, B.R. Global analysis of approaches for deriving total water storage changes from GRACE satellites. *Water Resour. Res.* **2015**, *51*, 2574–2594. [CrossRef]
13. Forootan, E.; Kusche, J.; Loth, I.; Schuh, W.D.; Eicker, A.; Awange, J.; Longuevergne, L.; Diekkrüger, B.; Schmidt, M.; Shum, C.K. Multivariate Prediction of Total Water Storage Changes Over West Africa from Multi-Satellite Data. *Surv. Geophys.* **2014**, *35*, 913–940. [CrossRef]
14. Castellazzi, P.; Longuevergne, L.; Martel, R.; Rivera, A.; Brouard, C.; Chaussard, E. Quantitative mapping of groundwater depletion at the water management scale using a combined GRACE/InSAR approach. *Remote Sens. Environ.* **2018**, *205*, 408–418. [CrossRef]
15. Hachborn, E.; Berg, A.; Levison, J.; Ambadan, J.T. Sensitivity of GRACE-derived estimates of groundwater-level changes in southern Ontario, Canada. *Hydrogeol. J.* **2017**, *25*, 2391–2402. [CrossRef]
16. Schumacher, M.; Forootan, E.; van Dijk, A.I.J.M.; Schmied, H.M.; Crosbie, R.S.; Kusche, J.; Döll, P. Improving drought simulations within the Murray-Darling Basin by combined calibration/assimilation of GRACE data into the WaterGAP Global Hydrology Model. *Remote Sens. Environ.* **2018**, *204*, 212–228. [CrossRef]
17. Zhou, H.; Luo, Z.; Tangdamrongsub, N.; Wang, L.; He, L.; Xu, C.; Li, Q. Characterizing drought and flood events over the Yangtze River Basin using the HUST-Grace2016 solution and ancillary data. *Remote Sens.* **2017**, *9*, 1100. [CrossRef]

18. Jin, S.; Zhang, T. Terrestrial Water Storage Anomalies Associated with Drought in Southwestern USA from GPS Observations. *Surv. Geophys.* **2016**, *37*, 1139–1156. [CrossRef]

19. Swenson, S.; Wahr, J. Monitoring the water balance of Lake Victoria, East Africa, from space. *J. Hydrol.* **2009**, *370*, 163–176. [CrossRef]

20. Reager, J.T.; Famiglietti, J.S. Global terrestrial water storage capacity and flood potential using GRACE. *Geophys. Res. Lett.* **2009**, *36*. [CrossRef]

21. Frappart, F.; Ramillien, G. Monitoring groundwater storage changes using the Gravity Recovery and Climate Experiment (GRACE) satellite mission: A review. *Remote Sens.* **2018**, *10*, 829. [CrossRef]

22. Rodell, M.; Houser, P.R.; Jambor, U.; Gottschalck, J.; Mitchell, K.; Meng, C.J.; Arsenault, K.; Cosgrove, B.; Radakovich, J.; Bosilovich, M.; et al. The Global Land Data Assimilation System. *Bull. Am. Meteorol. Soc.* **2004**, *85*, 381–394. [CrossRef]

23. Liang, X.; Lettenmaier, D.P.; Wood, E.F. One-dimensional statistical dynamic representation of subgrid spatial variability of precipitation in the two-layer variable infiltration capacity model. *J. Geophys. Res. Atmos.* **1996**, *101*, 21403–21422. [CrossRef]

24. Koren, V.; Schaake, J.; Mitchell, K.; Duan, Q.Y.; Chen, F.; Baker, J.M. A parameterization of snowpack and frozen ground intended for NCEP weather and climate models. *J. Geophys. Res. Atmos.* **1999**, *104*, 19569–19585. [CrossRef]

25. Koster, R.D.; Suarez, M.J. Energy and Water Balance Calculations in the Mosaic LSM. In *NASA Technical Memorandum*; National Aeronautics and Space Administration: Greenbelt, MD, USA, 1996.

26. Dai, Y.; Zeng, X.; Dickinson, R.E.; Baker, I.; Bonan, G.B.; Bosilovich, M.G.; Denning, A.S.; Dirmeyer, P.A.; Houser, P.R.; Niu, G.; et al. The common land model. *Bull. Am. Meteorol. Soc.* **2003**, *84*, 1013–1024. [CrossRef]

27. Fan, Y.; van den Dool, H. Climate Prediction Center global monthly soil moisture data set at 0.5° resolution for 1948 to present. *J. Geophys. Res. D Atmos.* **2004**, *109*. [CrossRef]

28. Milly, P.C.D.; Shmakin, A.B. Global Modeling of Land Water and Energy Balances. Part I: The Land Dynamics (LaD) Model. *J. Hydrometeorol.* **2002**, *3*, 283–299. [CrossRef]

29. Döll, P.; Kaspar, F.; Lehner, B. A global hydrological model for deriving water availability indicators: Model tuning and validation. *J. Hydrol.* **2003**, *3*, 283–299. [CrossRef]

30. Döll, P.; Müller Schmied, H.; Schuh, C.; Portmann, F.T.; Eicker, A. Global-scale assessment of groundwater depletion and related groundwater abstractions: Combining hydrological modeling with information from well observations and GRACE satellites. *Water Resour. Res.* **2014**, *50*, 5698–5720. [CrossRef]

31. Döll, P.; Fritsche, M.; Eicker, A.; Müller Schmied, H. Seasonal Water Storage Variations as Impacted by Water Abstractions: Comparing the Output of a Global Hydrological Model with GRACE and GPS Observations. *Surv. Geophys.* **2014**, *35*, 1311–1331. [CrossRef]

32. Ramillien, G.; Frappart, F.; Cazenave, A.; Güntner, A. Time variations of land water storage from an inversion of 2 years of GRACE geoids. *Earth Planet. Sci. Lett.* **2005**, *30*, 283–301. [CrossRef]

33. Güntner, A. Improvement of global hydrological models using GRACE data. *Surv. Geophys.* **2008**, *29*, 375–397. [CrossRef]

34. Atayer, E.S. Yeryuvarı Gravite Alanının Aylık GRACE Çözümleri İle İzlenmesi ve Duyarlılığı Üzerine Bir İnceleme. Ph.D. Thesis, Yıldız Teknik University, Istanbul, Turkey, 2012.

35. Lenk, O. Satellite based estimates of terrestrial water storage variations in Turkey. *J. Geodyn.* **2013**, *67*, 106–110. [CrossRef]

36. Caló, F.; Notti, D.; Galve, J.P.; Abdikan, S.; Görüm, T.; Pepe, A.; Şanli, F.B. DInSAR-based detection of land subsidence and correlation with groundwater depletion in konya plain, Turkey. *Remote Sens.* **2017**, *17*, 83. [CrossRef]

37. Yıldırım, Y.Ö. GRACE uydu verileri ile Türkiye'nin uzun dönemli su kütle değişiminin incelenmesi, Gebze Teknik University. 2015. Available online: https://bit.ly/1H1SX0L (accessed on 10 January 2019).

38. Eicker, A.; Forootan, E.; Springer, A.; Longuevergne, L.; Kusche, J. Does GRACE see the terrestrial water cycle "intensifying"? *J. Geophys. Res.* **2016**, *121*, 733–745. [CrossRef]

39. Günay, G. Türkiye'nin yüzey sulari ve yeralti sulari potansiyeli. *Bilim Aklın Aydınlığında Eğitim* **2011**, *132*, 56–60.

40. Türkeş, M. Influence of geopotential heights, cyclone frequency and Southern Oscillation on rainfall variations in Turkey. *Int. J. Climatol.* **1998**, *18*, 649–680. [CrossRef]

41. Türkeş; Serap, A.A.; Zerrin, D. Drought periods and severity over the Konya Sub-region of the Central Anatolia Region according to the Palmer Drought Index. *Coğrafi Bilim. Derg.* **2009**, *7*, 129–144.
42. Kapluhan, E. Türkiye'de Kuraklık Ve Kuraklığın Tarım Etkisi. *Marmara Coğrafya Derg.* **2013**, *27*, 487–510.
43. Marım, G.; Şensoy, A.; Şorman, A.; Şorman, A. Yukarı Fırat Havzası İçin Elde Edilen Kar Çekilme Eğrilerinin Zamansal Analizi ve Modelleme Çalışmaları. In *Kar Hidrolojisi Konferansı*; DSİ VIII.Regional Directorate, Atatürk University: Erzurum, Turkey, 2008.
44. Cheng, M.; Ries, J.C.; Tapley, B.D. Variations of the Earth's figure axis from satellite laser ranging and GRACE. *J. Geophys. Res. Solid Earth* **2011**, *116*. [CrossRef]
45. Swenson, S.; Chambers, D.; Wahr, J. Estimating geocenter variations from a combination of GRACE and ocean model output. *J. Geophys. Res. Solid Earth* **2008**, *113*. [CrossRef]
46. Swenson, S.; Wahr, J. Post-processing removal of correlated errors in GRACE data. *Geophys. Res. Lett.* **2006**, *33*. [CrossRef]
47. Swenson, S.; Wahr, J. Methods for inferring regional surface-mass anomalies from Gravity Recovery and Climate Experiment (GRACE) measurements of time-variable gravity. *J. Geophys. Res.* **2002**, *107*, ETG 3. [CrossRef]
48. Han, D.; Wahr, J. The viscoelastic relaxation of a realistically stratified earth, and a further analysis of postglacial rebound. *Geophys. J. Int.* **1995**, *120*, 287–311. [CrossRef]
49. Palmer, W.C. Meteorological Drought. In *US Weather Bureau Research Paper*; Department of Commerce, Weather Bureau: Washington, DC, USA, 1965.
50. Dai, A.; Lamb, P.J.; Trenberth, K.E.; Hulme, M.; Jones, P.D.; Xie, P. The recent Sahel drought is real. *Int. J. Climatol.* **2004**, *24*, 1323–1331. [CrossRef]
51. Heim, R.R. A review of twentieth-century drought indices used in the United States. *Bull. Am. Meteorol. Soc.* **2002**, *83*, 1149–1165. [CrossRef]
52. Wanders, N.; van Lanen, H.A.J.; van Loon, A.F. *Indicators for Drought Characterization on a Global Scale (Technical Report No. 24)*; Wageningen Universiteit: Wageningen, The Netherlands, 2010.
53. Kahya, E.; Marti, A.I. Do El Niño events modulate the statistical characteristics of Turkish streamflow? In Proceedings of the 27th Annual American Geophysical Union Hydrology Days, Fort Collins, CO, USA, 19–21 March 2007.
54. Hurrell, J.W.; Kushnir, Y.; Ottersen, G.; Visbeck, M. An overview of the north atlantic oscillation. In *Geophysical Monograph Series*; American Geophysical Union: Washington, DC, USA, 2003; ISBN 9781118669037.
55. Kurnaz, L. Drought in Turkey. Available online: https://bit.ly/2ADAp94 (accessed on 10 January 2019).
56. Türkeş, M.; Yıldız, D. Gözlenen Bugünkü ve Benzeş tirilen Gelecek Yağış Değişimleri ve Kuraklık Olayları Perspektifinde Türkiye'de Hidroelektrik Santrallerin Geleceği. Available online: https://bit.ly/2ShevsG (accessed on 10 January 2019).
57. Jin, S.; Park, J.U.; Cho, J.H.; Park, P.H. Seasonal variability of GPS-derived zenith tropospheric delay (1994-2006) and climate implications. *J. Geophys. Res. Atmos.* **2007**, *112*. [CrossRef]
58. Voss, K.A.; Famiglietti, J.S.; Lo, M.; De Linage, C.; Rodell, M.; Swenson, S.C. Groundwater depletion in the Middle East from GRACE with implications for transboundary water management in the Tigris-Euphrates-Western Iran region. *Water Resour. Res.* **2013**, *49*, 904–914. [CrossRef]
59. Longuevergne, L.; Wilson, C.R.; Scanlon, B.R.; Crétaux, J.F. GRACE water storage estimates for the middle east and other regions with significant reservoir and lake storage. *Hydrol. Earth Syst. Sci.* **2013**, *17*, 4817–4830. [CrossRef]
60. Rodell, M.; Famiglietti, J.S.; Chen, J.; Seneviratne, S.I.; Viterbo, P.; Holl, S.; Wilson, C.R. Basin scale estimates of evapotranspiration using GRACE and other observations. *Geophys. Res. Lett.* **2004**, *31*. [CrossRef]
61. Swenson, S.; Wahr, J. Estimating Large-Scale Precipitation Minus Evapotranspiration from GRACE Satellite Gravity Measurements. *J. Hydrometeorol.* **2006**, *7*, 252–270. [CrossRef]
62. Zhang, Z.; Chao, B.F.; Chen, J.; Wilson, C.R. Terrestrial water storage anomalies of yangtze river basin droughts observed by GRACE and connections with ENSO. *Glob. Planet. Chang.* **2015**, *126*, 35–45. [CrossRef]
63. Brönnimann, S. Impact of El Niño-Southern Oscillation on European climate. *Rev. Geophys.* **2007**, *45*. [CrossRef]

64. Fraedrich, K.; Müller, K. Climate anomalies in Europe associated with ENSO extremes. *Int. J. Climatol.* **1992**, *12*, 25–31. [CrossRef]
65. Jin, S.G.; Zhu, W.Y. Analysis of improving the precision of inSAR with GPS measurements. *J. Remote Sens. Info.* **2001**, *4*, 8–11.

 *remote sensing*

Article

GOCE-Derived Coseismic Gravity Gradient Changes Caused by the 2011 Tohoku-Oki Earthquake

Xinyu Xu [1,2], Hao Ding [1,*], Yongqi Zhao [1], Jin Li [3] and Minzhang Hu [4]

[1] School of Geodesy and Geomatics, Wuhan University, Wuhan 430079, China; xyxu@sgg.whu.edu.cn (X.X.); yqzhao@whu.edu.cn (Y.Z.)
[2] State Key Laboratory of Information Engineering in Surveying, Mapping and Remote Sensing, Wuhan University, Wuhan 430079, China
[3] Shanghai Astronomical Observatory, Chinese Academy of Sciences, Shanghai 200030, China; lijin@shao.ac.cn
[4] Key Laboratory of Earthquake Geodesy, Institute of Seismology, China Earthquake Administration, 40 Hongshan Celu, Wuhan 430071, China; huminzhang@126.com
* Correspondence: dhaosgg @sgg.whu.edu.cn; Tel.: +86-027-68778670

Received: 11 April 2019; Accepted: 25 May 2019; Published: 30 May 2019

Abstract: In contrast to most of the coseismic gravity change studies, which are generally based on data from the Gravity field Recovery and Climate Experiment (GRACE) satellite mission, we use observations from the Gravity field and steady-state Ocean Circulation Explorer (GOCE) Satellite Gravity Gradient (SGG) mission to estimate the coseismic gravity and gravity gradient changes caused by the 2011 Tohoku-Oki Mw 9.0 earthquake. We first construct two global gravity field models up to degree and order 220, before and after the earthquake, based on the least-squares method, with a bandpass Auto Regression Moving Average (ARMA) filter applied to the SGG data along the orbit. In addition, to reduce the influences of colored noise in the SGG data and the polar gap problem on the recovered model, we propose a tailored spherical harmonic (TSH) approach, which only uses the spherical harmonic (SH) coefficients with the degree range 30–95 to compute the coseismic gravity changes in the spatial domain. Then, both the results from the GOCE observations and the GRACE temporal gravity field models (with the same TSH degrees and orders) are simultaneously compared with the forward-modeled signals that are estimated based on the fault slip model of the earthquake event. Although there are considerable misfits between GOCE-derived and modeled gravity gradient changes (ΔV_{xx}, ΔV_{yy}, ΔV_{zz}, and ΔV_{xz}), we find analogous spatial patterns and a significant change (greater than 3σ) in gravity gradients before and after the earthquake. Moreover, we estimate the radial gravity gradient changes from the GOCE-derived monthly time-variable gravity field models before and after the earthquake, whose amplitudes are at a level over three times that of their corresponding uncertainties, and are thus significant. Additionally, the results show that the recovered coseismic gravity signals in the west-to-east direction from GOCE are closer to the modeled signals than those from GRACE in the TSH degree range 30–95. This indicates that the GOCE-derived gravity models might be used as additional observations to infer/explain some time-variable geophysical signals of interest.

Keywords: coseismic gravity gradient changes; gravity field model; GOCE; GRACE

---

1. Introduction

Because satellite gravity observations are not limited to Earth's surface conditions, and can cover the whole Earth quickly, they provide an independent way to detect the coseismic effects of large earthquakes, which is a good complement for other earthquake measurements (e.g., surface deformations), and are of great scientific significance.

A new generation of satellite gravimetry missions, including CHAMP (CHAllenging Minisatellite Payload) [1], GRACE (Gravity field Recovery and Climate Experiment) [2] and GOCE (Gravity field and steady state Ocean Circulation Explorer) [3], have been successfully implemented to detect global static and time-variable gravity signals. A large number of real observations and derived products from the CHAMP, GRACE and GOCE missions have been applied to Earth-science research, especially temporal gravity signals derived from the GRACE mission, which are used to monitor mass transport in the Earth system. The detection of coseismic gravity change signals using GRACE data has been widely studied. Generally, monthly gravity field models from GRACE before and after the earthquake were used to derive coseismic gravity or gravity gradient changes. Long-to-medium-wavelength coseismic and post-seismic gravity changes from large-scale earthquakes (e.g., the 2004 Sumatra-Andaman Mw = 9.1, 2010 Maule Mw = 8.8, 2011 Tohoku-Oki Mw = 9.0) have been adequately detected by the GRACE mission [4–17]. Wang et al. [13] and Li and Shen [17] also determined the coseismic gravity gradient changes caused by the 2011 Japan Tohoku-Oki earthquake. The GOCE mission has been proven successful for constructing regional geoid models combining with the EGM2008 and terrestrial gravity datasets [18,19], and can be used to study the lithospheric modeling, dynamic topography, and glacial isostatic adjustment [20–23]. However, a few studies have focused on inferring the coseismic gravity change signals from simulated data [24–26] or real GOCE observations [27–29].

The main scientific objective of the GOCE mission is to recover static gravity field models up to degree and order (d/o) 200 by using SGG (Satellite Gravity Gradient) observations [3]. Because of the lower sensitivity of the gradiometer on board of the GOCE satellite at lower frequencies, the gravity gradient observations are not sensitive to temporal signals, the main power of which is at long wavelengths (commonly, at spatial resolutions higher than 1500 km) [30,31]. However, the coseismic gravity change signals from the 2011 Japan Tohoku-Oki earthquake were still detected by the GOCE mission [27,28]. Garcia et al. [27] showed that GOCE's gradiometer, acting similarly to a seismometer in orbit, recorded the sound waves from the 2011 Tohoku earthquake. Fuchs et al. [28] combined a new vertical gravity gradient V'_{zz} from the diagonal components (V_{xx}, V_{yy}, and V_{zz}) along the orbit. The coseismic radial gravity gradient changes ΔV_{zz} in Fuchs et al. [28] were derived by subtracting the vertical gravity gradients computed by a reference model GOCO03s [32] from V'_{zz} with an along track and spatial filter applied to deal with the colored noise. Fuchs et al. [28] mentioned that they did not use the spatiospectral localization method like Han and Ditmar [26] to process geopotential coefficients, because there were no released global gravity field models (GGMs) from the post-earthquake GOCE data that could be used to detect coseismic gravity gradient changes.

In this paper, we choose a different approach than that proposed in Fuchs et al. [28]. First, we recover the GGMs (SH coefficients) before and after the earthquake from GOCE gravity gradient observations (V_{xx}, V_{yy}, and V_{zz}) along the orbit, based on the least-squares method. Note that, Fuchs et al. [28] did not estimate the SH coefficients of the pre- and post-earthquake global time-variable gravity field models. Then, we estimate the gravity changes (Δg) and the gravity gradient changes (ΔV_{xx}, ΔV_{yy}, ΔV_{zz}, and ΔV_{xz}) from the differences between two recovered global gravity models. In order to weaken the influences of colored noise and the polar gap problem from GOCE SGG observations on the recovered model, we use a tailored spherical harmonic (TSH) coefficients to recover the coseismic gravity change. Finally, we compare the results obtained from forward-modeled signals with the GRACE monthly gravity field models. Compared to Fuchs et al. [28], our approach can determine the coseismic gravity changes and the coseismic gravity gradient changes of other gravity gradient tensor (GGT) components in addition to ΔV_{zz}. The content of this manuscript is organized into four sections, beginning with a description of the methodology of GOCE data processing, forward modeling coseismic gravity gradient changes, and the computation of TSH coefficients in Section 2. The results and analysis are presented in Section 3. A summary and concluding remarks are provided in Section 4.

2. Methodology

2.1. Recovering Gravity Field Model from GOCE SGG Observations

In this paper, we derive the coseismic gravity and gravity gradient changes by evaluating the differences between a post-earthquake gravity field model and a pre-earthquake one from GOCE data; namely, we first recover gravity field models from the Gravity field and steady state Ocean Circulation Explorer Satellite Gravity Gradient (GOCE SGG) observations before and after the 2011 Tohoku-Oki earthquake. Here, we provide a brief description of the data processing strategies of recovering a gravity field model from GOCE SGG observations (more details can be found in Xu et al. [33]).

The component V_{ij} of the second-order gravity gradient tensor (GGT) is normally defined in the Local North-Oriented Frame (LNOF) with the x-axis pointing north, the y-axis pointing west and the z-axis up, as follows [33]:

$$V_{ij}(r,\theta,\lambda) = \sum_{m=0}^{\infty}\left[A_m^{ij}(r,\theta)\cos m\lambda + B_m^{ij}(r,\theta)\sin m\lambda\right], \quad \left.\begin{array}{c} A_m^{ij}(r,\theta) \\ B_m^{ij}(r,\theta) \end{array}\right\} = \sum_{n=m}^{\infty} H_{nm}^{ij}(r,\theta)\left\{\begin{array}{c} \overline{C}_{nm} \\ \overline{S}_{nm} \end{array}\right. \tag{1}$$

where the indices i and j define the gravitational gradient components (xx, yy, zz, xy, ...) with respect to the LNOF axes (x, y, z); the indices n and m are the degree and order, respectively, of the spherical expansion; and (r,θ,λ) are the spherical coordinates in the Earth's Fixed Reference Frame (EFRF), where r is the geocentric radius; and θ and λ are the spherical colatitude and longitude, respectively. $\overline{C}_{nm}$ and $\overline{S}_{nm}$ are the (fully normalized) geopotential cosine and sine coefficients. $A_m^{ij}(r,\theta)$, $B_m^{ij}(r,\theta)$, and $H_{nm}^{ij}(r,\theta)$ are the Fourier coefficients and transform coefficients, respectively; for more details, please refer to Koop [34].

Based on Equation (1), the functional and statistical models of the gravitational field recovered from the Satellite Gravity Gradient (SGG) data are defined as a standard Gauss-Markov model. Then, we can determine the geopotential coefficients $\overline{C}_{nm}$ and $\overline{S}_{nm}$, exploiting the least-squares (LS) method. However, the observed GGT of the GOCE mission is given in the Gradiometer Reference Frame (GRF), so we should transform the observations from the GRF to the LNOF by using the rotation matrix $\mathbf{R}_G^L$ [33]. However, the accuracies of the V_{xy} and V_{yz} components are lower than those of the other components, so they will contaminate the high-precision components (V_{xx}, V_{yy}, V_{zz}, and V_{xz}) in the transformation. To avoid this situation, we transform the base functions $(H_{nm}^{ij}(r,\theta)\left\{\begin{array}{c}\cos m\lambda \\ \sin m\lambda\end{array}\right\})$ in Equation (1) instead of transforming the GGT observations in GRF. In particular, we multiply the matrix $\mathbf{R}_G^L$ and its transposed matrix $\mathbf{R}_L^G$ on both sides of the base functions in Equation (1) at every epoch. Hence, we have

$$\mathbf{V}^{GRF} = \mathbf{R}_L^G \mathbf{V}^{LNOF} \mathbf{R}_G^L \tag{2}$$

where $\mathbf{V}^{GRF}$ and $\mathbf{V}^{LNOF}$ represent the gravitational gradient tensor in the GRF and LNOF, respectively.

Because the power spectral density (PSD) of the trace of the GGT represents the total error of the summation of the diagonal components (V_{xx}, V_{yy}, V_{zz}), the SGG observations are high-precision only within the designed measurement bandwidths (MBW) from 0.005 to 0.1 Hz according to Figure 1 [35]. The noise outside of this MBW increases with the decreasing frequency for frequencies below 0.005 Hz, and the increasing frequency for frequencies above 0.1 Hz, especially for the $1/f$ behavior at low frequencies, which show the character of the colored noise of the gradiometer [35]. To handle this colored noise in the SGG data, we apply a bandpass Auto Regression Moving Average (ARMA) filter with a pass-band of 0.005–0.041 Hz to both sides of the linear observation equation, these being Equation (2) and Equation (1) [36]. The maximum frequency (0.041 Hz) of the pass-band approximately corresponds to the maximum degree (220) of the recovered gravitational potential model based on the following formula:

$$f_{\max} = N_{\max}/T_r \tag{3}$$

where $T_r = 5383$ s is one satellite orbital revolution (cf. [37]).

According to Figure 1, the noise level of the components V_{xx} and V_{yy} is about 10 mE/$\sqrt{\text{Hz}}$ (1 mew $= 10^{-12}/\text{s}^{-2}$), that of V_{zz} is about 20 mE/$\sqrt{\text{Hz}}$, which is consistent with Rummel et al. [38]. Thus, to combine the diagonal components (V_{xx}, V_{yy}, and V_{zz}) in the LS, we set the ratio of the standard deviation factors of V_{xx}, V_{yy}, and V_{zz} to 1:1:2.

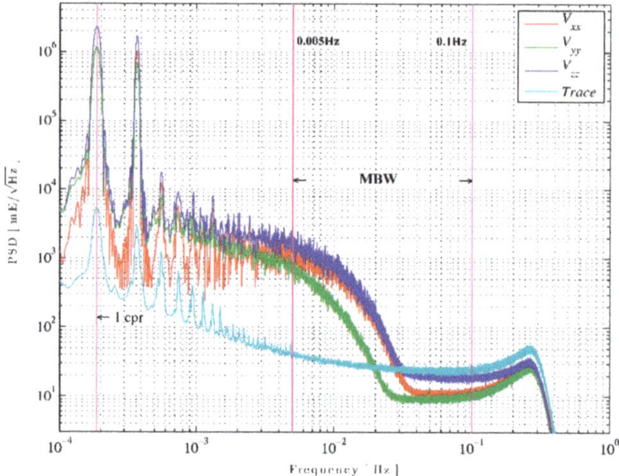

Figure 1. Power spectral densities of the diagonal components (V_{xx}, V_{yy}, and V_{zz}) and trace of the Gravity field and steady state Ocean Circulation Explorer (GOCE) gravity gradient tensor.

2.2. Tailored Spherical Harmonic Coefficients

Based on the data processing strategies in Section 2.1, the post-earthquake and pre-earthquake gravity field models can be recovered from GOCE SGG data. However, the lower and higher frequency bands of the recovered models are heavily influenced by the 0.005–0.041 Hz band-pass ARMA filter. Moreover, the polar gap problem of GOCE's orbit mainly affects the low-order spherical harmonic (SH) coefficients. We perform a numerical simulation to show how the recovered model's SH coefficients are affected by the band-pass filter and the polar gap problem. First, a 1 s sampling of 61-day GOCE satellite orbits are produced from the released reduced-dynamic orbit data (SST_PRD_2), with 10 s sampling by polynomial interpolation. Then, we simulate the GGT observations along the orbit by using the EGM2008 model up to d/o 220. We add simulated colored noise according to the prior PSD of the GGT's trace, as given by Cesare [35], to the simulated GGT data. Finally, by using these simulated observations, we recover the gravity field model based on the LS approach with the 0.005–0.041 Hz band-pass filter applied. The absolute errors of the recovered model's SH coefficients are obtained by calculating the difference between the model and the EGM2008 model. Figure 2 displays the error spectra of SH coefficients in log10 scale. According to this figure, SH coefficients up to d/o 160 with large errors are mainly located in the arrow-shaped region, which is framed by the red lines. Referring to the conclusions by Sneeuw [39], the errors of the near-zonal coefficients ($m \leq 10$) are mainly caused by the ill-posed problem occasioned by the 96.7° inclination of the GOCE satellite orbit. The other coefficients with large errors in the arrow-shaped region are due to the lower frequency limit of the band-pass filter. The coefficients outside the arrow region are called the tailored spherical harmonic (TSH) coefficients here, which correspond to two symmetrical quadrilaterals. These TSH coefficients will be used to show the coseismic gravity and gravity gradient changes.

As shown in Figure 2, the degree of the middle corner coefficients in the arrow-shaped region is approximately 30, which is very close to the degree 27 estimated by Equation (3), according to the minimum frequency of the 0.005–0.041 Hz band-pass filter. Thus, the minimum degree of the TSH coefficients is set to 30 herein. The vertices of the lower-left and lower-right corners of the arrow region approximately correspond to a degree of 80. Fuchs et al. [28] presented their results according to three bandwidths (0.005–0.05, 0.0039–0.03, and 0.00475–0.0175 Hz), and the bandwidth of 0.00475–0.0175 Hz was determined by the matched filter approach, which maximizes the expected signal. The degree range 30–95 approximately corresponds to the bandwidth of 0.005–0.0175 Hz [28]. So, we also choose 95 (see the red dashed line in Figure 2) for the maximum degree of the TSH coefficients. Moreover, the maximum degree of 95 is very close to the maximum d/o of the SH coefficients of Gravity field Recovery and Climate Experiment (GRACE)'s time-variable gravity field model, so we can compare the results from GRACE with those from GOCE at the same frequency band.

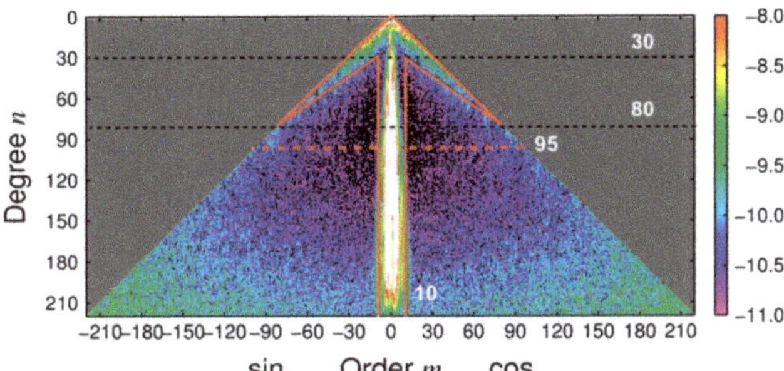

Figure 2. Error spectra of the recovered spherical harmonic (SH) coefficients in log10 scale compared to EGM2008.

2.3. Post-earthquake Gravity Changes from GOCE

Based on the theory and data processing strategies described above, we estimated two GGMs (the pre-earthquake model and the post-earthquake model) up to d/o 220 from the released GOCE's calibrated and corrected gravity gradients in the GRF (EGG_NOM_2 products) (GGT and IAQ) and precise science orbits with quality report (SST_PSO_2 products) (PRD and PRM) [40] before and after the 2011 Japan Tohoku-Oki earthquake. The IAQ products are the GRF to IRF attitude quaternions, and the PRM products are the Earth Fixed Reference Frame (EFRF) to Inertial Reference Frame (IRF) quaternions. We selected the data period from the 1st of November 2009 until the 28th of February 2011 for the pre-earthquake time span, and the period from the 15th of March 2011 until the 31st of May 2012 for the post-earthquake time span. Only the high-precision diagonal components (V_{xx}, V_{yy}, and V_{zz}) of the GGT are selected for modeling. Before forming the observation equation, some data preprocessing tasks were performed, such as data interpolation and outlier detection [33]. For a more detailed description of GGT data processing, please refer to Xu et al. [33]. To reduce the influence of high-frequency errors, a Gaussian filter with the smoothing radius of 210 km is also applied to the TSH coefficients. A radius of 210 km for the Gaussian filter is approximately determined with the formula 20000 km/N_{max}, corresponding to a maximum degree N_{max} of 95.

2.4. Post-earthquake Gravity Changes from GRACE

For comparison with GOCE, we also derived the coseismic gravity changes and gravity gradient changes from the Release05 (RL05) GRACE time-variable monthly gravity field models from the Center for Space Research (CSR), which were downloaded from the website http://icgem.gfz-potsdam.de/ICGEM/. The maximum d/o was 96.

The TSH coefficients that corresponded to the SH degree range 30–95 were chosen to maintain consistency with the processing of the GOCE models, and a Gaussian filter with the smoothing radius of 210 km was also applied to the TSH coefficients. The differences in the coefficients between the averages of the monthly models for one year before and after the earthquake were used to calculate the coseismic gravity changes and gravity gradient changes on a sphere with a height of 260 km.

2.5. Forward Modeling Coseismic Gravity and Gravity Gradient Changes

The PSGRN/PSCMP code for the modeling co- and post-seismic response of the Earth's crust to earthquakes from Wang et al. [41] is used to compute the coseismic gravity changes of the 2011 Tohoku-Oki earthquake based on a five-layer half-space Earth model and the fault slip model from Caltech provided by Wei and Sladen [42]. The layer depths, densities, and seismic velocities are extracted from the CRUST2.0 global tomography model [43], which are derived from the epicenter (38.1°N, 142.8°E) grid cell layer parameters, and are shown in Table 1. The upper four crustal layers in the model are treated as elastic materials. The bottom half-space mantle layer is treated as biviscous materials, i.e., a Burgers body with a transient (Kelvin) viscosity and a steady state (Maxwell) viscosity (Details about the Burgers viscosities will be described in Section 2.6). The free-air corrections, as calculated by the vertical surface displacements, are added to the calculated coseismic gravity changes from the PSGRN/PSCMP program [9]. A Bouguer layer seawater compensation effect is also corrected according to the vertical displacements with the consideration of the land-ocean differentiation [44]. Since Broerse et al. [44] pointed out that simply considering a global ocean for the seawater correction would lead to non-negligible errors (up to a few 10% for the 2011 Tohoku-Oki earthquake), using a practical land-ocean mask ensures a more reliable seawater effect correction for the model prediction in our study.

Table 1. The 5-layer half-space Earth model used in the prediction of co- and post-seismic gravity changes.

Depth (km)	Density (ρ) (10^3 kg/m^3)	V_P(km/s)	V_S(km/s)	Material Type
0–1	2.10	2.10	1.00	Elastic
1–8	2.70	6.00	3.40	Elastic
8–15	2.90	6.60	3.70	Elastic
15–22	3.05	7.20	4.00	Elastic
22–∞	3.40	8.20	4.70	Biviscous (Burgers body)

The computation region is from 28°N to 48°N latitude and from 132°E to 152°E longitude. The epicenter (38.1°N, 142.8°E) is located at the center of this region. The cell size of the grid is set to 0.1° × 0.1°. The global gridded coseismic gravity changes on a sphere are formed by the forward-modeled coseismic gravity changes, as well as filling in zero values outside the computational region. Based on the global gridded coseismic gravity changes, we estimate the gravitational potential spherical harmonic (SH) coefficients up to d/o 250, by using the classical spherical harmonic analysis approach [45]. Then, based on the estimated gravitational potential SH coefficients, we calculate changes of both the coseismic gravity and radial gravity gradient on the Earth's surface and on a sphere with a height of 260 km above the WGS84 reference ellipsoid, which are shown in Figures 3 and 4. According to the maximum d/o 250 of the SH coefficients, we use a Gaussian filter with a radius of 110 km here. The epicenter is also shown as a black star in these figures. The units of the gravity and the gravity gradient are mGal (1 mGal = 10^{-5} ms^{-2}) and mE, respectively. According to Figures 3 and 4, the spatial patterns of the coseismic gravity changes and gravity gradient changes show classic dipole characteristics in the nearly west-to-east direction. Additionally, the attenuation effect of the coseismic gravity change signal at the satellite height is very clear.

The amplitudes of the coseismic gravity changes and coseismic radial gravity gradient changes are reduced by more than a factor of 10 and 20, respectively.

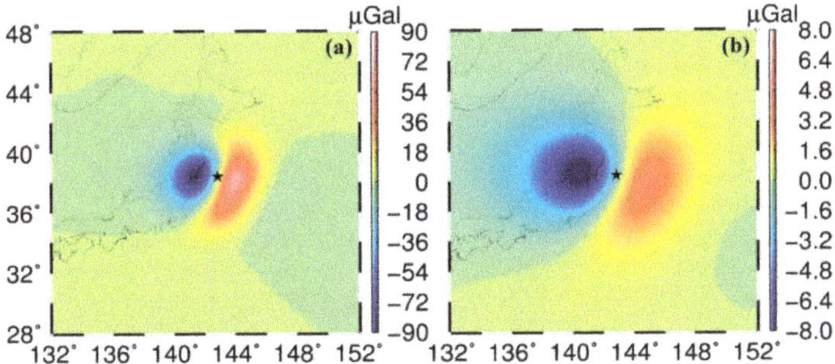

Figure 3. Coseismic gravity changes (**a**) on the Earth's surface and (**b**) on a sphere with a height of 260 km, when using forward-modeled spherical harmonic coefficients up to d/o 250 with a Gaussian filter applied (radius of 110 km). The units are µGal.

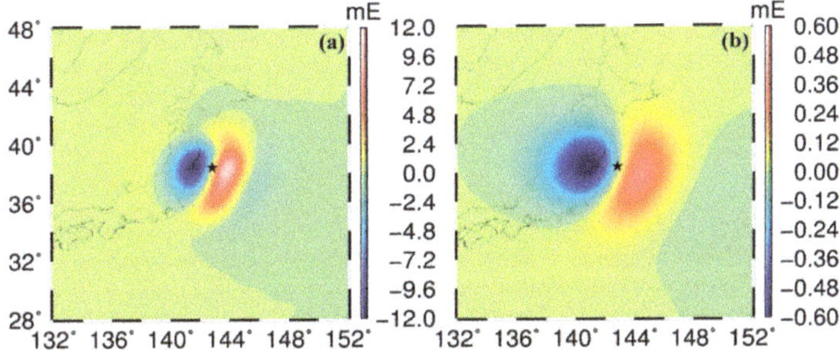

Figure 4. Coseismic radial gravity gradient change (**a**) on the Earth's surface and (**b**) on a sphere with a height of 260 km when using spherical harmonic coefficients up to d/o 250 with a Gaussian filter applied (radius of 110 km). The units are mE.

2.6. Modeling of Post-seismic Gravity Changes

For the GOCE and GRACE observations, the coseismic gravity field changes are estimated by using the differences in coefficients between the pre-earthquake model and the post-earthquake model. Here we note that taking the one-year mean field differences as the coseismic signals will inevitably bring in the impact of the post-seismic effects, which are likely to affect the peak-to-peak range, as well as the spatial pattern of the extracted signals [46]. Han et al. [46] revealed that a dominant post-seismic positive gravity change signal is visible surrounding the epicenter within a couple of years after the 2011 Tohoku-Oki earthquake, with the amplitude up to around +6 µGal under a 500 km spatial scale comparable as GRACE observations. Moreover, they gave the results from aviscoelastic relaxation model and an afterslip model, respectively, and showed that the observed vertical deformation at the coast and offshore agrees better with the viscoelastic relaxation model than the afterslip model. Therefore, we choose the viscoelastic relaxation model to estimate the post-seismic gravity change effect here. Based on our 5-layer half-space model in Table 1, we use Burgers viscosities including a transient viscosity of 10^{17} Pa s, and a steady state viscosity of 10^{18} Pa s, for the biviscous mantle (with a ratio value of 1 between Kelvin and Maxwell rigidity). Monthly gravity field changes after the Tohoku-Oki earthquake were then calculated according to our biviscous post-seismic model. The model predicted post-seismic gravity changes are comparable with those reported in Han et al. [46] in both amplitude and spatial pattern under a 500 km spatial scale, even though we use different viscosities for the biviscous mantle. Our Burgers viscosities used in the post-seismic model are smaller than those from Han et al. [46] (a transient viscosity of 10^{18} Pa s and a steady state viscosity of 10^{19} Pa), which might be due to the distinct differences in layer depths between our model and the one used in Han et al. [46].

We converted the gridded gravity changes into geopotential coefficients up to d/o 96, and derived the 1-year mean field after the earthquake to estimate the impact from post-seismic changes.

2.7. Computation of Coseismic Gravity Changes from the Hydrological and Oceanic Mass Redistributions

According to [40,47], the temporal corrections (direct tide, solid Earth tide, ocean tide, pole tide and non-tidal correction) have already been removed from the released GGT observations at each epoch. But only SH coefficients up to d/o 20, estimated from atmospheric and oceanic mass variations, and from seasonal variations of hydrology, are used to model the non-tidal signals [47]. Since the coefficients with the d/o lower than 30 are not included in TSH coefficients, the non-tidal GGT corrections certainly have no contribution to the derived gravity gradients in the manuscript. In order to evaluate the influences of the hydrological and oceanic mass redistributions on coseismic gravity changes, we calculated the pre-earthquake models (data from November 2009 to February 2011) and post-earthquake models (data from March 2011 to May 2012) up to d/o 96 from the ECCO-OBP (Ocean Bottom Pressure from Estimating the Circulation and Climate of the Ocean) [48,49] and GLDAS (Global Land Data Assimilation Systems) [50] models, respectively.

3. Results

3.1. Coseismic Gravity and Gravity Gradient Changes from the forward-modeled TSH Coefficients

Based on the forward-modeled TSH coefficients, the coseismic gravity changes Δg and gravity gradient changes of the components (ΔV_{xx}, ΔV_{yy}, ΔV_{zz}, and ΔV_{xz}) in the LNOF on a sphere with a height of 260 km above the WGS84 reference ellipsoid were computed, which are shown in Figures 5 and 6. Figure 5 shows that the magnitude of the forward-modeled coseismic gravity changes is approximately 1.2 µGal. The maximum magnitude of the components (ΔV_{xx}, ΔV_{yy}, ΔV_{zz}, and ΔV_{xz}) is approximately 0.1 mE, which corresponds to the radial component. Compared to Figures 3 and 4, the spatial patterns of Figures 5 and 6 are no longer dipole patterns, and instead show multiple extrema. Additionally, the gravity changes Δg and the changes in the gravity gradient components ΔV_{yy} and ΔV_{zz} show multiple rings. The spatial patterns of Δg and ΔV_{zz} are very similar and nearly isotropic. According to Figure 6, the components (ΔV_{xx}, ΔV_{yy}, ΔV_{zz}, and ΔV_{xz}) have different spatial patterns. The components ΔV_{xx} and ΔV_{xz} have similar spatial patterns, namely, nearly west-to-east stripes and multiple poles in the North-South direction. The ΔV_{yy} component has a spatial pattern with nearly North-South stripes and multiple poles in the west-to-east direction. Moreover, we also use two other fault slip models, provided by the GSI (Geospatial Information Authority of Japan) and the USGS (United States Geological Survey), to compute the coseismic gravity gradient changes from the TSH coefficients(see the supplementary materials). Compared with the Caltech fault slip model (as mentioned in Section 2.5), the GSI and USGS models have different depth extends. The slips extend to 70 km and 58 km in depth for the GSI and USGS models, respectively, while for the Caltech model, the maximum depth of slip is 47 km. The dip angles of the three slip models are very close, all around 10°. Although calculated with different fault slip models, the spatial patterns of the gravity and gravity gradient changes are similar to the signals plotted in Figures 5 and 6.

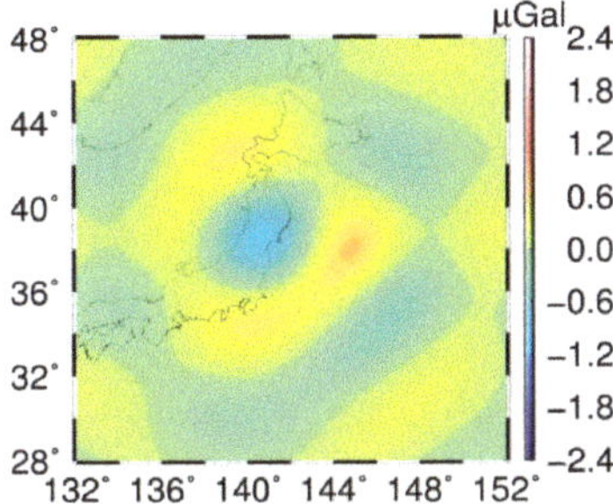

Figure 5. Gravity changes on a sphere with a height of 260 km computed by the tailored spherical harmonic coefficients from the forward model with a Gaussian filter applied. The SH degree range is 30–95 and the radius of the filter is 210 km. The unit is μGal.

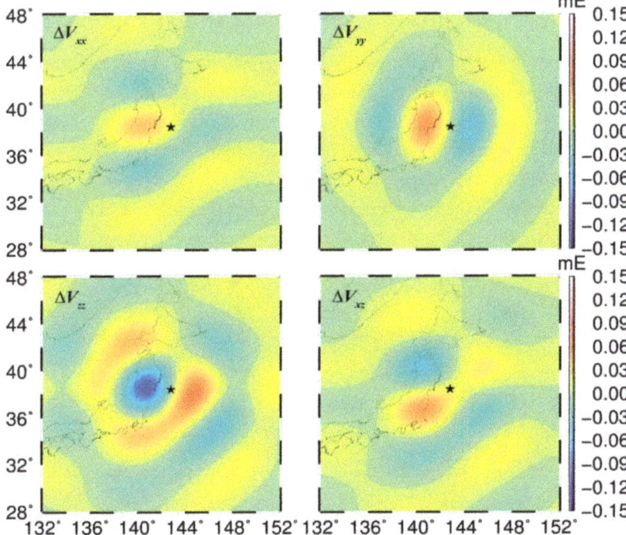

Figure 6. Gravity gradient changes (ΔV_{xx}, ΔV_{yy}, ΔV_{zz}, and ΔV_{xz}) in Local North-Oriented Frame on a sphere with a height of 260 km computed by the tailored spherical harmonic coefficients from the forward-modeled coseismic signals with a Gaussian filter applied. The SH degree range is 30–95 and the radius of the filter is 210 km. The units are mE.

3.2. Post-seismic Gravity Changes from the Viscoelastic Model

We compute the 1-year mean post-seismic radial gravity gradient changes (ΔV_{zz}) based on the TSH coefficients method from the computed viscoelastic relaxation models in Section 2.6, which are shown in Figure 7. A Gaussian filter with the smoothing radius 210 km is applied to the TSH coefficients. The maximum and minimum values of the radial gravity gradient changes from the viscoelastic relaxation model are 0.0095 mE, and -0.0047 mE, respectively. This 1-year mean field will be used in removing the post-seismic effect, and deriving the coseismic gravity signals both from the GOCE and GRACE data (in Sections 3.4 and 3.5). Comparison between our viscoelastic relaxation results and those from Han et al. [46] under the SH degree 60 truncation has also been provided in the supplementary of the manuscript.

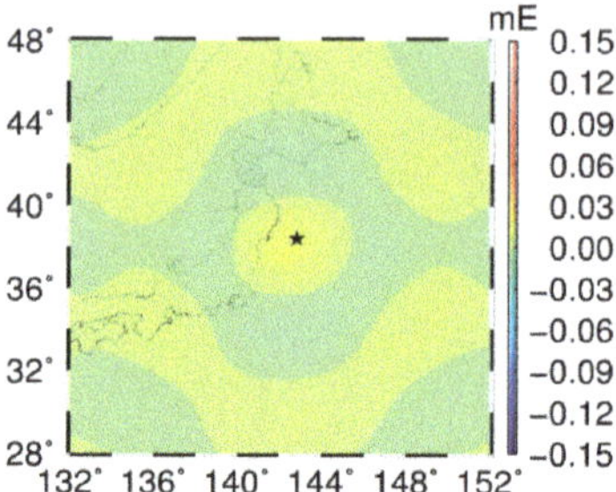

Figure 7. Radial gravity gradient changes (ΔV_{zz}) in Local North-Oriented Frame on a sphere with a height of 260 km computed by the tailored spherical harmonic coefficients from the viscoelastic relaxation model with a Gaussian filter applied. The SH degree range is 30–95 and the radius of the filter is 210 km. The units are mE.

3.3. Coseismic Gravity Changes from the Hydrological and Oceanic Mass Redistributions

Based on the TSH coefficients method, we compute the coseismic radial gravity gradient changes (ΔV_{zz}), which are shown in Figure 8. According to Figure 8, the radial gravity gradient changes derived from the hydrological and oceanic mass redistributions are very small. The maximum and minimum value of the radial gravity gradient changes from both of GLDAS and ECCO-OBP are 0.0022 mE, and −0.0027 mE, respectively. Nevertheless, the contribution from the hydrological and oceanic mass redistributions will be removed in deriving the coseismic gravity signals both from GOCE and GRACE data.

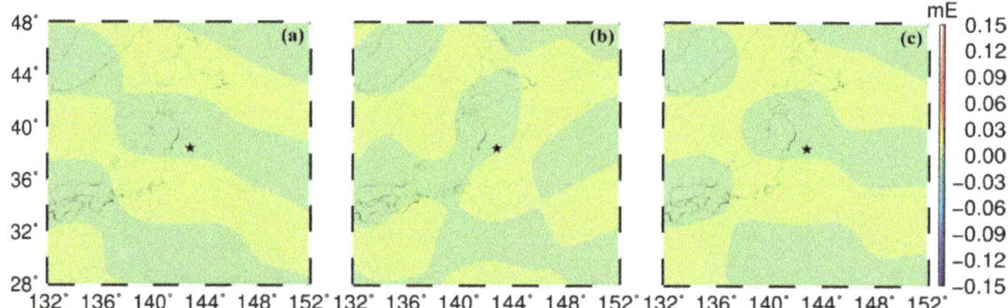

Figure 8. Radial gravity gradient changes (ΔV_{zz}) in Local North-Oriented Frame on a sphere with a height of 260 km computed by the tailored spherical harmonic coefficients from GLDAS (**a**), ECCO-OBP (**b**) and both of GLDAS and ECCO-OBP (**c**) with a Gaussian filter applied. The SH degree range is 30–95 and the radius of the filter is 210 km. The units are mE.

3.4. GRACE-derived Coseismic Gravity Changes and Gravity Gradient Changes

The coseismic gravity changes and gravity gradient changes from GRACE are shown in Figures 9 and 10. It should be noted that before computing the coseismic gravity changes, we removed the post-seismic effects from the one-year mean post-seismic model, and the contribution from the hydrological and oceanic mass redistributions. According to Figures 9 and 10, the coseismic gravity and gravity gradient changes from GRACE have nearly the same spatial pattern, showing only clear multi-pole characteristics in the North-South direction alongside west-to-east stripes. Only the gravity gradient changes ΔV_{xx} and ΔV_{xz} have similar spatial patterns compared to the forward-modeled coseismic signals plotted in Figure 6. Additionally, the gravity changes and the gravity gradient components ΔV_{yy} and ΔV_{zz} have very different spatial patterns compared to those from the forward-modeled coseismic signals. The spatial stripes for the component ΔV_{yy} are oriented west-to-east, which is completely different from the North-South stripes of the forward-modeled coseismic signal. The reason for this situation should be that GRACE's satellite is in a polar orbit with an inclination of 89.5°, so the inter-satellite range-rate observations along the orbit are almost in the North-South direction. Thus, these inter-satellite range-rate observations are not sensitive to signals in the west-to-east direction.

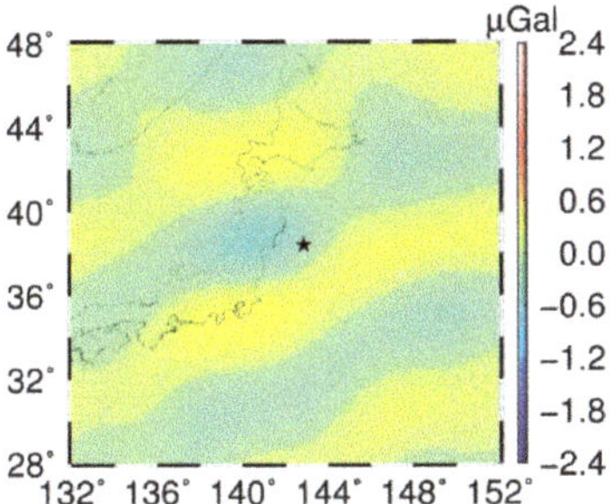

Figure 9. Gravity changes on a sphere with a height of 260 km computed by the tailored spherical harmonic coefficients from GRACE (the Gravity field Recovery and Climate Experiment). The SH degree range is 30–95. The unit is μGal.

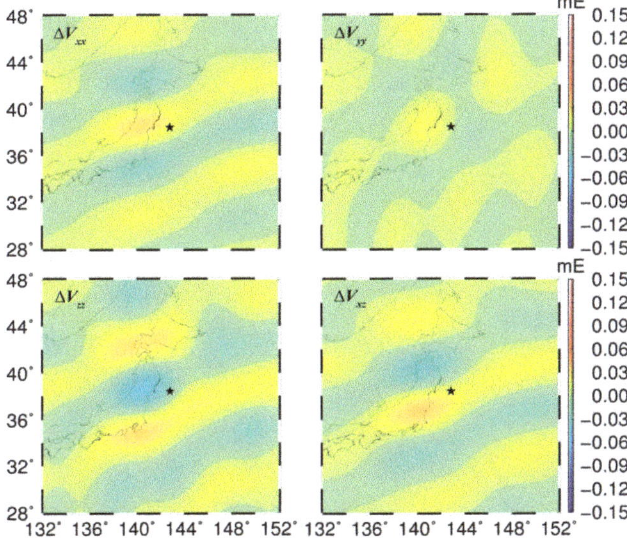

Figure 10. Gravity gradient changes (ΔV_{xx}, ΔV_{yy}, ΔV_{zz}, and ΔV_{xz}) in Local North-Oriented Frame on a sphere with a height of 260 km computed by the tailored spherical harmonic coefficients from GRACE. The SH degree range is 30–95. The units are mE.

Furthermore, when comparing the forward-modeled coseismic results in Figures 5 and 6 with those from the GRACE data in Figures 9 and 10, the magnitudes of the coseismic gravity changes from GRACE are smaller than those from the forward model.

3.5. GOCE-derived Coseismic Gravity Changes and Gravity Gradient Changes

We use the same method of processing the forward-modeled coseismic signals to compute the coseismic gravity and gravity gradient changes from GOCE, which are shown in Figures 11 and 12. The post-seismic effects and the contribution from the hydrological and oceanic mass redistributions are also removed like GRACE-derived coseismic gravity signals. A Gaussian filter with a radius of

210 km was also applied. Figure 11 shows that the magnitude of the forward-modeled coseismic gravity changes is approximately 2.4 µGal. The maximum magnitude of the components (ΔV_{xx}, ΔV_{yy}, ΔV_{zz}, and ΔV_{xz}) is approximately 0.15 mE, which corresponds to the radial component.

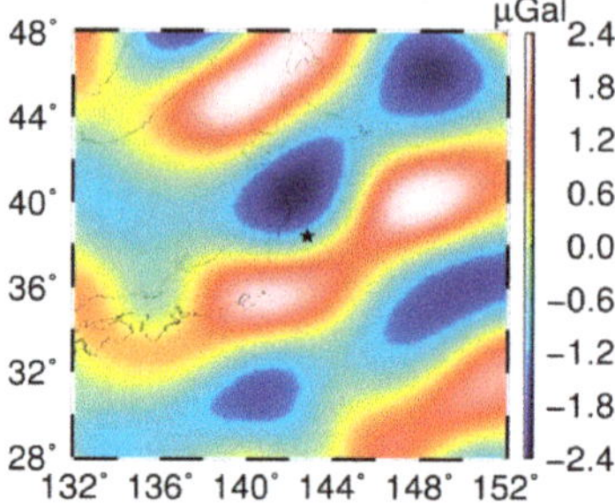

Figure 11. Gravity changes on a sphere with a height of 260 km computed by the tailored spherical harmonic coefficients from the GOCE observations with a Gaussian filter applied. The SH degree range is 30–95, and the radius of the filter is 210 km. The unit is µGal.

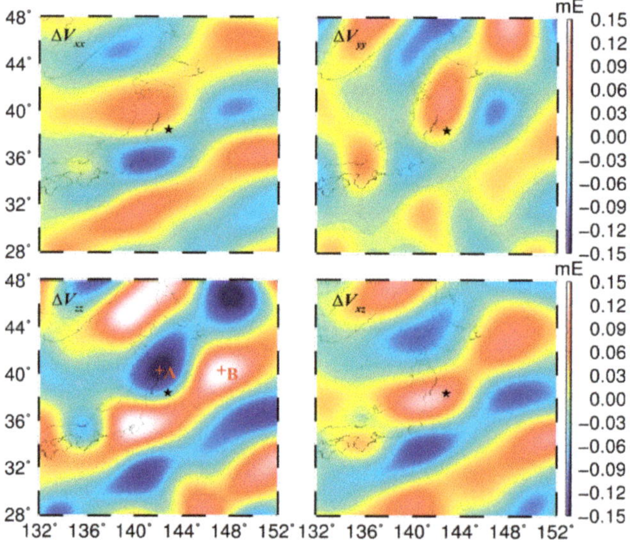

Figure 12. Gravity gradient changes (ΔV_{xx}, ΔV_{yy}, ΔV_{zz}, and ΔV_{xz}) in Local North-Oriented Frame on a sphere with a height of 260 km computed by the tailored spherical harmonic coefficients from the GOCE observations with a Gaussian filter applied. The SH degree range is 30–95, and the radius of the filter is 210 km. The units are mE.

4. Discussion

When compared to the forward-modeled coseismic signals in Figures 7 and 8, the spatial patterns of the coseismic signals in Figures 11 and 12 from the GOCE observations, although noisy, are somewhat similar, exhibiting multiple extrema. The spatial patterns of the gravity changes Δg and gravity gradient changes of the radial component ΔV_{zz} exhibit multiple rings. According to Figure 12, the components ΔV_{xx} and ΔV_{xz} show similar spatial patterns, i.e., the west-to-east stripes and multiple poles in the North-South direction. Note that, there are more gravity gradient change peaks compared to the modeled gravity gradient changes in the area of interest, which is the same situation with Figure 8 in Fuchs et al. [28]. Referring to Fuchs et al. [28], we also use the averaged accuracy of gravity gradients before the 2011 earthquake to represent the accuracy of the derived coseismic gravity gradients. The deviations of the mean (1σ) in 0.5° grid cells for the differences of gravity gradients (ΔV_{xx}, ΔV_{yy}, ΔV_{zz} and ΔV_{xz}) between the GOCE-derived pre-earthquake model and the GOCO03S [32] are computed, which are 0.018, 0.022, 0.036 and 0.024 mE, respectively. According to Figure 12, we could see that there is a significant change (greater than 3σ) in gravity gradients between pre- and post-earthquake gravity gradients.

We note that the uncertainties (σ) might be underestimated because the two models are not completely independent and the errors are not stationary, but we will not do further assessment here. Additionally, according to Figures 5, 6, 11 and 12, the amplitudes of the coseismic gravity gradient changes are larger than the forward-modeled coseismic signals, and the geographical positions of the multi-poles are different, which is similar to what is seen in Fuchs et al. [28]. The geographic positions of these multi-poles from the GOCE data are located approximately 200 km to the northeast of the modeled results. The reasons for these discrepancies are still not very clear. According to Fuchs et al. [28], Heki et al. [51] and Feng et al. [52], the differences between the results from the forward model and GOCE data might be caused by systematic errors in the SGG observations and the weak sensitivity of SGG observations to the location of the earthquake. In addition, the large uncertainties of fault slip models also contribute to the discrepancies. Dai et al. [16] show that the fault slip model still leads to around 40% relative difference of gravity changes compared to the GRACE observations, even inverted with a multiple data source, which is likely, because the current commonly-used dislocation models are not sophisticated enough. The noticeable differences between model prediction and observation need to be investigated by further studies. According to Figures 7, 10 and 12, the maximum and minimum values from the viscoelastic relaxation model are nearly 15–21% of the ones (max: 0.031 mE, min: −0.046 mE) derived from the GRACE one-year mean field differences, and nearly 3–6% of the ones (max: 0.161 mE, min: −0.167 mE) derived from the GOCE one-year mean field differences. According to Figures 8 and 12, the maximum and minimum value of the radial gravity gradient changes from both of GLDAS and ECCO-OBP are nearly two orders of magnitude smaller than the GOCE-derived coseismic gravity gradient signals.

In order to show the gravity change time series, we processed the GOCE SGG data to obtain the monthly gravity field models with the time period from November 2009 to May 2012. The data processing strategies are the same with that determining the pre-earthquake and post-earthquake models above. According to the GOCE daily reports EGG [53], there are a lot of special events, several data gaps and frequent calibrating operations (about once a month), from November 2009 to May 2012. Therefore, we are unlikely to be able to derive continuous monthly time-variable gravity field models due to the practical data quality. In addition, the solutions after the earthquake have larger oscillations, which agree with the fact that there are more special events in SGG data after the earthquake than those before the earthquake for most months [53]. Figure 13a,b show the radial gravity gradient changes at A (40.6°, 141.2°) and B (40.2°, 147.2°) points on Figure 12 derived from the monthly time-variable gravity field models before and after the earthquake, while Figure 13c,d shows the average radial gravity gradient changes at a negative area [(40°~41°), (140.5°~142.5°)] around A and a positive area [(39°~40.5°), (147~148.5°)] around B.

According to Figure 13, although the radial gravity gradient change time series contain some significant oscillations, they are still comparable with the co-seismic gravity change time series from GRACE for some smaller earthquake (Mw $\approx$ 8.5) given by some previous studies (see as Figure 2 of Han et al. [54], and Figures 6 and 7 of Chao and Liau [55]). In order to show the changes more clearly, we use a 1D median filter (with 5th order) to process those four time series, and the obtained filtered results denoted by the red curves in Figure 13. Those filtered curves show obvious steps before and after (gray areas in Figure 13) the earthquake. We further use step functions to fit the original observations (the obtained fitting curves are also plotted in Figure 13; blue curves), and the step values can be estimated at the same time (see Figure 13). Here we use a bootstrap procedure (a Monte Carlo process, see Efron and Tibshirani [56]) to estimate the uncertainties for those step values. This method has been widely used for the uncertainty estimations in different geophysical research [57–62]. Here we give a simple example to explain the error estimation process (more details can be found in Shen and Ding [60]):

(1)   A step function $s(t)$ is used to fit the original time series (the earthquake time is used as a priori information in the way some previous studies have done), then the step value s is obtained at the same time.

(2)   300 different random Gaussian white noise time series $n_i(t)$ are synthesized, which have the same length, and with the same mean power in the frequency domain (i.e., almost the same noise level; here we use mean power $P = 1.3 \times 10^{-2}$ mE^2/cpy; see the supplementary Figure S3). Note that here we ignore the small differences of the oscillation amplitudes in the radial gravity gradient change time series (see Figure 13), because the time points are limited, the 300 different noise time series can almost reproduce all possible different oscillations.

(3)   The fitted step function $s(t)$ is added in the 300 noise time series $n_i(t)$, and 300 new noisy time series $f_i(t) = s(t) + n_i(t)$ are obtained.

(4)   For the 300 noisy time series $f_i(t)$, repeating the process 1) and 2) for all of them, then the 300 new step values s_i can be estimated. The standard deviation of those different S_i is used as the final error estimation for the step value **s**.

From Figure 13, we can see that the step values are clearly over three times their corresponding uncertainties (the double uncertainties are denoted by the blue areas in Figure 13 as the corresponding two standard deviations). Statistically, we may suggest that those steps represent the coseismic gravity gradient change signals caused by the 2011 Tohoku-Oki earthquake.

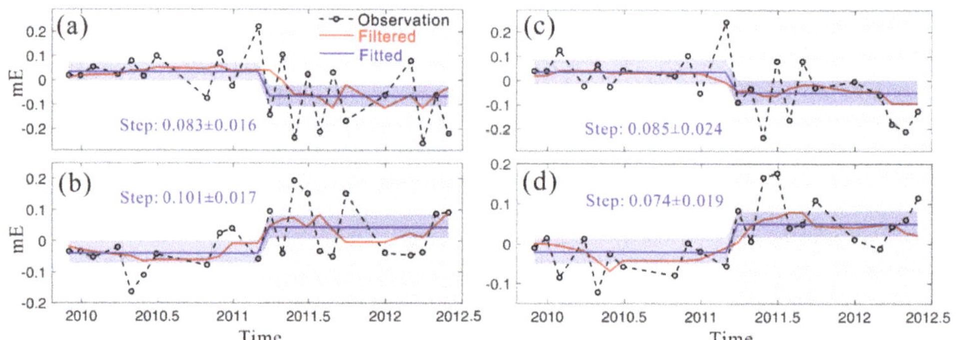

Figure 13. Radial gravity gradient changes at A (**a**), B (**b**) points and two selected areas around A and B (**c,d**) before and after earthquake. The red curves denote the filtered results after using a 1D median filter to the original time series, the blue curves are the fitted step functions. The double uncertainties are denoted by the blue areas.

To reveal how the spatial patterns of the GOCE and modeled gravity gradient changes agree with each other, we computed the correlation coefficients (see Table 2) of the gravity changes and gravity gradient changes (ΔV_{xx}, ΔV_{yy}, ΔV_{zz}, ΔV_{xz} and Δg) between the observed models (GOCE and GRACE) and the forward model for the TSH coefficients (see Figure 2). The Root Mean Square (RMS) of observed (ΔV_{xx}, ΔV_{yy}, ΔV_{zz}, ΔV_{xz} and Δg) of GOCE/GRACE, and the differences between the observed and forward modeling signals, are also computed and shown in Table 2. According to Table 2, both of GOCE and GRACE do not perform well, because all the correlation coefficients are less than or equal to 0.85. Although most correlation coefficients corresponding to GOCE are lower than those of the GRACE time-variable gravity field models, the correlation coefficient corresponding to the

ΔV_{yy} component from GOCE is slightly larger than the one of GRACE. Thus, for the SH degree range 30–95, the GOCE mission could reveal a greater signal of coseismic gravity gradient changes in the west-to-east direction than GRACE, because GRACE has sensitive observations along the North-South direction. The coefficient for ΔV_{zz} is very close to the value in Fuchs et al. [28]. All RMS of observed signals are reduced after subtracting the forward-modeled coseismic signals from the GOCE-derived results. However, because the amplitude of the results from GRACE is only about half of the one from the forward model, the RMS of GRACE's ΔV_{yy} increases when the modeled signal is subtracted from the observed signal and the RMS of ΔV_{yy} doesn't change. Of course, compared to GOCE, GRACE performs relatively well in the SH degree range 30–95, especially for the ΔV_{xx} component, which has the maximum correlation coefficient because only inter-satellite range-rates are observed along its orbit.

Table 2. Correlation coefficients, RMS of observed gravity and gravity gradients (ΔV_{xx}, ΔV_{yy}, ΔV_{zz}, ΔV_{xz} and Δg), RMS of the differences between the observed and forward modeling signals. The observed signals are from GRACE and GOCE. The positions for the computation are located in the region of 28°N–48°N, 132°E–152°E. The unit of ΔV_{xx}, ΔV_{yy}, ΔV_{zz}, and ΔV_{xz} is mE and the unit of Δg is µGal.

	GOCE					GRACE				
Quantities	ΔV_{xx}	ΔV_{yy}	ΔV_{zz}	ΔV_{xz}	Δg	ΔV_{xx}	ΔV_{yy}	ΔV_{zz}	ΔV_{xz}	Δg
Correlation coefficients	0.68	0.56	0.61	0.68	0.63	0.85	0.53	0.68	0.80	0.74
RMS of observed signals	0.044	0.041	0.078	0.056	1.179	0.010	0.005	0.015	0.012	0.179
RMS of differences	0.038	0.036	0.066	0.047	1.043	0.007	0.012	0.015	0.008	0.158

5. Conclusions

We employed the least squares method with a bandpass auto regression moving average filter, to recover two global gravity field models up to degree and order 220, from GOCE satellite gravity gradient data from the 15 March 2011 to the 31 May 2012 after the 2011 Japan Tohoku-Oki earthquake event. Then, we used the recovered models to estimate the coseismic gravity and gravity gradient changes that were caused by the earthquake by subtracting the pre-earthquake model from the post-earthquake model. This approach is different from that in Fuchs et al. [28]. They used the diagonal components (V_{xx}, V_{yy}, and V_{zz}) along the orbit to construct the new vertical gravity gradient V_{zz}, which was used to present the coseismic gravity gradient changes. To extract the coseismic gravity signals from the recovered global gravity field models, we proposed TSH coefficients according to the influences of colored noise in the SGG data and the polar gap problem on the recovered models.

The gravity changes Δg and the gravity gradient changes (ΔV_{xx}, ΔV_{yy}, ΔV_{zz}, and ΔV_{xz}) were computed by the GOCE-derived tailored spherical harmonic (TSH) coefficients, GRACE-derived TSH coefficients and the forward-modeled TSH coefficients. In data processing, the degree range 30–95 was used to obtain the TSH coefficients. A Gaussian filter with a radius of 210 km was also applied to the TSH coefficients according to the maximum frequency of 95. When comparing the coseismic gravity signals from GOCE, the forward model, and GRACE, the spatial patterns of the coseismic gravity and gravity gradient changes from GOCE were analogous to those from the forward model. There is a significant change (greater than 3σ) in the gravity gradients between the pre- and post-earthquake gravity gradients, which is associated with the earthquake. Moreover, the radial gravity gradient changes from the derived monthly time-variable gravity field models before and after the earthquake show obvious steps before and after the earthquake, whose amplitudes are at a level over three times that of their corresponding uncertainties, and thus significant. Statistically, it is reasonable to infer that these steps represent the coseismic gravity gradient change signals caused by the 2011 Tohoku-Oki earthquake. For the GGT components (ΔV_{xx}, ΔV_{zz}, and ΔV_{xz}), the correlation between the GOCE-derived coseismic gravity signals and the forward-modeled coseismic results was weaker than that from GRACE. However, the correlation coefficient that corresponded to the ΔV_{yy} component from GOCE is larger than that from GRACE. The component ΔV_{yy} had spatial patterns that included North-South stripes and multiple poles in the west-to-east direction. This situation means that the GOCE mission might reveal more coseismic gravity signals in the West-East direction than GRACE in the SH degree range 30–95. The estimated time series of radial gravity gradient changes (see Figure 13) suggest that the coseismic gravity changes can be reproduced from GOCE observations through the proposed method.

Supplementary Materials: The following are available online at http://www.mdpi.com/2072-4292/11/11/1295/s1, Figure S1: Gravity gradient changes (ΔV_{xx}, ΔV_{yy}, ΔV_{zz}, and ΔV_{xz}) in LNOF on a sphere with a height of 260 km computed by the TSH coefficients from the forward-modeled signals of the GSI fault slip model with a Gaussian filter applied, Figure S2: Gravity gradient changes (ΔV_{xx}, ΔV_{yy}, ΔV_{zz}, and ΔV_{xz}) in LNOF on a sphere with a height of 260 km computed by the TSH coefficients from the forward-modeled signals of the USGS fault slip model with a Gaussian filter applied, Figure S3: The observation (the fitted step has been removed) and the inputted noise (a), and their Fourier power spectra (b) (logarithmic scale in dB), Figure S4: Postseismic gravity changes (Δg) on the ground from the spherical harmonic coefficients up to d/o 60 of the viscoelastic relaxation model computed in the paper (left) and provided by Han (right).

Author Contributions: Conceptualization X.X.; methodology, X.X. and H.D.; investigation, X.X. and H.D.; resources, J.L. and M.H.; data curation, Y.Z.; writing—original draft preparation, X.X.; writing—review and editing, X.X., H.D., and J.L.

Funding: This research was financially supported by the National Natural Science Foundation of China (Grant No. 41574019, 41774020, 11873075). DAAD Thematic Network Project (57421148). The Major Project of High resolution Earth Observation System. The Natural Science Foundation of Shanghai (17ZR1435600).

Acknowledgments: The authors thank Shin-Chan Han for the help in our post-seismic effect evaluation and providing the viscoelastic and afterslip models of the Tohuku-Oki earthquake for the verification of our post-seismic models. We thank Wenbin Shen for useful discussion. We also acknowledge the European Space Agency for providing the GOCE data. We are also grateful to the three anomalous reviewers who have provided constructive comments and suggestions to improve our work.

Conflicts of Interest: The authors declare no conflict of interest.

References

1. Reigber, C.; Schwintzer, P.; Neumayer, K.H.; Barthelmes, F.; König, R.; Förste, C.; Balmino, G.; Biancale, R.; Lemoine, J.-M.; Loyer, S.; et al. The CHAMP-only earth gravity field model EIGEN-2. *Adv. Space Res.* **2003**, *31*, 1883–1888. [CrossRef]

2. Tapley, B.D.; Bettadpur, S.; Watkins, M.; Reigber, C. The gravity recovery and climate experiment: Mission overview and early results. *Geophys. Res. Lett.* **2004**, *31*, L09607. [CrossRef]

3. Drinkwater, M.; Haagmans, R.; Muzi, D. The GOCE gravity mission: ESA's first core explorer. In Proceedings of the Third GOCE User Workshop, Frascati, Italy, 6–9 November 2006; pp. 1–7.

4. Han, S.C.; Shum, C.K.; Bevis, M.; Ji, C.; Kuo, C.Y. Crustal dilatation observed by GRACE after the 2004 Sumatra-Andaman earthquake. *Science* **2006**, *313*, 658–662. [CrossRef]

5. Chen, J.L.; Wilson, C.R.; Tapley, B.D.; Grand, S. GRACE detects coseismic and postseismic deformation from the Sumatra-Andaman earthquake. *Geophys. Res. Lett.* **2007**, *34*, L13302. [CrossRef]

6. Panet, I.; Pollitz, F.; Mikhailov, V.; Diament, M.; Banerjee, P.; Grijalva, K. Upper mantle rheology from GRACE and GPS postseismic deformation after the 2004 Sumatra-Andaman earthquake. *Geochem. Geophys. Geosyst.* **2010**, *11*, Q06008. [CrossRef]

7. De Linage, C.; Rivera, L.; Hinderer, J.; Boy, J.P.; Rogister, Y.; Lambotte, S.; Biancale, R. Separation of coseismic and postseismic gravity changes for the 2004 Sumatra-Andaman earthquake from 4.6 yr of GRACE observations and modeling of the coseismic change by normal modes summation. *Geophys. J. Int.* **2009**, *176*, 695–714. [CrossRef]

8. Han, S.C.; Sauber, J.; Luthcke, S. Regional gravity decrease after the 2010 Maule (Chile) earthquake indicates large-scale mass redistribution. *Geophys. Res. Lett.* **2010**, *37*, L23307. [CrossRef]

9. Heki, K.; Matsuo, K. Coseismic gravity changes of the 2010 earthquake in central Chile from satellite gravimetry. *Geophys. Res. Lett.* **2010**, *37*, L24306. [CrossRef]

10. Broerse, D.; Vermeersen, L.; Riva, R.; van der Wal, W. Ocean contribution to co-seismic crustal deformation and geoid anomalies: Application to the 2004 December 26 Sumatra-Andaman earthquake. *Earth Planet. Sci. Lett.* **2011**, *305*, 341–349. [CrossRef]

11. Matsuo, K.; Heki, K. Coseismic gravity changes of the 2011 Tohoku-Oki earthquake from satellite gravimetry. *Geophys. Res. Lett.* **2011**, *38*, L00G12. [CrossRef]

12. Han, S.C.; Sauber, J.; Riva., R. Contribution of satellite gravimetry to understanding seismic source processes of the 2011 Tohoku-Oki earthquake. *Geophys. Res. Lett.* **2011**, *38*, L24312. [CrossRef]

13. Wang, L.; Shum, C.K.; Simons, F.J.; Tapley, B.; Dai, C. Coseismic and postseismic deformation of the 2011 Tohoku-Oki earthquake constrained by GRACE gravimetry. *Geophys. Res. Lett.* **2012**, *39*, L07301. [CrossRef]

14. Zhou, X.; Sun, W.; Zhao, B.; Fu, G.; Dong, J.; Nie, Z. Geodetic observations detecting coseismic displacements and gravity changes caused by the Mw = 9.0 Tohoku-Oki earthquake. *J. Geophys. Res.* **2012**, *117*, B05408. [CrossRef]

15. Cambiotti, G.; Sabadini, R. A source model for the great 2011 Tohoku earthquake (Mw = 9.1) from inversion of GRACE gravity data. *Earth Planet Sci. Lett.* **2012**, *335–336*, 72–79. [CrossRef]

16. Dai, C.; Shum, C.K.; Wang, R.; Wang, L.; Guo, J.; Shang, K.; Tapley, B.D. Improved constraints on seismic source parameters of the 2011 Tohoku earthquake from GRACE gravity and gravity gradient changes. *Geophys. Res. Lett.* **2014**, *41*, 1929–1936. [CrossRef]

17. Li, J.; Shen, W.B. Monthly GRACE detection of coseismic gravity change associated with 2011 Tohoku-Oki earthquake using northern gradient approach. *Earth Planets Space* **2015**, *67*, 29. [CrossRef]

18. Odera, P.A.; Fukuda, Y. Evaluation of GOCE-based global gravity field models over Japan after the full mission using free-air gravity anomalies and geoid undulations. *Earth Planets Space* **2017**, *69*, 135. [CrossRef]

19. Gilardoni, M.; Reguzzoni, M.; Sampietro, D.; Sansò, F. Combining EGM2008 with GOCE gravity models. *Bollettino di Geofisica Teorica ed Applicata* **2013**, *54*, 285–302.

20. Bouman, J.; Ebbing, J.; Martin Fuchs, M.J.; Sebera, J.; Lieb, V.; Szwillus, W.; Haagmans, R.; Novak, P. Satellite gravity gradient grids for geophysics. *Sci. Rep.* **2016**, *6*, 21050. [CrossRef]

21. Bouman, J.; Ebbing, J.; Meekes, S.; Fattah, R.A.; Fuchs, M.; Gradmann, S.; Haagmans, R.; Lieb, V.; Schmidt, M.; Dettmering, D.; et al. GOCE gravity gradient data for lithospheric modeling. *Int. J. Appl. Earth Obs.* **2015**, *35*, 16–30. [CrossRef]

22. Reguzzoni, M.; Sampietro, D. GEMMA: An Earth crustal model based on GOCE satellite data. *Int. J. Appl. Earth Obs.* **2015**, *35*, 31–43. [CrossRef]

23. Bingham, R.J.; Haines, K.; Lea, D. A comparison of GOCE and drifter-based estimates of the North Atlantic steady-state surface circulation. *Int. J. Appl. Earth Obs.* **2015**, *35*, 140–150. [CrossRef]

24. Mikhailov, V.; Tikhotsky, S.; Diament, M.; Panet, I.; Ballu, V. Can tectonic processes be recovered from new gravity satellite data? *Earth Planet Sci. Lett.* **2004**, *228*, 281–297. [CrossRef]

25. Migliaccio, F.; Reguzzoni, M.; Sansò, F.; Dalla Via, G.; Sabadini, R. Detecting geophysical signals in gravity satellite missions. *Geophys. J. Int.* **2008**, *172*, 56–66. [CrossRef]

26. Han, S.-C.; Ditmar, P. Localized spectral analysis of global satellite gravity fields for recovering time-variable mass redistributions. *J. Geod.* **2008**, *82*, 423–430. [CrossRef]

27. Garcia, R.F.; Bruinsma, S.; Lognonné, P.; Doornbos, E.; Cachoux, F. GOCE: The First Seismometer in Orbit around the Earth. *Geophys. Res. Lett.* **2013**. [CrossRef]

28. Fuchs, M.J.; Bouman, J.; Broerse, T.; Visser, P.; Vermeersen, B. Observing coseismic gravity change from the Japan Tohoku-Oki 2011 earthquake with GOCE gravity gradiometry. *J. Geophys. Res. Solid Earth.* **2013**, *118*, 5712–5721. [CrossRef]

29. Fuchs, M.J.; Hooper, A.; Broerse, T.; Bouman, J. Distributed fault slip model for the 2011 Tohoku-Oki earthquake from GNSS and GRACE/GOCE satellite gravimetry. *J. Geophys. Res. Solid Earth.* **2016**, *121*, 1114–1130. [CrossRef]

30. Han, S.C.; Shum, C.K.; Ditmar, P.; Visser, P.; van Beelen, C.; Schrama, E.J.O. Effect of high-frequency mass variations on GOCE recovery of the Earth's gravity field. *J. Geodyn.* **2006**, *41*, 69–76. [CrossRef]

31. Bouman, J.; Fiorot, S.; Fuchs, M.; Gruber, T.; Schrama, E.; Tscherning, C.C.; Veicherts, M.; Visser, P. GOCE gravity gradients along the orbit. *J. Geod.* **2011**, *85*, 791–805. [CrossRef]

32. Mayer-Gürr, T.; Rieser, D.; Hoeck, E.; Brockmann, J.M.; Schuh, W.-D.; Krasbutter, I.; Kusche, J.; Maier, A.; Krauss, S.; Hausleitner, W.; et al. The new combined satellite only model GOCO03s. Presented at the GGHS2012, Venice, Italy, 9–12 November 2012.

33. Xu, X.; Zhao, Y.; Reubelt, T.; Tenzer, R. A GOCE only gravity model GOSG01S and the validation of GOCE related satellite gravity models. *Geod. Geodyn.* **2017**, *8*, 260–272. [CrossRef]

34. Koop, R. *Global Gravity Field Modeling Using Satellite Gravity Gradiometry*; Publications on Geodesy, New Series, Number 38; Netherlands Geodetic Commission: Delft, The Netherlands, 1993.

35. Cesare, S. *Performance Requirements and Budgets for the Gradiometric Mission*; Technical Note, GOC-TN-AI-0027; Alenia Spazio: Turin, Italy, 2008.

36. Schuh, W.D. Improved modelling of SGG-data sets by advanced filter strategies. In *ESA-Project "From Eötvös to mGal+"*; Final Report, ESA/ESTEC Contract 14287/00/NL/DC, WP 2; ESA: Noordwijk, The Netherlands, 2002; pp. 113–181.

37. Fuchs, M.; Bouman, J. Rotation of GOCE gravity gradients to local frames. *Geophys. J. Int.* **2011**, *187*, 743–753. [CrossRef]

38. Rummel, R.; Yi, W.; Stummer, C. GOCE gravitational gradiometry. *J. Geod.* **2011**, *85*, 777–790. [CrossRef]

39. Sneeuw, N.; van Gelderen, M. The polar gap. In *Geodetic Boundary Value Problems in View of the One Centimeter Geoid*; Sansò, F., Rummel, R., Eds.; Lecture Notes in Earth Sciences; Springer: Berlin, Germany, 1997; Volume 65, pp. 559–568.

40. EGG-C. *GOCE Level 2 Product Data Handbook; GO-MA-HPF-GS-0110.* 2010. Available online: https://earth.esa.int/web/guest/document-library/browse-document-library/-/article/goce-level-2-product-data-handbook-5713 (accessed on 12 May 2017).

41. Wang, R.; Lorenzo-Martiín, F.; Roth, F. PSGRN/PSCMP—A new code for calculating co- and post-seismic deformation, geoid and gravity changes based on the viscoelastic-gravitational dislocation theory. *Comput. Geosci.* **2006**, *32*, 527–541. [CrossRef]

42. Wei, S.; Sladen, A.; The ARIA Group. Updated Result: 3/11/2011 (Mw 9.0), Tohoku-Oki, Japan. 2011. Available online: http://www.tectonics.caltech.edu/slip_history/2011_taiheiyo-oki/ (accessed on 18 June 2015).

43. Bassin, C.; Laske, G.; Masters, G.G. The current limits of resolution for surface wave tomography in North America. *EOS Trans.* **2000**, *81*, F897.

44. Broerse, T.; Riva, R.; Vermeersen, B. Ocean contribution to seismic gravity changes: The sea level equation for seismic perturbations revisited. *Geophys. J. Int.* **2014**, *199*, 1094–1109. [CrossRef]

45. Colombo, O. *Numerical Methods for Harmonic Analysis on the Sphere*; Report No. 310; Dept. of Geodetic Science and Surveying, Ohio State University: Columbus, OH, USA, 1981.

46. Han, S.-C.; Sauber, J.; Pollitz, F. Broadscale postseismic gravity change following the 2011 Tohoku-Oki earthquake and implication for deformation by viscoelastic relaxation and afterslip. *Geophys. Res. Lett.* **2014**, *41*, 5797–5805. [CrossRef]

47. Bouman, J.; Rispens, S.; Gruber, T.; Koop, R.; Schrama, E.; Visser, P.; Tscherning, C.C.; Veicherts, M. Preprocessing of gravity gradients at the GOCE high-level processing facility. *J. Geod.* **2009**, *83*, 659–678. [CrossRef]

48. Fukumori, I. A partitioned Kalman filter and smoother. *Mon. Weather Rev.* **2002**, *130*, 1370–1383. [CrossRef]

49. Kim, S.B.; Lee, T.; Fukumori, I. Mechanisms Controlling the Interannual Variation of Mixed Layer Temperature Averaged over the Niño-3 Region. *J. Clim.* **2007**, *20*, 3822–3843. [CrossRef]

50. Rodell, M.; Houser, P.R.; Jambor, U.; Gottschalck, J.; Mitchell, K.; Meng, C.-J.; Arsenault, K.; Cosgrove, B.; Radakovich, J.; Bosilovich, M.; et al. The Global Land Data Assimilation System. *Bull. Am. Meteorol. Soc.* **2004**, *85*, 381–394. [CrossRef]

51. Heki, K.; Mitsui, Y.; Matsuo, K.; Tanaka, Y. Accelerated subduction of the Pacific Plate after mega-thrust earthquakes: Evidence from GPS and GRACE. Presented at the AGU Fall Meeting 2012, San Francisco, CA, USA, 3–7 December 2012.

52. Feng, W.; Li, Q.; Li, Z.; Hoey, T. Slip distribution of 2011 Japan Mw 9.0 earthquake determined by GPS-coseismic displacement and GRACE gravity changes. Presented at the EGU Scientific Assembly, Vienna, Austria, 7–12 April 2013.

53. ESA. Available online: https://earth.esa.int/web/sppa/mission-performance/esa-missions/goce/quality-control-reports/daily-egg-quality-overview (accessed on 20 January 2019).

54. Han, S.-C.; Sauber, J.; Pollitz, F. Coseismic compression/dilatation and viscoelastic uplift/subsidence following the 2012 Indian Ocean earthquakes quantified from satellite gravity observations. *Geophys. Res. Lett.* **2015**, *42*, 3764–3772. [CrossRef]

55. Chao, B.F.; Liau, J.R. Gravity changes due to large earthquakes detected in GRACE satellite data via empirical orthogonal function analysis. *J. Geophys. Res. Solid Earth* **2019**. [CrossRef]

56. Efron, B.; Tibshirani, R. Bootstrap methods for standard errors, confidence intervals, and other measures of statistical accuracy. *Stat. Sci.* **1986**, *1*, 54–75. [CrossRef]

57. Widmer-Schnidrig, R.; Masters, G.; Gilbert, F. Observably split multiplets-data analysis and interpretation in terms of large-scale aspherical structure. *Geophys. J. Int.* **1992**, *111*, 559–576. [CrossRef]
58. Häfner, R.; Widmer-Schnidrig, R. Signature of 3-D density structure in spectra of the spheroidal free oscillation 0S2. *Geophys. J. Int.* **2013**, *192*, 285–294. [CrossRef]
59. Shen, W.B.; Ding, H. Observation of spheroidal normal mode multiplets below 1 mHz using ensemble empirical mode decomposition. *Geophys. J. Int.* **2014**, *196*, 1631–1642. [CrossRef]
60. Ding, H.; Chao, B.F. Data stacking methods for isolation of singlets of the Earth's normal modes: Extensions comparisons and applications. *J. Geophys. Res. Solid Earth* **2015**, *120*, 5034–5050. [CrossRef]
61. Chen, J.L.; Wilson, C.R.; Tapley, B.D. Contribution of ice sheet and mountain glacier melt to recent sea level rise. *Nat. Geosci.* **2013**, *6*, 549–552. [CrossRef]
62. Han, J.; Tangdamrongsub, N.; Hwang, C.; Abidin, H.Z. Intensified water storage loss by biomass burning in Kalimantan: Detection by GRACE. *J. Geophys. Res. Solid Earth* **2017**, *122*, 2409–2430. [CrossRef]

MDPI
St. Alban-Anlage 66
4052 Basel
Switzerland
Tel. +41 61 683 77 34
Fax +41 61 302 89 18
www.mdpi.com

Remote Sensing Editorial Office
E-mail: remotesensing@mdpi.com
www.mdpi.com/journal/remotesensing